FUTURE INDUSTRIAL PROSPECTS
OF MEMBRANE PROCESSES

Proceedings of a symposium organised by the Commission of the European Communities, Directorate-General for Science, Research and Development, in the framework of the R & D programmes on 'Basic Research in Industrial Technologies for Europe' and 'Radioactive Waste Management and Disposal', held in Brussels, Belgium, 6–7 December 1988.

Scientific Committee of the Commission of the European Communities:

Chairmen: S. FINZI
A. GARCIA ARROYO
W. VAN DER EIJK
R. SIMON

Secretaries: L. CECILLE and J.-C. TOUSSAINT

FUTURE INDUSTRIAL PROSPECTS OF MEMBRANE PROCESSES

Edited by

L. CECILLE and J.-C. TOUSSAINT

Commission of the European Communities, Brussels, Belgium

ELSEVIER APPLIED SCIENCE
LONDON and NEW YORK

ELSEVIER SCIENCE PUBLISHERS LTD
Crown House, Linton Road, Barking, Essex IG11 8JU, England

Sole Distributor in the USA and Canada
ELSEVIER SCIENCE PUBLISHING CO., INC.
655 Avenue of the Americas, New York, NY 10010, USA

WITH 51 TABLES AND 158 ILLUSTRATIONS

British Library Cataloguing in Publication Data

Future industrial prospects of membrane processes.
1. Synthetic membranes
I. Commission of the European Communities   II.
Cécille, L.   III. Toussaint, J.-C.
660.2'842

ISBN 1-85166-412-2

Library of Congress CIP data applied for

Publication arrangements by Commission of the European Communities, Directorate-General Telecommunications, Information Industries and Innovation, Scientific and Technical Communication Unit, Luxembourg

EUR 12114

LEGAL NOTICE

Printed in Northern Ireland by The Universities Press (Belfast) Ltd.

PREFACE

This first international symposium on "Future Industrial Prospects of Membrane Processes" organised by the Commission of the European Communities provided an excellent opportunity to survey the present status of the membrane technologies in Europe by presenting the main results achieved within the two European Community R&D programmes "BRITE" (Basic Research in Industrial Technologies for Europe) and "Radioactive Waste Management and Disposal".

By bringing together for the first time experts from both sectors this symposium has also allowed a comparison to be made of the results obtained in nuclear and non-nuclear applications demonstrating that technological innovations developed in one sector could be used in the other. This emphasized the common problems which exist for the industrialisation of membrane processes.

Early and wide diffusion of scientific and technical results has always been one of the primary concerns of the Commission of the European Communities and we hope through these proceedings to have demonstrated the importance of the cross frontier collaboration as well as cross sectorial collaboration.

These proceedings included all of the papers presented at the symposium, together with the opening address and a summary of the discussions.

The podium and the audience during the opening session

CONTENTS

OPENING SESSION

Opening address

The importance of membrane processes within the
BRITE programme

The C.E.C. research activities on membrane
processes dealing with radioactive waste
management

OPENING ADDRESS

A. Garcia Arroyo
Director for Technological Research (XII/C)

S. Finzi
Director for Nuclear Safety Research (XII/D)

G.A. On behalf of the Commission of the European Communities, it is a
great pleasure for Mr. Finzi and myself to welcome you to the
Symposium on Future Industrial Prospects of Membrane Processes. We
consider that this symposium is an important event in the membrane
science and technology for the following reasons. First, it will
present the objectives currently aimed at and the main results
achieved and in matter of membrane applications in two well distinct
Community's research programmes, namely the BRITE and the
Radioactive Waste Management programmes.

Then, this symposium will provide a good opportunity for experts
involved in the nuclear and conventional industries to exchange
their respective experience gained in different specific membrane
applications.

Before addressing, in more detail, the specific actions of the
Community on membrane processes, we think it would be worthwhile to
let you know how the BRITE and the Radioactive Waste Management are
fitting within the overall Community's efforts in the field of
research and technology. For that, I give the floor to Mr. Finzi.

S.F. Thank you, Mr. Garcia Arroyo, Ladies, Gentlemen. Owing to the
growing importance attached by the Community to its action in the
field of research and technology, the Commission decided from 1984
onwards to coordinate all its R&D activities within a broad general
structure called framework programme.

In accordance with the research priorities which are bound to change
with time, this programme is revised and possibly re-elaborated
every five years. The present framework programme, which is the
second of its kind, spans the 1987-1991 period. This comprises R&D
works on eight main items.

The first : "quality of life", covers research in the fields of
health, radiation protection and the environment. The second deals
with information technology and telecommunications. Information
technology basically means the ESPRIT programme which probably most
of you know, at least by name. As you can see on this transparency,
about 40% of the Community fundings for research and technology are
allocated to these projects.

The third item of the framework programme concerns industrial
technologies in which the BRITE programme plays an important part.
Mr. van der Eijk will outline the main contents of this programme in
a few minutes. It must be pointed out that the BRITE programme will
be associated with the EURAM programme which focuses on materials
technology.

The fourth item relates to biological resources and covers the applications of biotechnology to agriculture and the agri-business.

Energy, the fifth item of the framework programme, is a field in which the Community has been involved since its inception. This item comprises research activities in the field of thermonuclear fusion, nuclear safety which includes actions on radioactive waste management and also non-nuclear energy. With respect to the radioactive waste management programme, Mr. Simon, replacing Mr Orlowski who is unable to attend due to other commitments, will shortly give an overview about its content.

Under the heading Science and Technology for development, the Community has set up a specialised programme to promote the use of science and technology to tackle Third World problems.

The seventh item of the framework programme is entitled marine resources. This is a new area which covers both research to increase the knowledge of the ocean environment and work on the exploitation of the sea-bed and agriculture.

Lastly, the eighth item, under the heading of European Scientific and Technical Cooperation, covers a number of programmes whose common purpose is to improve the liaison between the scientific institutions in all the countries of the Community and stimulate common initiatives.

For the period 1987-1991, the total funding of the framework programme amounts to approximately 5.4 milliards ECUS.

I want to add that the management of the different research programmes is assured by two general directorates : DG XII - Science, research and development, J.R.C. - and DG XIII - Telecommunications, information industries and innovation.

DG XII is in fact managing all the part of the tarte at the left of the drawing. This means that, in spite of the large difference between the aims and the scope of the individual programmes, a liaison between them is assured, resulting not only in a similar managing approach, but also in the attention which is put in finding occasions for a cross fertilisation between the programmes. This symposium is a relatively modest, but interesting example of this liaison.

G.A. Thank you, Mr. Finzi. Before giving the floor to Messrs. van der Eijk and Simon, who will outline how membrane processes fit in their respective programmes, Mr. Finzi and I would like to thank you for the interest you show in this symposium - we have been told that more than 150 participants have so far registered - and we hope that by the technical interest of the presentations as well as by the in-depth discussions on the different applications considered here, this symposium will bring tangible benefits in your future work.

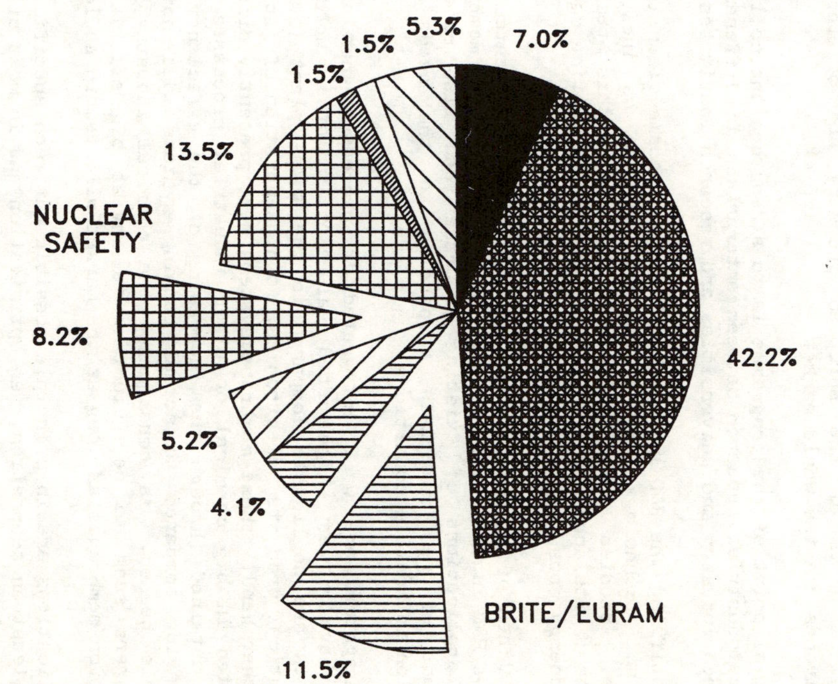

FRAMEWORK PROGRAMME (1987-1991)
7.0%
5.3%
1.5%
1.5%
13.5%
NUCLEAR SAFETY
8.2%
5.2%
4.1%
11.5%
BRITE/EURAM
42.2%
Quality of life
Information technol.
Industrial Technol.
Biological resources
Energy
S/T for Development
Marine resources
European S/T cooper.
Total funding: 5396 MECU

The importance of membrane processes within the BRITE programme

---

W. van der Eijk, J-C. Toussaint

**Commission of the European Communities
Directorate General Science, Research and Development
BRITE programme**

The BRITE programme of the Commission of the European Community has been developed in collaboration with industry in view of improving the technological basis of the European industry and of increasing its competitiveness on the world market.

An essential point of the programme is to strengthen the collaboration in strategic industrial research between enterprises in different countries and between industry and universities and research institutes.

When identifying the R&D needs of industry it became clear that membrane science and technology was an obvious area to be included in the programme. Studies showed at that time that this technology had 2 benefits : its specificity and efficiency in separating components and its low energy consumption.

The main lines of action of BRITE in this area are oriented towards the innovative R&D in the development of new membranes, new membrane systems and new applications of existing processes and also related to the development of new materials responding to the needs imposed by application in hostile conditions.

Membrane systems are well integrated in processes when the products produced have a very high added value or in other words when the investments are negligable compared to the total plant (for example : pharmaceutical products, electronics, food, separation of isotopes, etc). On the other hand, these membrane systems meet presently difficulties to be inserted in big chemical or other industrial processes because the benefits obtained through the increasing of the effeciency or through energy saving compared to actual running systems will not cover the investments needed. As consequence the chemical industry and polymer manufacturers seem to be reluctant to invest big efforts in R&D of polymers for membranes as long as the market will remain so low.

Many applications are in principle possible but each specific application needs at least an adaptation, an optimization and in most of the cases a specific development.

For the two calls for proposals published in the J.O. in 1985 and 1987, 39 proposals have been submitted regrouping 140 partners. From these 39 proposals, 8 have been selected regrouping 34 partners from eight different countries (Germany, Denmark, France, Italy, Ireland, the Netherlands, Spain, United Kingdom). The total budget is of 17.8 Mio Ecu of which 8.9 Mio Ecu are funded by E.C. and corresponding to about 5% of the total budget of the BRITE programme. 4 to 5 projects are running. The 3 others will start very soon.

The subdivision by subject is :

2 projects in Gas separation (purification of natural gas)
1 project in Pervaporation
1 project in food and water industry
2 projects in the treatment of effluents
1 project in the development of inorganic membrane
1 project in microfiltration

The subdivision of partners by type of organisations demonstrate the importance of the industry in this area :

18 industrial enterprises	53%	
7 universities	21%	
9 research centres	26%	

From the experience of the projects already running we could notice that the membrane processes are very nice processes but in many cases :

- the actual primary energy cost is so low that conventional separation systems are cheaper
- usual systems in place are in many cases already depreciated
- the investment costs are too high.

Nevertheless the introduction of pollution regulations will have as effect that in some cases only with membrane processes the aimed results could be attained. The use of membrane allows not only to reject friendly liquids for the environment but also to recuperate products and energy which can be reused in the process or in other applications.

These new trends will be a stimulus for continuing R&D in the field of membrane technology.

THE C.E.C. RESEARCH ACTIVITIES ON MEMBRANE PROCESSES DEALING WITH RADIOACTIVE WASTE MANAGEMENT

R. Simon and L. Cécille
Commission of the European Communities, Brussels

1. The C.E.C. research programme on radioactive waste management

Research and development on management of radioactive waste is one of the tasks, to which the Community is committed in the treaties of Rome in 1957. Since 1975, the Commission has carried out R&D programmes in the field of radioactive waste management and disposal on the basis of cost-sharing contracts. The current, third programme has a 5-year budget of 62 Million ECU. As this programme covers a wide range of research areas from waste characterisation to the safety analysis of disposal options, about 13% of the funds are devoted to the development of treatment processes and only 2% to the specific topic of membrane processes.(See Figure 1)

The overriding objective of our research programme is to assure the health and safety of the public in the management and disposal of the waste in an economically viable way.

For treatment and conditioning processes the following aims or performance requirements are derived from this general objective:

. Discharges of radionuclides to the human environment must be minimized. We are therefore developing processes with higher decontamination factors for the cleaning of liquid and gaseous effluents, taking due account of the different toxicity and life time of the respective nuclides.

. The volume of the wastes conditioned for disposal should be reduced. The present disposal philosophy relies upon a multiple barrier concept for the final containment of the nuclides. Any reduction of the waste volume can significantly contribute to the efficiency of these barriers. It will also facilitate the handling, transport and intermediate storage of the conditioned waste.

. The cost-effectiveness of certain management steps should be increased.

2. Treatment of radioactive liquid waste

The treatment processes traditionally applied to the purification of waste solutions are   - ion exchange
                          - chemical precipitation   and
                          - evaporation.

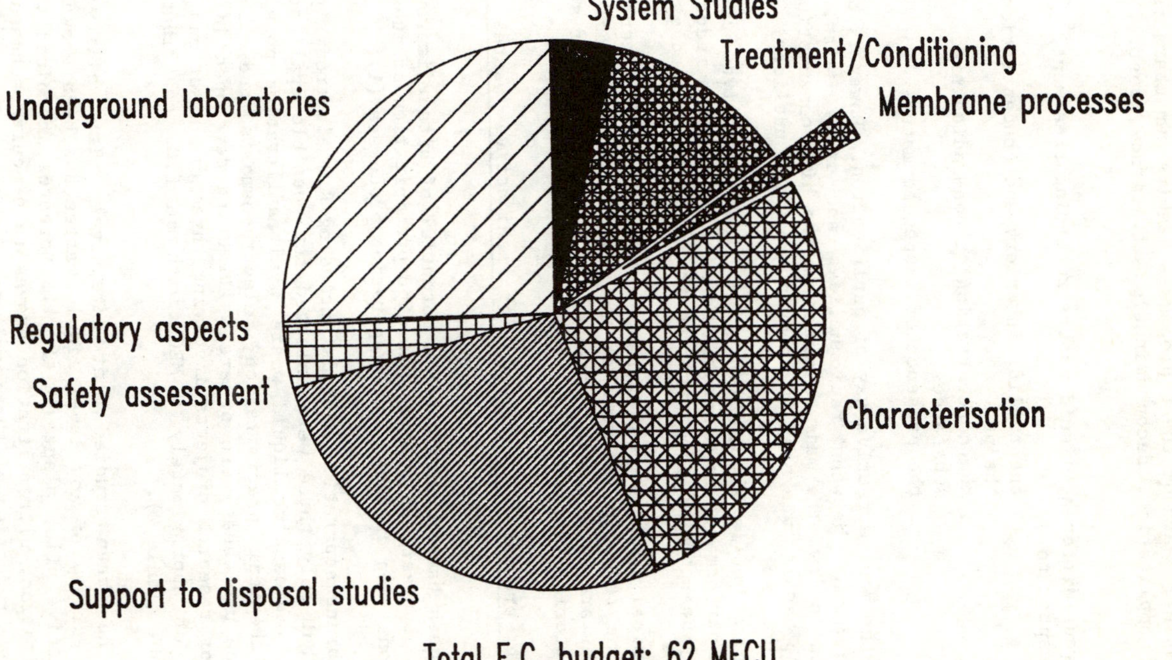

FIG 1: BREAKDOWN OF FUNDS BETWEEN THE DIFFERENT ITEMS INVESTIGATED WITHIN THE THIRD EC RESEARCH PROGRAMME ON RADIOACTIVE WASTE MANAGEMENT

(1985–1989)

As quoted in Table 1, each of these processes has its own characteristics (range of application, decontamination efficiency, cost...) and limitations.

There are numerous different liquid wastes requiring treatment. They vary in respect to

- the nuclide inventory and concentration
- the pH
- the concentration of non-radioactive substances
- the presence of suspended material.

Evaporation is the most effective way of purification. However, as it does not discriminate between radionuclides and inert salts, its volume reduction is low. The main drawback is the high consumption of energy, but also has its weakness in removing volatile nuclides.

Chemical precipitation is about 20-50 times less expensive but its decontamination performance is moderate. Satisfactory volume reduction factors can only be achieved by dewatering the precipitate sludges.

The third classical option, ion exchange, exhibits high decontamination and volume reduction factors at reasonable cost, but only if the radionuclides are under ionic forms and the concentration of salts and suspended material is low.

3. <u>The membrane processes developed for radioactive wastes and effluents</u>

Membrane techniques have a considerable potential of either improving the performance of treatment systems based on precipitation or substituting conventional ion exchange and evaporation processes (1).

A number of membrane separation methods including R.O electrodyalysis were explored, but only three were selected for possible application to treatment of radioactive liquid waste : ultrafiltration possibly combined with chemical precipitation, electro-osmosis and liquid membranes. These processes were or are still being developed in the framework of four research projects involving three organisations and a total budget of approximately 4 MECU with an E.C. contribution between 40 and 50% (Table 2).

Ultrafiltration has been studied most extensively. Its ability to decontaminate a variety of low level liquid wastes has been shown at Harwell in an active pilot plant (2, 3, 4). Moreover, combined with chemical precipitation, ultrafiltration proves to be quite successful in terms of volume reduction of the resulting sludges as demonstrated at Harwell and also to a lesser extent at Hanau for the processing of low level liquid wastes generated during the fabrication of uranium fuels (5).

TABLE 1 : Main features of conventional treatment processes for radioactive liquid waste

PROCESS	RANGE OF APPLICATION	Decontamination Factor DF	Volume reduction Factor VRF	COST	REMARKS
Ion-exchange	- Liquid wastes with low salt content - Radionuclides under ionic forms	$10 - 10^4$	$10^2 - 10^4$	Relatively expensive	Regeneration gives rise to secondary wastes
Chemical precipitation	Very wide	$10 - 10^3$	$10 - 10^2$	Cheap	VRF is highly dependent on the dewatering system applied to the chemic.precip.sludges
Evaporation	Very wide	$10^4 - 10^5$	Salt content in liquid waste is the limiting factor for the VRF	Expensive (energy consuming)	

TABLE 2 : __Membrane processes developed for treatment of radioactive liquid waste__
__within the E.C. research programmes on Radioactive Waste Management__

PROCESS	RANGE OF APPLICATION	ORGANISATION	DURATION	TOTAL COST (MECU)
Ultrafiltration possibly combined with chemical precipitation	Various reprocessing liquid wastes and Harwell waste	AERE – Harwell	1981 – 1985 1986 – 1989	2.9
	Liquid waste streams generated during uranium fuel fabrication	NUKEM – Hanau	1981 – 1983	0.2
Electro–osmosis	Concentration of chemical precipitation sludges	AERE – Harwell	1981 – 1985	0.4
Liquid membranes	Exhaustive decontamination of Marcoule type reprocessing concentrate	CEN – Cadarache	1987 – 1989	0.5

TOTAL   4.0

Another important application of UF is its use as a dewatering process for precipitation sludges, concentrating the precipitate to a water content which is suitable for cementation without having recourse to drying, centrifugation or evaporation (6).

Electro-osmosis is another method that has shown promise in the dewatering of sludges. First results obtained in Harwell indicate, that precipitation sludges with about 2-5% solids can be concentrated to 40% solids content using only a fraction (1.5-7%) of the energy needed to achieve the same performance by evaporation (7, 8).

At last, as a mean of improving selectivity in the ion-exchange type operation, the utilisation of the so-called liquid membrane techniques for the exhaustive decontamination of the Marcoule type reprocessing concentrate (9) (which is expected to decrease by a factor 40 the volumes of waste products to be disposed of in geological formations) is also investigated in the Community programme. By means of this technique, it becomes possible to use low quantities of very selective and expensive extractants like bidentates and crown-ethers. Work is in progress to demonstrate, at the laboratory-scale the process feasibility and performances on fully active samples of reprocessing concentrate.

4. Conclusions

For a number of low and medium level liquid waste streams, the implementation of membrane processes like ultrafiltration, electro-osmosis and liquid membranes in conjunction or as substitutes to conventional treatment processes was shown quite effective or at least promising to definitely improve current management practices both for decreasing still further the discharges of radioactive effluents into the environment and then to reduce the volumes of waste products to be disposed of in geological formations. Likewise for some specific applications, membrane processes could advantageously compete, in terms of cost, with energy consuming processes like evaporation.

Obviously, on account of the current stage of development of some of these novel processes, important R & D efforts are still needed to demonstrate, at a significant scale, their good prospects for subsequent industrial application in the nuclear field.

Another important finding of these research projects lies on the fact that most of the problems encountered in the setting-up of membrane processes for the treatment of radioactive liquid waste were not so much linked to the presence of radionuclides in the waste streams as to the limitations of the membrane processes themselves (membrane fouling, pressure build-up, wear of the equipment, circuit...). Therefore, it is expected that the new technologies developed to cope with are equally valuable for the non-nuclear applications and conversely.

REFERENCES

(1) K.W. CARLEY-MACAULY et al. "Radioactive waste advanced management methods for medium active liquid waste", EUR 7037 Harwood academic publishers (1981)

(2) R.G. GUTMAN et al. "Active liquid treatment by a combination of precipitation and membrane processes", EUR 10822 EN (1986)

(3) I.W. CUMMING and A.D. TURNER "Optimisation of an UF pilot plant for treatment of radioactive waste" (This symposium)

(4) I.W. CUMMING and A.D. TURNER "Recent developments in testing new membrane systems at AERE-Harwell for nuclear applications" (This symposium)

(5) H.M. MULLER "Dekontamination durch ultrafiltration von Schwach-aktiven abwässern aus der brennelementfertigung", EUR 9281 DE (1984)

(6) M. HOWDEN and T.L.J. MOULDING "Progress in the reduction of liquid radioactive discharges from the Sellafield site", International conference on nuclear fuel reprocessing and waste management "RECOD 87" (Paris)

(7) A.D. TURNER et al. "Electrical processes for the treatment of medium-active liquid wastes", EUR 10565 EN (1986)

(8) A.D. TURNER "Application of electro-osmosis to radioactive waste processing" (This symposium)

(9) J.F. DOZOL "Use of liquid membranes for treatment of nuclear effluents" (This symposium)

FUNDAMENTALS OF MEMBRANE SYSTEMS AND ENGINEERING ASPECTS

Review of nuclear and non-nuclear applications of membrane processes - present problems and future R&D work

Production and characterisation of membranes

Economical evaluation of the membrane technology

Studies on transport phenomena in hollow fibre membranes by means of dimensionless groups

REVIEW OF NUCLEAR AND NON-NUCLEAR APPLICATIONS OF MEMBRANE PROCESSES -
PRESENT PROBLEMS AND FUTURE R & D WORK

R.G.Gutman, Pall Europe, Portsmouth, U.K.
R.H.Knibbs, UKAEA, Harwell Laboratory, U.K.

Summary

This paper describes those membrane processes that are of industrial
significance in the fluid phase separations. The review covers
pressure driven, cross-flow processes (reverse osmosis,
ultrafiltration and microfiltration) and electrically driven membrane
processes (electro-dialysis and electro-osmosis). A brief
description of the mechanism of each of the different types of
membrane process is given. The most common types of module design,
spiral wound, hollow fibre and tubular are illustrated and compared
and the operating limitations of temperature, pressure and pH are
discussed. A review of membrane processes already finding large
scale industrial applications is given and the paper concludes with a
brief discussion of possible avenues of future R and D that might
help to alleviate the problems of concentration polarisation and
fouling of membranes.

1.   INTRODUCTION
     A membrane has been defined by the European Society of Membrane
Science and Technology as "an intervening phase separating two phases
and/or acting as an active or passive barrier to the transport of matter
between phases." This broad definition covers a wide variety of different
types of membrane, and hence a whole host of different separation
processes. In this paper, attention will be focussed more narrowly, on a
number of membrane processes that are used for the industrial scale fluid
phase separations. The membrane processes that are considered are the
cross-flow processes, in which the driving force for transport across the
membrane is pressure (i.e. reverse osmosis, ultrafiltration and
microfiltration) and electrically driven processes.

2.   PROCESSES
2.1 Reverse Osmosis
     There is a superficial similarity between the three pressure driven
membrane processes, which all involve the permeation of solvent (usually
water) through a "semi-permeable" membrane, and the rejection of a second
suspended or dissolved component. The term reverse osmosis is used to
describe membranes and processes in which the separation is of solvent
from solutes with similar molecular dimensions (within say one order of
magnitude of the solvent). Reverse osmosis membranes are capable of
removing from aqueous solution ionic solutes and also low molecular weight
organic compounds.
     Separation is understood to take place by a solution-diffusion
mechanism. Both the solute and solvent are thought to dissolve in the

dense membrane phase, and then pass through it by diffusion. In the case of many of the more successful reverse osmosis membranes, their high bound water content both assists the transport of water across the membrane and also tends to exclude ions from the membrane phase. In fact the best reverse osmosis membranes now available are capable of better than 99% removal of salts like sodium chloride, and of better than 97% removal of small organic compounds like ethanol.

Process performance is, however, strongly influenced by the solution's osmotic pressure, which reduces the effective driving force across the membrane. The effective upper pressure limit for reverse osmosis is currently about 70 bar, and this restricts the process to the treatment of dilute saline solutions (e.g. up to about 5% NaCl solutions). Concentration polarisation, that is the tendency of rejected solutes to accumulate at or near the membrane surface, also has an important effect on reverse osmosis performance, since it reduces membrane flux and removal efficiency. Polarisation can be controlled by the use of cross-flow, which is the direction of the feedstream tangentially over the surface of the membrane. Cross-flow provides an effective mechanism for the transport of accumulated solutes away from the surface of the membrane, and is utilised in virtually all commercial reverse osmosis devices.

Four different types of device have become established industrially over the last twenty years. In the desalination industry, hollow fibre and spiral wound equipment predominates. Hollow fibre modules, like that shown in Figure 1, contain many thousands of individual hollow fibre membranes, often less than 100μm in outside diameter, potted into header blocks and then mounted in a pressure vessel. The fibres are externally pressurised, and operate in laminar cross-flow. The high packing density leads to hollow fibre modules having the best membrane area/volume ratio (up to ~ 8000m^{-1}) of all systems, but this compactness results in sensitivity to feed channel plugging.

Spiral wound modules achieve an area/volume ratio of only one-tenth that of hollow fibre modules, but they nevertheless often achieve comparable performance in desalination applications. A typical spiral module design is shown in Figure 2. The module is constructed from flat sheets of membrane that are coiled up together with retentate and permeate drainage layers around a central permeate mandrel. Spiral modules normally operate in laminar flow, with a degree of turbulence induced by the flow disrupting effect of the retentate drainage mesh.

Because of their compactness, both hollow fibre and spiral wound plants have to be equipped with extensive pretreatment to reduce the risk of plugging. The systems therefore do not lend themselves to fouling applications, for which tubular or flat plate systems tend to be used. Tubular modules, like the design shown in Figure 3, generally operate in highly turbulent flow and have wide, unobstructed feed channels. Flat plate systems, like that shown in Figure 4, generally have narrower channels – often containing a mesh separator – and do not always operate in turbulent flow. However, both types of system have performed well in difficult applications, such as the processing of high viscosity materials in the food and dairy industry and where "clean-in-place" techniques are applied.

Although some success has been achieved in the development of mineral membrane systems, such as glass membranes and dynamic hydrous zirconium oxide based membranes, most reverse osmosis membranes are plastic and are made from a relatively narrow range of polymers. In many cases the

Figure 1. End view of Amicon hollow fibre ultrafiltration module.
(Courtesy of Amicon)

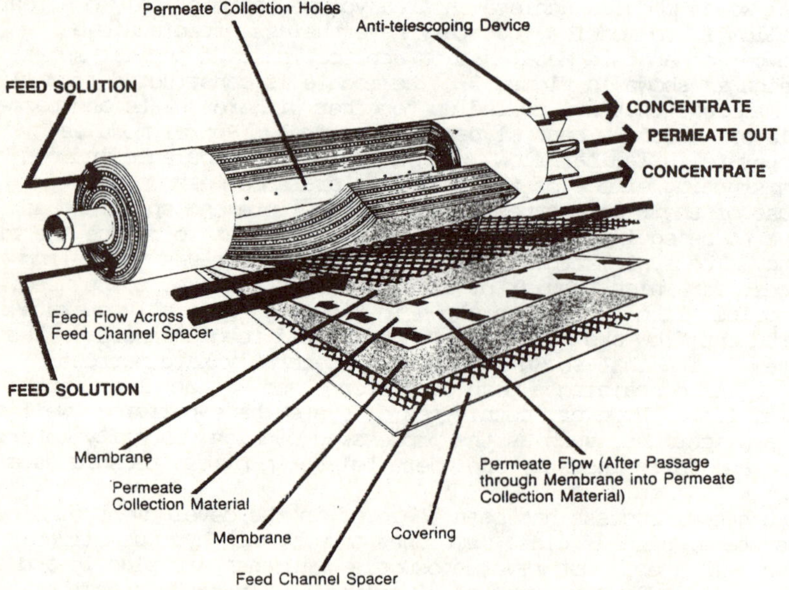

Figure 2. Construction of a spiral-wound module (Courtesy of Kock
Membrane Systems Inc)

<u>B1 R.O. MODULE ASSEMBLY</u>

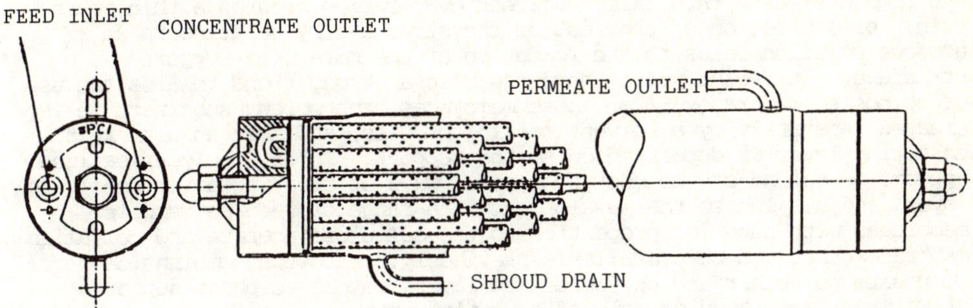

Figure 3. Tubular reverse osmosis module (Courtesy of PCI)

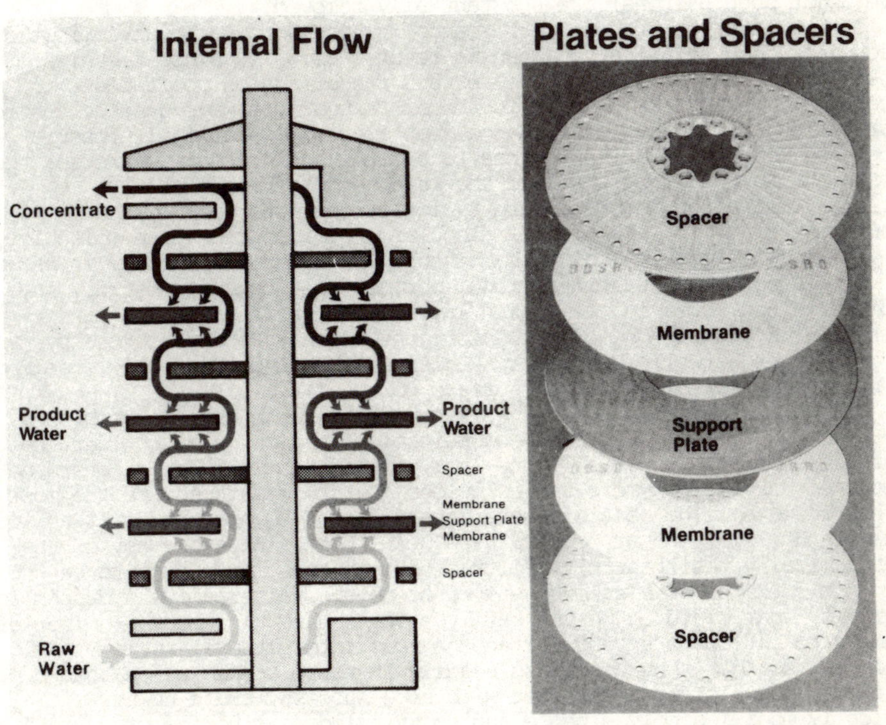

Figure 4. DDS Series reverse osmosis module (Courtesy of DDS)

plastic membrane has an asymmetric structure, and is produced by a phase inversion process, in which the polymer is dissolved in organic solvents and then cast as a thin film. Solvent evaporation creates a thin dense skin; precipitation of the rest of the structure by gelation in an aqueous solution leads to the formation of the more open support structure. In recent years there has been a strong trend towards the use of composite membranes. For these membranes, the porous substructure is created separately by a solvent casting process, and then an ultrathin desalting layer is deposited on to the support. In-situ polymerisation appears to be one of the most successful ways of creating a top desalting layer, and has led to the development of a range of reverse osmosis membranes with superior properties to the cellulose acetate and polyamide membranes that had been available previously. With the combination of composite polyether and polyamide top layers on polysulphone support structures, membranes capable of operation at pH 3-11 and temperatures up to 50°C have been produced for the first time.

2.2 Ultrafiltration

The term ultrafiltration is used to describe the membrane separation of macromolecular species from solution. Dissolved salts and low molecular weight organic compounds permeate freely through ultrafiltration membranes, but large molecular weight compounds (with a molecular weight above about 5000 Daltons) are substantially removed.

Ultrafiltration process performance is even more seriously affected by concentration polarisation than is reverse osmosis, since the size of species to be removed is much greater and its diffusion coefficient is consequently lower. Cross-flow is therefore nearly always used to control polarisation and improve performance, but even so operational pressure (and hence membrane flux) often has to be limited in order to prevent the membrane being coated by a gel of the rejected material.

Ultrafiltration membranes are typically characterised by the measurement of their rejection of globular proteins or dextran molecules. From the known dimensions of such molecules in aqueous solution, it has been concluded that ultrafiltration membranes, although asymmetric, have nanometer sized pores in their skin layers, and their mechanism of filtration is usually one of mechanical retention. Recent microscopy, for example at Harwell, has partly confirmed this hypothesis by revealing the presence of nanometer sized surface pores (1).

Most ultrafiltration membranes are, like reverse osmosis membranes, made from plastics but the range of suitable plastics is much wider than for reverse osmosis. Polysulphone, polyvinylidene fluoride, regenerated cellulose, acrylic polymers and cellulose acetate membranes are the most common products. The best of these materials can be operated over a wide range of pH (1-13) and at temperatures up to 80°C. Because many of the more chemically resistant ultrafiltration membranes were found to be hydrophobic there has been much effort to develop alternative hydrophilic membranes, that could be wetted easily. There have also been a number of attempts to introduce strong surface charges into ultrafiltration membranes, so that they foul less readily in the presence of charged colloids. These efforts have met with some success, and a number of chemically modified polysulphone and polyvinylidene fluoride products are now available.

Despite these recent advances in polymer membranes, further improvements are still being sought. Current plastic ultrafiltration membranes have, for example, very poor resistance to steam sterilization and limited compatability with organic solvents. Some of these problems can be overcome by the use of ceramic membranes, although these have their own disadvantages, particularly their sensitivity to mechanical shock. At the moment composite zirconia/carbon and composite γ/α alumina membrane systems are available. Both systems are tubular, and in the former case the membrane can be operated at over 140°C and at pH's between 0-14. Ceramic membranes are also capable of operation at higher pressures than most plastic systems, and they can therefore be operated at very high cross-flows (and hence pressure drops). In some applications this operation at high cross-flow has been reported as giving higher membrane flux.

Ultrafiltration modules are in many cases similar in design to their reverse osmosis counterparts. The greatest difference is in the design of hollow fibre modules, where, because of the much lower operational pressures, ultrafiltration systems utilise internally pressurised, larger bore fibres. This gives better hydrodynamics, and enables turbulent flow to be achieved in some cases.

2.3 Microfiltration

The term microfiltration is generally understood to describe the separation of sub-micron or micron sized particles from a feed suspension. SEM examination of microfiltration membranes shows that they have pores with dimenstions in the micron range, and indicates that the principal removal mechanism is mechanical retention. In many cases, however, removal by adsorptive mechanisms is also very significant, and under such conditions microfilters may remove species (e.g. proteins and pyrogens) well below the membrane's rating.

The major applications of microfiltration are for the clarification of lightly contaminated fluid streams, and for such applications disposable filters are successfully used. Such microfiltration applications, which constitute a major industry in themselves, fall outside the scope of the present paper, which has been limited to cross-flow devices. Cross-flow microfiltration technology is still in an embryonic stage, but there do appear to be some applications particularly in th separation of streams with high solids loading, where the use of cross-flow is advantageous.

Most microfiltration membranes are polymeric. Many of them have a microporous structure and are made by a phase inversion process; other microfiltration media are fibrous in construction. Microfilters are available with ratings from 0.1μm to greater than 10μm, and in a range of polymers including nylon, polypropylene, polyester, fluorinated polymers, polysulphone, polyacrylonitrile, regenerated cellulose and polyvinyl chloride. Modification of membrane surfaces to influence wetting characteristics and surface charge, as described earlier for ultrafiltration membranes, is practised to an even greater degree with microfilters.

A number of plastic membranes have been produced by more unusual routes. For example, PTFE and polypropylene membranes have been produced by stretching of dense films, and other polypropylene membranes have been produced by a thermally induced phase inversion process. Track etched membranes, which have extremely uniform, cylindrical pores, have been produced by a process in which a dense polymer film is irradiated with alpha particles or neutrons, and the damaged regions of polymer are then etched away to form the cylindrical pores (Figure 5).

The physical and chemical resistance of many plastic microfilters is

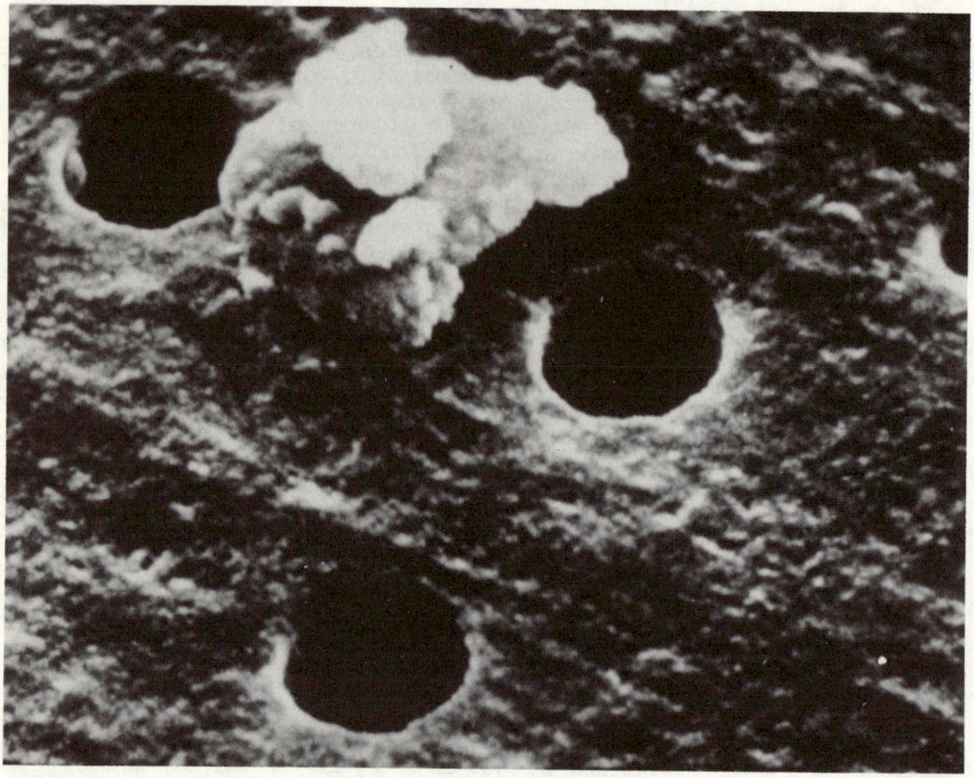

Figure 5. Scanning electron micrograph of the surface of a Nuclepore
membrane. (Courtesy of Nuclepore)

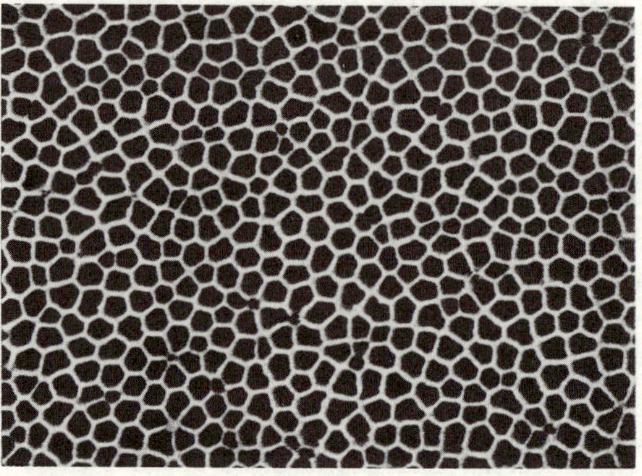

Figure 6. View of an anopore membrane

very good; some can be steam sterilized repeatedly. A range of ceramic microfilters (zirconia/carbon composites and alumina) is now available. One of the most recent ceramic developments is an alumina membrane with a regular "honeycomb" structure (Figure 6), that is produced by an electrolytic process. There has also been much development of metal media in recent years, and with the introduction, for example, of the Pall porous metal membrane, stainless steel microfilters with 99% removal efficiency at $0.5\mu m$ are now available.

Because cross-flow microfiltration is a relatively new concept, designs of cross-flow module are still far from optimal. Systems derived from flat plate, tubular and hollow fibre ultrafiltration modules are available, together with some novel microfilter systems, such as a rotating device developed by Sulzer.

2.4 Electrical Processes

Electrical driving forces can be used advantageously in a number of ways to assist or create a membrane process. The most successful development to date has been electrodialysis, in which ionic solutes are driven across ion selective membranes. By applying a large potential difference across an electrodialysis stack (an array of alternate cation and anion permeable membranes), salts can be concentrated and depleted in alternate chambers of the stack. Electrodialysis is used extensively for the desalination of brackish waters, and also for the concentation of sea water, and the dewatering of protein solutions.

In electro-osmosis a potential difference is applied across a membrane and a mobile layer of water and associated ions (with opposite charge to that of the membrane surface) are dragged through the membrane. Dewatering of suspensions can be achieved by such a technique, particularly where the conductivity of the liquid is high.

It is also possible, by the application of a potential difference, to attract colloidal fouling species away from the surface of a membrane, and also to clean a membrane by the generation of electrolysis gases within the membranes pores. Both these novel techniques have been the subject of considerable development at Harwell, and will be described in more detail in other presentations at this meeting.

3.   APPLICATIONS

A number of industries have found large scale applications for membrane processes. In the time available it is only possible to give a brief reveiw of the major applications. It is convenient to start with water treatment by reverse osmosis.

3.1 Potable Water from Brackish Water Sources

The treatment of brackish borehole waters or river waters with salinities up to $10,000g/m^3$, and osmotic pressures of ~ 7 bar, have been the main target for reverse osmosis. It is estimated that the world capacity of reverse osmosis plant producing potable water from brackish water is now over 3 million m^3 per day and it is an expanding market. Membrane life times for such plant are usually over 3 years and costs of water production compare favourably with alternative techniques such as evaporation, electrodyalisis and ion exchange. Table 1 gives an indication of costs for brackish water desalting (1). The costs of electrodyalisis are comparable on feeds with low salinities but become less competitive at salinities over $5000g/m^3$. Cellulose acetate was the first material used commercially for reverse osmosis applications and modules were produced in flat sheet, spirally wound and tubular form.

<u>Table I</u>

COSTS FOR BRACKISH WATER DESALTING

Source	Plant		Capital Cost £/m³/d	Operating Cost £/m³
	Type	Size m³/d		
Applegate (7) (1984)	R.O.	3,800	226	0.29
Applegate (7) (1984)	R.O.	95,000	155	0.21
Wojcik (8) (1983)	R.O.	~ 40,000	180	−
Porter (9) (1979)	R.O.	1,900	90–100	0.08–0.12
Lloyd et al (10) (1978)	R.O.	100	−	0.08
Gluekstern (11) (1984)	R.O.	−	150–190	0.23
Crossley (12) (1983	R.O.	240	300	−
Crossley (12) (1983	R.O.	24,000	100	−
Caracciolo (13) (1977)	R.O.	1,900	100	0.12
Applegate (7) (1984)	E.D.	95,000	209–276	0.19–0.26
Gluekstern (11) (1984)	E.D.	−	190–320	0.23–0.40

<u>Table II</u>

COSTS FOR SEA-WATER DESALTING

Source	Plant		Capital Cost £/m³/d	Operating Cost £/m³
	Type	Size m³/d		
Applegate (7) (1984)	R.O.	400	1,600	1.71
Applegate (7) (1984)	R.O.	19,000	1,000	−
Ericsson & Hallmas (14) (1985)	R.O.	19,000	1,300	1.41
Wade and Hornsby (15) (1982)	R.O.	12,000	−	1.01*
Ericsson & Hallmas (14) (1985)	M S.F.	−	−	1.5–3.3

*With energy recovery.

However, it has limited chemical resistance and hydrolysis causing loss of acetyl groups is a major cause of failure due to a reduction in desalting efficiency. It also has pressure limitations in that it compacts and suffers subsequent reduction in flux. These limitations made cellulose acetate membranes unsuitable for desalinating sea water where pressures of up to 60 bar are required to overcome the osmotic pressure and give a reasonable flux level. By the early seventies Dupont were marketing B9, RO modules containing aromatic polyamide hollow fibres. These polyamide membranes gave salt rejections similar to the cellulose acetate membrane. They had a lower flux but this was compensated for by the ability to obtain a very large surface area per unit volume. Unfortunately the B9 membranes have little tolerance to free chlorine and in addition were not suitable for desalting sea water.

3.2 Potable Water from Sea Water

The treatment of sea water with salinities of 35,000 to 60,000g/m³, and osmotic pressures of ~ 40 bar, was not feasible on a commercial scale with the early membranes. During the seventies a range of new membranes came on to the market. The B10 polyamide membrane could be used for treating sea water, but more recently composite RO membranes have been introduced. UOP produced the PA300 membrane which had a polysulphone support layer and a polyamide desalting layer, this membrane was more chemically resistant and could operate up to 55 bar. Film-Tech then marketed the FT30 membrane which is reported to be a cross-linked polyamide composite membrane and Toray produced the PEC1000 composite membrane which was also suitable for sea water desalting. It is estimated that world capacity of sea water RO plant producing potable water is now about 0.3 million m³ per day, and growing rapidly. Costs of water production are now comparable with multiple stage flash distillation and Table 2 gives some cost data for sea water desalting.

3.3 Ultra Pure Water

Large volumes of silica-free water are used in power generation for feeds to steam turbines, the electronics industry also use large volumes of ultra pure water for washing printed circuits and then there are the medical and pharmaceutical uses. Ultrafiltration is used upstream of reverse osmosis plants to remove suspended solids in many applications but the major ultrafiltration applications are in the food industry.

3.4 Food and Dairy Industries

Reverse osmosis equipment used in the food and dairy industries is of hygenic design and costs are of the same order as sea water capital and operating costs i.e. ~ £1000/m³ per day capital costs and £1-1.4/m³ operating costs. The plants are much smaller but the high value of the product make it commercially attractive to use RO. Whey processing from cheese manufacture is a major application. Some protein remains in the whey, by using RO the solids in the whey can be increased four-fold and the permeate, having a low BOD, can be discharged to the drain. The retentate is converted to a powder by evaporation and drying and is then used as additives to various alimentary products (2). RO has also found an application in concentrating tomato juice without imparing flavour. The fibrous nature of the tomato juice is such that flux decline is relatively small. Other fruit juices can be concentrated by RO but flux decline is greater and the rate of expansion of these applications is much slower. Egg white can also be concentrated by reverse osmosis but ultrafiltration gives considerably higher flux rates. Over 1 million tons per year of egg white is produced by the Western nations and about 60% of

this is processed to a concentrated form. The major process problems
are:

    (a) "Browning" caused by the reaction of glucose and amino acids
    resulting in a decrease in protein solubility, and
    (b) the denaturation of protein.

Ultrafiltration can overcome both these difficulties in that it can remove
glucose while concentrating the proteins; is an ambient temperature
process and hence does not denature the protein; it also provides less
shear with higher flows than an RO system and hence does not affect the
foam stability properties of the reconstituted product. (3) UF and RO are
also used to obtain vegetable extracts from soya bean whey, wheat starch
factories and to remove the protein from cotton seed.

3.5 Effluent Treatment

3.5.1 Nuclear Industry

Membrane processes are gradually becoming common in the treatment of
nuclear wastes. Due to their limited radiation stability RO membranes can
only be used for low and intermediate level waste treatment. It has been
shown that RO membranes can remove trace elements if the correct membrane
is chosen. Experiments on simulated waste have shown that a 97% salt
rejection membrane (based on sodium chloride) removed 97.5% of ^{60}Co (4).
Laundry waste is difficult to treat by evaporation because of foaming. As
well as removing activity, the removal of phosphates and the biological
oxygen demand may be important if discharge is to be to river or lake. If
laundry waste is treated using reverse osmosis the permeate can be
discharged without foam treatment and the concentrate can be drummed. It
is possible to concentrate up to 5% wt solids. The degree of
concentration is limited by the fouling of the membranes. This occurs
once the concentration of dissolved salts reaches saturation and begins to
precipitate out on the membrane.

The first RO plant to treat laundry waste at the Ginna Nuclear plant
(5), comprised 18 modules and handled 11m³/day of laundry and hot shower
effluent. Inlet water to the laundry was pre-treated to minimise fouling
of the membranes, both tubular and sprial wound membranes were used.

The removal of radioactive species by co-precipitation followed by
ultrafiltration has also been used to substantially reduce effluent
discharge limits and work on this topic will be reported to this
conference. British Nuclear Fuels is currently building a large-scale
plant for the removal of activity from a range of wastes containing ferric
hydroxide. In this case UF is used to reduce the volume of waste by a
factor of about 200 whilst retaining the majority of the α-activity and
substantially reducing βγ activity (6). Inorganic membranes have been
used for these dewatering type applications and the plant usually consists
of a two-stage process, as shown in Figure 7. Plants of this type are
operating under fully automatic control at Harwell using Carbosep UF
membranes which consist of a carbon tube internally coated with a zirconia
membrane and have a molecular weight cut off of 20,000. Figure 8 shows
such plants and Figure 9 shows a secondary dewatering plant containing a
membrane area of 11m³.

3.5.2 Non-nuclear Applications

Water borne Electro-paints - The use of UF in electrophoretic
painting has grown rapidly during the last 15 years and it is now standard
practise to incorporate UF in all electro-painting systems. Most
electro-paints are aqueous solutions of resin to which pigment and organic
solvents have been added. The solutions contain 10-15% solids. These
paints are used for priming bare metal which is dipped into the paint bath

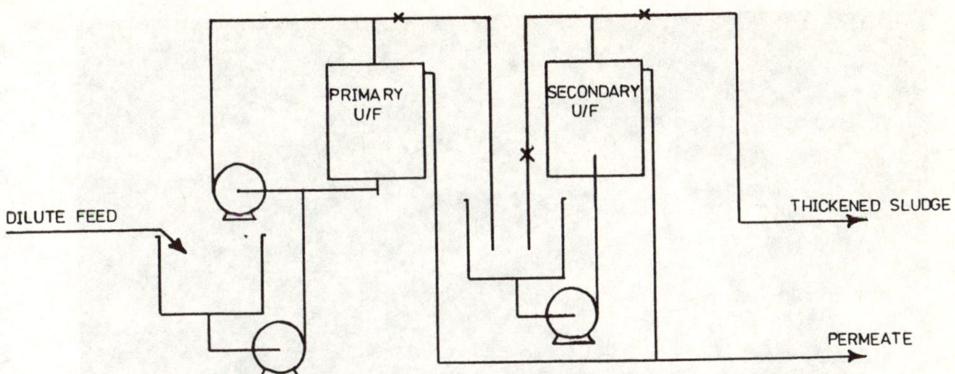

Figure 7. Schematic representation of a two-stage ultrafiltration -
dewatering plant

Figure 8. Fully automated, two stage ultrafiltration - dewatering plant

FIGURE 9. Secondary ultrafiltration – dewatering plant containing 11m^2 membrane area.

and forms the cathode while the bath forms the anode. Positively charged
UF membranes which repel paint particles are used and membrane fluxes of
about 1m/day can be maintained for many months.

<u>Machining and Cutting Oils</u> - Oil/water emulsions are used as
lubricants in machining operations. Considerable volumes of waste are
generated due to contamination by solids, loss of lubricating properties
and bacterial growth. In most cases disposal is by incineration;
however, if the wastes are untreated they are of high volume and negative
heating value and hence pose disposal problems. Ultrafiltration can
reduce the cost of disposal by concentrating the oils, grease and dirt and
at the same time providing an effluent with a positive heating value.

<u>Pulp and Paper Industries</u> - UF has been used on spent sulphite
liquors, to separate wood sugars from lignosulphonates. Monosaccharides
have a molecular weight of about 180 whereas lignin or lignosulphonic
acids have a molecular weight of 18,000. The concentrate, purified
lignosulphonic acids, can be used for a number of organic syntheses and
the permeate contains xylose and furfural. The natural temperature of the
waste liquor is about 80°C and has a pH of 2.5 so that cellulose acetate
membranes are not suitable, however polysulphone membranes have been
successfully used.

3.6 Applications in Biotechnology

Membranes are now used in biotechnology both for the primary
separation of cells from fermenter broths and also for product recovery.
At least one large-scale ultrafiltration plant has been built to separate
a filamentous organism (Nocardia sp) from an antibiotic product. In
comparison with the traditional rotary drum vacuum filtration process, the
membrane route has enabled higher product recovery to be achieved, and has
eliminated the problems associated with disposal of filter aid. Further
downstream, the concentration of antibiotic solutions by reverse osmosis
has also found favour over conventional techniques (vacuum evaporation),
since the energy requirements of the membrane process, which involves no
change of phase, are much less.

Currently, the major use of membrane processes is for the
concentration/purification of enzymes. The advantages of ultrafiltration
compared to vacuum evaporation for such applications are considerable,
since concentration can be followed by a diafiltration step, during which
dissolved salts and low molecular weight products can be washed out prior
to final drying of the enzyme.

Much attention is also being focussed on the development of various
types of membrane reactor. In many cases the membrane process has been
integrated with an enzyme reactor or fermenter, so that broth can be
recirculated through the separator with product being withdrawn and the
retentate returned to the reactor. In other applications, attempts have
been made to immobilise enzymes or whole cells onto a membrane, so that
the membrane itself becomes the site of the reaction process.

3.7 Other Applications Under Development

Feasibility studies have been carried out on the dewatering of
dyestuff, recovery of insoluble catalyst metal powders, recovery of
valuable soluble products and a wide range of other applications too
numerous to mention here, except to say that the potential for membrane
processes is still growing rapidly.

4.   CONCLUSIONS

Cross-flow membrane processes are already used extensively in a variety of non-nuclear industries, and are now finding applications in the nuclear industry also. More widespread use of the technology would almost certainly result if some of the current process problems could be alleviated. In particular there is a requirement for new equipment designs and improved processing techniques, so that actual membrane performance can be brought closer to the level that should be theoretically achievable. In many applications, concentration polarisation and fouling are a serious limitation to the performance that can be achieved. The use of Taylor vortices to reduce fouling, e.g. in pulsatile flow or rotating devices, is one particularly promising approach to this problem.

There is also a continuing need for improved membranes, for example reverse osmosis membranes that can be steam sterilized or operated over a wider pH range than at present. In some cases alleviation of the fouling problems described above will require the development of special "antifouling" membranes. In many ultrafiltration applications, for example, protein and antifoam agent fouling can be severe, and improved, hydrophilic ultrafiltration membranes may be needed to prevent such problems.

REFERENCES

(1) GUTMAN, R.G., (1987). Membrane Filtration. Adam Hilger.
(2) LANG, F. and LANG, A., (1976). The Milk Industry 78, 9, 16.
(3) PORTER, M.C. and MICHAELS, A.S., (1971). Chem.Tech. April, 248.
(4) WILLIAMS, M.K. et al. (1982) MLM-2899.
(5) MAMET, V.A. et al, (1978). Taploenergetika (1978).
(6) HOWDEN, M., (1988). Progress in the Production of Radioactive Discharges from Sellafield. BNFL plc publication.
(7) APPLEGATE, L.E., (1984). Chem.Eng. 91, 64.
(8) WOJCIK, C.K., (1983). Desalination 46, 17.
(9) PORTER, M.C., (1972). Ind.Eng.Chem.Prod.Res.Dev. 11, 234.
(10) LLOYD, A.I., THEODOSIOU, S.A. and KNIBBS, R.H., (1978). Proc. 6th Int.Symp. Fresh Water from the Sea 3, 327.
(11) GLUEKSTERN, P. and ARAS, N., (1984). In Synthetic Membrane Processes, ed. G. BELFORT (New York: Academia) ch.12.
(12) CROSSLEY, I.A., (1983). In "Desalination Technology – Development and Practice" ed. A.PORTEOUS (New York: Applied Science).
(13) CARACCIOLO, V.P. et al. (1977). In "Reverse Osmosis and Synthetic Membranes" ed. S. Sourirajan (Canada: NRC) ch.16.
(14) ERICSSON, B. and HALLMAS, B., (1985). Desalination 55, 441.
(15) WADE, N.M. and HORNSBY, M.R., (1982). Desalination 40, 245.

PRODUCTION AND CHARACTERISATION OF MEMBRANES

A.RENARD - G. CUEILLE - B. MIRABEL
TECH-SEP
Rue Penberton - Saint-Maurice de Beynost -
BP 347 - 01703 MIRIBEL CEDEX - FRANCE -

Summary

Membrane separation processes are emerging as highly relevant unit operation in various industries. Developments of well adapted membranes came out of the application of physical chemistry, surface and polymer science. Modern methods for textural analysis especially electron microscopy, showed that membranes are made of a very thin barrier supported by a less dense substrate which is of little importance in the mass transfert properties of the membrane. Such an asymmetric structure can be obtained by adequate treatment of polymer solutions or by special methods of sintering in the case of mineral membranes. The present efforts are aimed at improving selectivity and stability as well as higher operational fluxes. These qualities are required by the development of sophisticated uses of membranes, with stringent technical and economic conditions, particularly in the field of food and pharmaceutical industries. The manufacture of membranes is now based on a mixture of science and practical experience closely related to results obtained in applications. Much effort is also being devoted to the development of characterization methods.
This effort has to be continued in conjunction with users and process designers in order to improve the actual performances of membranes.

1.INTRODUCTION

Membranes are not sold by manufacturers as ordinary chemicals because what the customer buy is a level of performance in order to achieve a particular task.

This is the reason why the manufacture of membranes is an intricate mix of pure science (polymer science, physical chemistry...) and pratical experience acquired by processing the often complex products of customers.

The manufacturer has to improve constantly his methods of production and develop new ones in order to answer the demand of the market. He has also to develop proper characterization methods for several reasons :

to choose the best available membrane for a particular application,
to study the effects of changing a parameter during membrane manufacturing,
to get a range of significant methods of quality control.

More and more sophisticated methods are now in use for the characterization of membranes (1). But the intrinsic properties of membranes, as finely known as the actual technology permits, dot not allow the engineer to get the best results when designing or operating a membrane system. Some characteristics (water flux, cut-off etc. ...) are indicated by manufacturers but since they use different conditions, this is of limited help for the user. Moreover, methods such as cut-off determination are established with model compounds (PEG, dextrans,

proteins, etc. ...).
    The results are hardly correlated with retention of components in complex feed-stocks where polarization and fouling can change the retention considerably.
    Membrane characterization, always based on more or less realistic assumptions, can be separated in several categories of methods :
. Pore size distribution by physical methods and subsequent tests such water and air flux determination.
. Separation characteristics with model compounds.
. Separation characteristics with reconstituted or real feed-stocks.

    The interactions between manufacture and characterization are summarized in the following diagram.

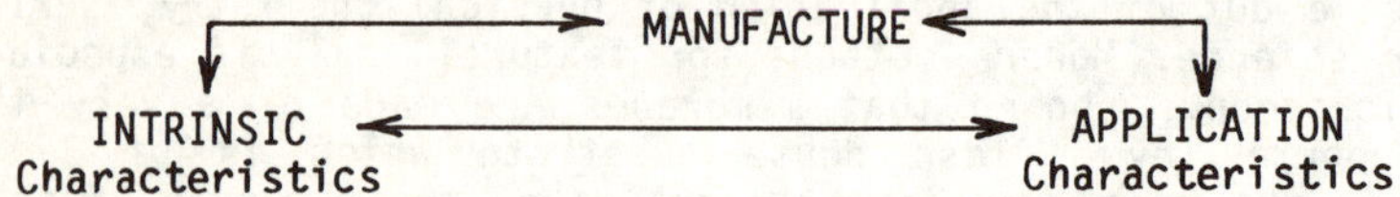

INTRINSIC Characteristics	APPLICATION Characteristics
. Chemical composition	. Selectivity for the desired compounds with real feed-stocks
. Pore distribution	
. Surface properties	. Flux of product
. Electrical charge	. Thermal and chemical stability
. Texture (multilayer, asymmetry)	. Resistance to fouling
. Retention of model compounds	. Ability for cleaning
. Air flux	

    The purpose of this paper is to give an introduction to the methods currently in use for membrane manufacturing and characterization. From two practical examples, it will be shown how the manufacturer can meet the requirements of various industries by a proper combination of field tests and laboratory methods.

2.MANUFACTURE OF MEMBRANES , AN OVERVIEW.

    Membranes with different physical structures can be used for reverse osmosis, ultrafiltration and microfiltration. The most important is the asymmetric type in which the pores are much larger on one side than on the other. The bulk of the membrane acts essentially as a support for the ultrathin filtering layer. Polymeric membranes are produced by dissolution of a polymer (or blends) in a solvent and precipitation by a non-solvent in the procees known as "phase-inversion". Composite membranes are also asymmetric but the support and the filtering thin layer are produced in different stages. Some polymeric microfiltration membranes can also be manufactured through special processes such as track etching. They have usually no or little asymmetry.
    Mineral membranes are also of asymmetric structure. They are made by coating porous carbon tubes with a layer of metal oxide (TECH-SEP) or by sintering alumina in a multichannel monolith (CERAVER, NORTON).

2.1.Preparation of polymeric membranes

    The technique known as phase inversion is the most important method for the preparation of reverse osmosis, ultrafiltration and some microfiltration membranes (2). Very different structures can be obtained by phase inversion according to the conditions of operation. The correlation between structure and conditions of membrane formation were extensively studied by Strathman (3,4). General rules are available in the litterature but the practical recipes used by the manufacturers in order to get reproducibly large quantities of membranes remain undisclosed.

The classical description of phase inversion uses a ternary diagram :

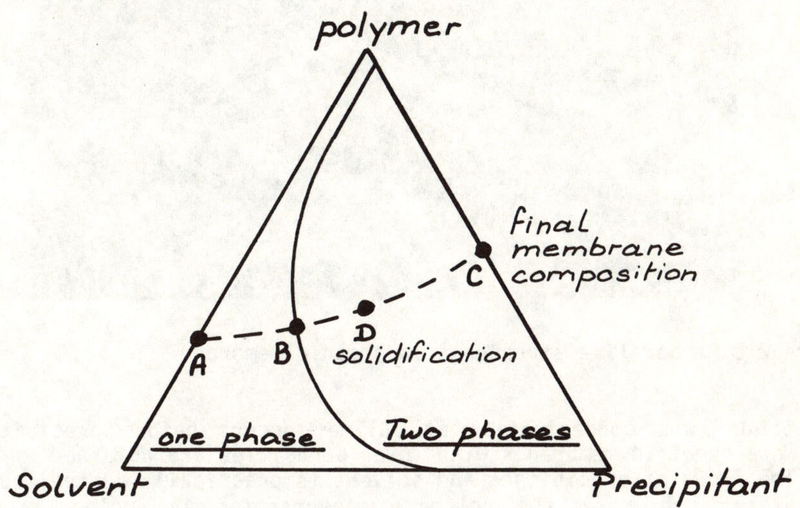

Figure 1 : displacement in a ternary phase diagram during formation of organic membranes.

When the homogenous solution of polymer (point A) is quenched in the precipitant, the composition moves along ABC to the two phase region : a solid porous phase, the membrane, and a liquid phase filling the pores.

Point B is the transition from the one phase to the two phases region where the mixture splits in a polymer rich phase and a polymer depleted region.

At point D, the polymer rich phase becomes solid.

This kind of diagram is only indicative because the practical pathway is far from equilibrium and mainly determined by the kinetic of demixion. This kinetic is related to the flow Jp of precipitant entering the polymer solution and the flow Js of solvent out.

When Jp/Js is high, the rate of precipitation is high and leads to very porous membranes with large voids of finger - like structure in the membrane sublayer.

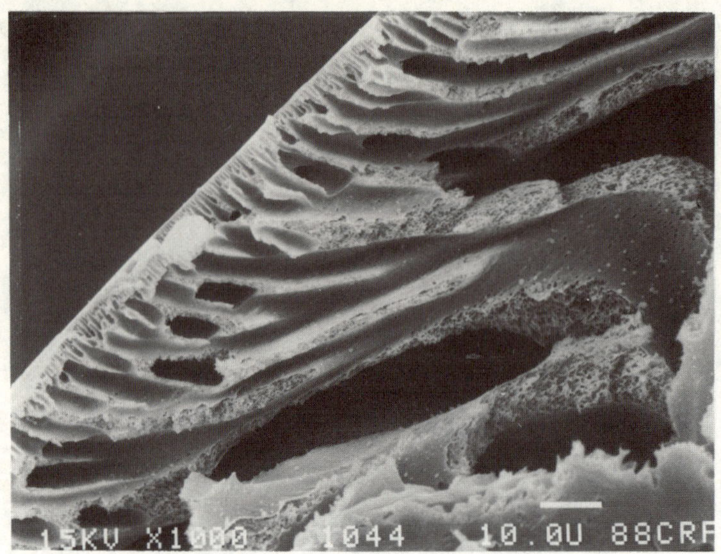

Figure 2 : Finger-like structures in organic membrane

If Js $>$ Jp, concentration of polymer occurs before precipitation. More symetric membranes with lower porosities are obtained.

The flows of precipitant and solvent is practically controlled by parameters such as : the polymer concentration and molecular weight distribution, the solubility parameters, the presence of additives in the solvent or in the precipitant, the temperature etc. For RO and UF membranes, the early mechanism is the formation of the dense ultrathin-layer at the interface between the solution and the precipitant. The evolution of the structure is then governed by the diffusion of the precipitant and the solvent. By a proper selection of conditions, one can obtain finger-like cavities extending for the whole membranes with an asymetric structure.

Membrane formation depends on so many parameters that quantitative statements of individual variables are very difficult to establish. A long experience made possible some semi-quantitative statements :
. Strong solvents of the polymers lead to dense membranes
. Strong precipitants (rapid phase separation) give a dense active layer and a finger-like structure of the membrane
. Strong viscosity of the polymer solution reduce the formation of cavities
. Weak solvents or precipitants lead to a small gradient of porosity across the membrane.

Membranes have to withstand high operating pressures in modules. Consequently they are cast onto various supports such as paper sheet, nylon or polyester fabric , sintered plastic. TECH-SEP manufactures flat sheet membranes by continuous casting on a moving belt.

Due to the small size of their lumen, hollow fibers are self sup-porting. For reverse osmosis, with pressure up to 70 bars, the fibers are very fine filaments with outside diameter of about 0.1 mm.

2.2. Preparation of mineral membranes :

Because of their enhanced chemical and thermal stability, mineral membranes are gaining importance in ultrafiltration and microfiltration.

The common characteristic of mineral membrane is the tubular shape with internal diameters from 4 to 15 mm. They have an asymmetric "skin", internal or external with pores from several nanometers to several microns.

Manufacturer	Matérials	Techniques	Pore Size
TECH-SEP	Carbon/ZrO$_2$	UF, MF	4 nm to 0.2 μ
ALCOA	Alumina	UF, MF	5 nm to 5 μ
NORTON	Alumina	MF	0.2 to 1 μ
Carbone-Lorraine	Carbon/Carbon	MF	0.2 to 1 μ

Table 1 : The main manufacturers of mineral membranes.

The properties of the mineral membranes result from the properties of the constituting materials :

- Chemical inertia (carbon, zircone) allows work in hard conditions : acids, alkalis, oxydants, regeneration and cleaning by agressive chemicals.
- The mechanical strength allows the work at high pressure.
- The thermal resistance allows the work at temperatures up to 350°C
- The techniques of preparation promotes the formation of narrow pore size distribution, allowing better separations.

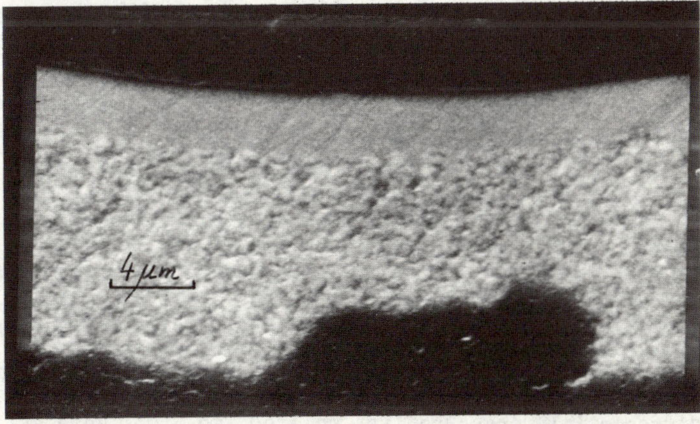

Figure 3 : Scanning electron micrograph of a TECH-SEP mineral membrane

TECH-SEP CARBOSEP inorganic membranes are made by coating a porous support tube with a layer of metal oxide. The coating is applied by circulating an aqueous suspension of metal oxide in the lumen of a porous carbon tube which is 6 mm in diameter. The metal oxide is filtered from the suspension and deposited on the porous carbon support covering up pores in the tube. By successive deposits of particles ranging from 0.1 to 0.01   m, one can create an asymmetric layer covering the carbon support. The tubes are then baked to assure a proper consolidation of the layer by sintering.

3.CHARACTERIZATION OF MEMBRANES :

Numerous methods are used to characterize membranes. They can be divided in three main groups :
.Pore size and pore size distribution determinations
  Microscopy (SEM, TEM)
  Bubble point method
  Liquid displacement porometry
  Liquid-vapour equilibrium technique (BJH method)
  Gas-liquid equilibrium technique (permporometry)
  Liquid-solid equilibrium technique (thermoporometry).
.Determination of electrical properties : net charge, Zeta potential
.Determination of retention : observed and intrinsic retention

3.1 Pore size distribution :
Microscopy
 Scanning electron microscopy (SEM) is extensively used to study the cross-sectionnal structure of membranes. Due to its limited resolution (50-100 A) it can not be used to detect the pores in the superficial layer of ultrafiltration membranes. Transmisssion electron microscopy (TEM) can detect pores but the difficult preparation of samples and interpretation of results restricts its use.
 Bubble point method :
 The pore size is evaluated by measuring the pressure drop necessary to blow a gas through a water filled membrane. This pressure is related to the pore radius by the Young and Laplace equation

$$\Delta p = \frac{4\gamma \cos\theta}{d}$$

p : air pressure across the membrane
γ : surface tension of the liquid/gas interface
θ : contact angle
d : pore size

Gas pass through the largest pore first and the bubble point is a measure of the largest pore. This method is currently used to check the integrity of membranes.

Liquid displacement porometry :
The gas/liquid interface is replaced by a liquid/liquid interface of low surface tension (eg butanol/water). The flow of liquid under a differential pressure is related to the pore size by the Hagen-Poiseuille equation :

$$J = n\frac{\pi . r^4 . \Delta P}{\eta . \lambda . 8}$$

with :

n = number of pores
η = viscosity
r = radius of pores
λ = thickness of the membrane

A low interfacial tension allows the measurements of very small pores :
5 to 10 A.

Liquid-vapour equilibrium :
This method uses the well known phenomenon of capillary condensation and evaporation of a gaz into narrow pores. Nitrogen or argon are often used. The total pore volume is measured from the quantity of gaz absorbed near the saturation pressure. A hysteresis loop is observed when a full isotherm is determined. The shape of the pores is indicated by the shape of the hysteresis loop and the volume of a particular pore size is deduced from the hysteresis loop.

The method applies for pores from 17 A to 500 A. Limitations are on the assumptions of pores shape and the use of dry membranes whose structure can be dammaged by drying.

Permporometry :
This method is also based on capillary condensation. An organic vapor is diluted in an inert gas and condensed in the pores by changing the relative vapor pressure from 0 to 1. The gas transport is measured as a function of the vapor pressure.

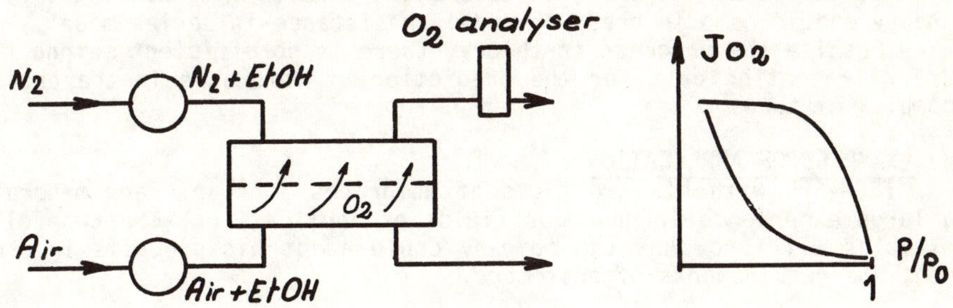

Figure 4 : schematic diagram of permporometry

Pore size distribution is calculated from the hysteresis loop of the isotherm JO2 vs P/P0. Some problems can also be encountered with the drying ot the membrane by ethanol.

Thermoporometry :
This method is based on the calorimetry of the solid/liquid transfor mation of a capillary condensate. The solidification temperature depends of the liquid interface curvature and thus of the size of the pore. The curve of thermal flux versus temperature (differential scanning calorimetry) can be converted into a pore size distribution. The method is used for pore radii from 15 to 1500 A and is preferably applied to structure unmodified by drying.

Many unexplained discrepancies remain between the various methods. In spite of their limitations, permporometry and thermoporometry are the easier and more current methods.

3.2 <u>Electrical properties</u> :

The main electrical properties are the ion exchange capacity, measured by chemical titration, and the zeta potential. These parameters are very important in the study of fouling.

3.3 <u>Retention properties</u> :

Separation characteristics of a membrane are dependant of the dimensions, shape, electrical charge, adsorption properties of the solute molecule. A single parameter such as molecular weight give only a rought estimation of the rejection capacity of a membrane. manufacturers use the cut-off value determined with proteins or other polymers of known molecular weight. The main interest of such determinations is that they are simple to carry out and well adapted for quality control.

Because of its low adsorption on membranes and known characteristics, dextrane is widely used but problems are associated with the broad molecular weight distribution and non spherical shape of the molecule.

The other widely used substances are polyethylene glycol, vitamin B12 and proteins : cytochrome C, β lactoglobulin, bovine serum albumin, immunoglobulines.

The main problem is to establish usable relations between laboratory tests with defined substances and membrane performance with complex media.

Boundary layer phenomena and fouling change dramatically the performances. Adsorption of products onto the membrane is probably much more important than the molecular size.

Various theories tried to take these effects into account : film theory model, osmotic pressure model, resistance-in series model.

Despite the progress in theory, there is no efficient method for the use of retention data for the prediction of membranes in the process of complex mixtures.

4. <u>EXAMPLES OF APPLICATION</u> :

TECH-SEP with his two types of membranes , organic and mineral, has a large experience in numerous fields of applications. The two following examples will show how the company could adapt his products in order to meet the requirements of customers.

4.1. <u>Ultrafiltration of electrophoretic paint</u> :

Electrophoretic paints are widely used for the protective coating of cars and household appliances. Ultrafiltration is used for concentration and recycle of the paint which is diluted by rinsing of the parts after coating. Due to the characteristics of these industries, the UF system must work continuously at constant flux and the cleaning and regeneration operations must be reduced to a minimum.

Typically, a cleaning operation is requested one or two times a year in contrast with the daily cleaning of the food industry.

Fouling may occur by accident (mainly pollution of the paint bath) or by interaction of the paint with the membranes. This interaction must be reduced to a minimum to insure long term operations. A test rig was developped for simulation of the long term fouling. The time scale was reduced by operating the membranes at high pressure and low tangential velocity. After the occurence of fouling (flux reduced by 5) the observed parameter was the time requested to recover the flux after application of normal pressure and velocity conditions.

Membranes with no charge or of sign opposed to the electric charge of the membrane (eq cationic membrane and anionic paint) hardly recovered their flux. This confirmed the opinion that membranes had to be of the same sign than the paint in order to repell the suspendend pigments.

Optimization of the electrical charge was carried out by this fouling simulation method in combination with determination of rejection using dextranes. The last method was
used to adjust the cut-off in order to assure the lowest possible rejection for solvents, salts and low MW compounds of the paint.

This methodology led to ionic membranes well adapted to the paints :
IRIS 3050 for cationic paints
IRIS 3042 for anionic paints

4.2. Concentration of human albumin :

Human albumin is a very high value product issued from blood fractionation. At the final step, albumin is concentrated up to 200 g/l by ultrafiltration prior to sterile packaging.

The manufacturers are always seeking for an UF unit showing the following characteristics :
- Low dead volume : limitation of losses
- Very high rejection :
- High flux to reduce the time of operation (fragile product)

A low dead volume can be achieved by using cassettes or plate and frame to reduce the time of operation.

A low dead volume can be achieved by using cassettes or plate and frame UFP10 system.

The very high rejection (99,8 %) can be only checked by using the real product but it was possible to combine the rejection curve for model compounds and air flux measurements to guide the preparation of adequate polysulfone membranes.

A statistical analysis of rejection for HSA as a function of the ratio rejection for bovine serum albumin/rejection for β lactoglobulin, of water and air flux was established. Rejection for HSA increased with the ratio and with a decrease of air flux. This sharpening of the cut off (see following figure) was obtained by manipulating the preparation variables.

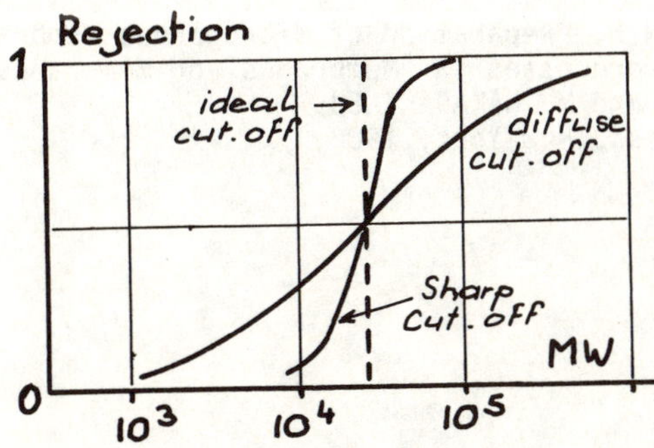

Figure 5 : Typical rejection curves

TECH-SEP is now able to prepare UF membranes responding to the high quality standards of the bio-industries. This can only be achieved by an extensive field and laboratory experience.

5.<u>CONCLUSIONS</u> :
     Despite large efforts in the past two decades, manufacture of membranes remains much of an art than of a science. Preparation of high quality products requires very skilled people and a lot of practical experience.
     The need for a wider range of techniques for characterization is clear for the manufacturer and the process designer. The current methods give insufficient data on pore shape and size distribution. There is also a need of unification of method as has been done for microfiltration membranes under the pressure of the pharmaceutical industry.
     A better understanding of the structure of the membranes will made possible more precise relations between structure, properties and the manufacturing conditions. A large part of the cost of membranes is associated with quality control and a more efficient feed back from the application to the manufacturing process is requested in order to cut those costs.
     A first requirement for advances in membranes technology is to esblish a more basic understanding of fouling mechanism in relation with characterization and manufacturing practices.
     Through numerous international collaborations, TECH-SEP is currently doing major efforts to improve his products and the performances he provides to the industry.

<u>REFERENCES</u>

(1) Characterization of ultrafiltration membranes.
    Proceedings of a workshop arranged by department of food engineering.
    Lund University, SWEDEN 1987

(2) R.E. KESTING - Synthetic polymeric membranes
                   Mc Graw Hill - New-York 1971

(3) STRATHMANN H., KOCK K. The formation mechanism of phase inversion membranes.
    Desalination, 21, 3 , p 241. 1977.

(4) STRATHMANN H. Preparation of microporous membranes by phase inversion processes in "Membranes and Membranes processes"
    E. DRIOLI and M. NAKAGAKI Eds
    Plenum Press, New-York, 1986.

ECONOMICAL EVALUATION OF THE MEMBRANE TECHNOLOGY

H . STRATHMANN
Fraunhofer-Institut für Grenzflächen- und Bioverfahrenstechnik

Summary

In recent years membranes and membrane processes have become industrial products of substantial technical and commercial importance. The worldwide sales of synthetic membranes in 1988 will be in excess of 1.2×10^9 US \$, with an annual increase of 12 to 14 %. In spite of already impressive sales figures and growth rates the use of membranes in industrial scale separation processes is not without problems and market forecasts are difficult because of the multitude of different products, processes, and applications, and because of the rapid development of new products opening up new applications. In this paper the present market for synthetic membranes has been analyzed in terms of its regional distribution and industrial areas of application. Various market segments such as the water and wastewater industry, the chemical and pharmaceutical industry, as well as the food industry, biotechnology and the environmental protection industry have been identified. Problems effecting the future industrial scale utilization of membranes and membrane processes are discussed. Various membrane-related research projects carried out in industry and government-supported insitutions are analyzed in terms of their impact on the development of the membrane industry. Several research and development areas have been identified as being of prime importance for the future growth of the membrane-based industry.

I. INTRODUCTION

Membranes and membrane processes have become industrial products of substantial technical and commercial importance. Even if the optimism of the sixties and seventies concerning the membrane market has been replaced by a more realistic view, there is no doubt about the significance of membranes in industrial scale separation processes. Membranes are used today to produce potable water from seawater, to treat industrial effluents, to recover hydrogen from off-gases, or to fractionate, concentrate, and purify molecular solutions in the chemical and pharmaceutical industry. Membranes are also key elements in artificial kidneys and controlled drug delivery systems. The worldwide sales of synthetic membranes in 1988 will be in excess of 1.2×10^9 US \$, with an annual increase of 12 to 14 %. Taking into account that membranes or membrane modules are only 20 - 30 % of the total cost of a separation apparatus the total sales of the membrane-based industry is close to 5×10^9 US \$. In spite of the rather impressive sales figures the use of membranes in industrial scale separation processes is rather problematic. First of all, membranes and membrane processes are extremely heterogeneous and include a multitude of different products and processes in a large variety of applications. This often leads to rather small market segments for a given product in a given application. Furthermore, the state of development of membrane processes in different applications is also very different. For instance, reverse osmosis, microfiltration, or electrodialysis can be considered mature processes in the production of ultrapure or potable water. In biotechnology or in the chemical and pharmaceutical process industry they are hardly used until today, because of certain inefficiencies of todays commercial membranes and problems related to the chemical engineering aspects of the process design. Other membrane processes such as pervaporation, and to a large extent gas separation, are still in the development stage. In those applications, where membranes are well-established, such as in artificial kidneys or in chlorine-alkaline electrolysis the market size and potential growth are well-known and served by a small number of relatively large companies operating on an international scale. In other applications, where membranes are not yet established, such as in biotechnology, gas separation, waste air and wastewater treatment systems, it is more difficult to make any predictions about the future development. Although, presently sales of membranes in these applications are low there is a huge speculative potential market with accordingly high growth rates.

Besides these processes, which are today the acknowledged state-of-the-art or used on a pilot plant scale, there are various membrane processes, where an industrial application is extremely interesting, which, however, are only conceptually available today because of the lack of suited membranes.

The entire problematic of market forecasts for membranes and membrane separation processes is illustrated in the schematic graph of Figure 1, which lists various membrane processes and industrial applications in the order of their state of technical development, thus identifying processes, which in certain applications can be regarded as "status-of-the-art. In other applications the same processes are still only used on a pilot plant scale, and in others, again, they are just tested in the laboratory. Other processes are still under development and have not yet found applications where they could be considered as status-of-the-art. Finally, there are membranes and membrane processes, which are only conceptionally available. All these processes could reach great technical significance in the future, if the problems that concern the membranes as well as the processes can be solved.

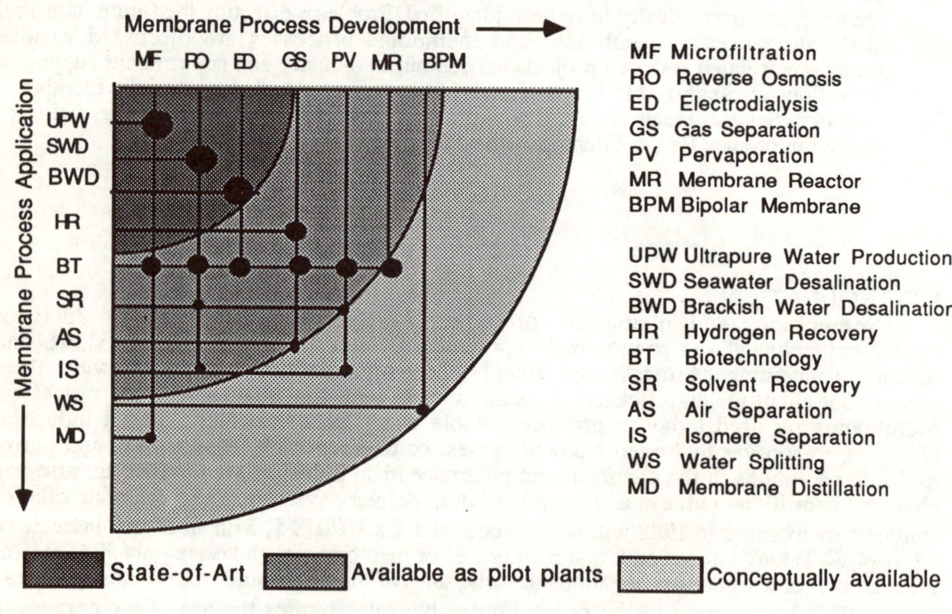

FIGURE 1   Schematic diagram showing membrane separation processes and their application as a function of status of development.

The development of membrane technology on an industrial scale is rather difficult, in general, and especially when new applications are to be covered with new processes. Significant investments for an efficient membrane production and application-oriented process engineering are necessary. This know-how can only be obtained through a close and interdisciplinary collaboration. This has often hindered a quick and widespread implementation of membrane technology.

2. <u>MEMBRANE PROCESSES AS UNIT OPERATIONS</u>

For a better understanding of the technical and economical significance of membrane technology the basis of membranes and membrane processes shall be briefly reviewed at this point.

2.1 <u>Definition of a Membrane, its Structures and Functions in Separation Processes</u>

A multitude of very different structures is summarized under the term "membrane". All these structures have in common that they affect the permeation of different chemical components in a very specific way. Membranes can be liquid, homogeneous or heterogeneous, symmetric or asymmetric, and they can consist of organic or inorganic materials.

Membranes achieve the separation of molecular mixtures by restricting the permeation of certain components while others may pass unhindered. Membranes may operate by different separation mechanisms. For instance, separation can be achieved by a pure "sieving-effect" like with porous membranes, where the components of a mixture are sorted out according to their size. Separation can also be based on different solubilities and diffusion coefficients of the single components in the membrane matrix. Technically relevant membranes, their properties, structures, and applications are summarized in Table I. Structures of various membranes are shown in the scanning electron micrographs in Figure 2.

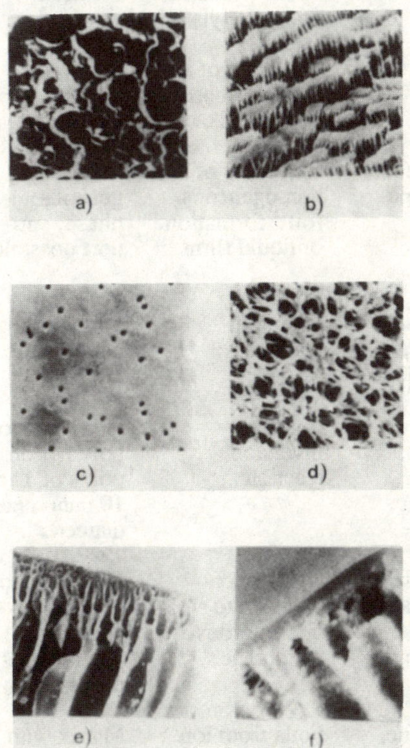

FIGURE 2   Scanning electron micrographs of membrane structures: a) PTFE-sintered membrane; b) stretched PTFE-membrane; c) capillary pore membrane; d) symmetric MF-membrane; e) asymmetric UF-membrane; f) composite membrane.

TABLE I   Properties and applications of technically relevant synthetic membranes

Membranes	Basic Materials	Manufacturing Procedures	Structures	Applications
Ceramic membranes	Clay, silicate, aluminiumoxide, graphite, metal powder	Pressing and Sintering of fine powders	Pores from 0.1 to 10 micron diameter	Filtering of suspensions, gas separations, separation of isotopes
Polymeric sinter membranes	Polytetrafluoro-ethylene, polyethylene, polypropylene	Pressing and sintering of fine powders	Pores from 0.1 to 50 micron diameter	Coarse filtration of aggressive media, introduction of gases, cleaning of air
Stretched membranes	Polytetrafluoro-ethylene, polyethylene, polypropylene	Stretching of partially crystalline foil perpendicular to the orientation of crystallyts	Pores of 0.1 to 1 micron diameter	Filtration of aggressive media, cleaning of air, sterile filtration, medical technology
Etched polymer films	Polycarbonate	Radiation of a foil and subsequent acid etching	Pores of 0.51 micron diameter	Analytical and medical chemistry, sterile filtration
Homogeneous membranes	Silicone rubber, hydrophobic liquids	Extruding of homogeneous foils, formation of liquid films	Homogeneous phase, support possible	Gas separations, carrier-mediated transport
Symmetrical microporous membranes	Cellulose derivatives, polyamide, polypropylene	Phase inversion reaction	Pores of 50 to 5000 nonometers diameter	Sterile filtration, dialysis, membrane distillation
Integral asymmetric membranes	Cellulose derivatives, polyamide, polysulfone, etc.	Phase inversion reaction	Homogeneous polymer or pores of 1 to 10 nanometers diameter	Ultrafiltration, hyperfiltration, gas separations, pervaporation
Composite asymmetric membranes	Cellulose derivatives, polyamide, polysulfone, polydimethyl-siloxane	Application of a film to a microporous membrane	Homogeneous polymer or pores from 1 to 5 nanometers diameter	Ultrafiltration, hyperfiltration, gas separations, pervaporation
Ion exchange membranes	Polyethylene, polysulfone polyvinyl-chloride etc.	Foils from ion exchange resins or sulfonation of homogeneous polymers	Matrix with positive or negative charges	Electrodialysis, electrolysis

The most simple form of a synthetic membrane are porous plate or foils. They are produced by pressing and sintering a polymeric, ceramic or a metal powder. These membranes have relatively large pores, a wide pore size distribution and a low porosity. The symmetric or asymmetric phase inversion membranes are more complex in their production as well as in their structure. They are produced by precipitation of a polymer solution. These membranes are used today in micro- and ultrafiltration was well as in gas separation and pervaporation. Composite membranes are more and more employed in the last three processes mentioned above. The selective layer and the support structure of these membranes consist of different materials. The membranes with functional groups, like the simple ion exchange membranes are used in electrolysis and electrodialysis, or the liquid membranes that are used in coupled transport with selective complexing agents and chelates, are gaining increasing importance.

2.2 Technical Relevant Membrane Separation Processes

Membrane processes are just as heterogeneous as the membranes. Significant differences occur in the membranes that are used, the driving forces for the mass transport, the applications and also in their technical and economical significance. Technically relevant membrane processes and their most important characteristics are given in Table II. In some of the processes membranes as well as the processes have reached such a level that a completely new development can not be expected. Examples are the micro- and ultrafiltration, reverse osmosis, dialysis or electrodialysis. In these processes improvements and optimization of existing systems and their adaptation to special applications are to be expected. Other processes are in the very beginning of their industrial application and they give the possibility for totally new developments for membranes and modules. Examples are pervaporation and gas separation.

2.3 Technical and Economical Problems Related to the Use of Membrane Processes

The first step in solving a specific separation problem is the development of suited membranes and membrane modules. Then, the process has to be developed and adapted to the problem. Frequently, membrane processes like ultra- or microfiltration and reverse osmosis, which have been very successful in the production of ultrapure water, have failed at first in other applications, like biotechnology. The reason is that they were not adapted to the specific needs of this application in terms of the chemical engineering aspects.

The use of a membrane process may be questionable, even if all technical problems are solved, for economical reasons and the economics are often determined by the membrane costs which, in turn, depend on market size and the expected market development. Membranes and membrane separation processes have the advantage, that they can be tailored to many separation problems. This leads often to small market segments for a particular product. Small market segments, however, mean high development and production costs. An example is the artificial kidney which is produced today as a disposable unit in large quantities. The production costs for an artificial kidney of 1 m^2 membrane area are about 20 - 40 DM including the demanding quality control, sterilization and packaging. Membranes with similar properties are also used in technical ultrafiltration. Here the price is 300 - 1000 DM per m^2, although the larger module units should lead to lower production costs. Relatively small markets often lead to less advantageous cost structures. The membrane module development is closely related to the membrane process and its technical application. It is performed according to economical considerations, where the costs per installed membrane area are an important factor and to other chemical engineering criteria, such as the flow of the feed solution at the membrane surface and the flow path of the permeate. The goal is to obtain an optimum flow to and from the membrane while minimizing pressure loss, concentration polarization and membrane fouling. For the different applications, different types of modules have been developed which generally fulfill the given demands. The main types of membrane modules used today on an industrial scale are shown in Table III. It is obvious that there are substantial differences in the specific costs and the practical applications of the different membrane module concepts.

TABLE II Technically relevant membrane processes

Membrane separation process	Driving force for mass transport	Type of membrane employed	Separation mechanism of the membrane	Application
Micro-filtration	Hydrostatic pressure difference 50-100 kPa	Symmetrical porous membrane with a pore radius of 0.1 to 20 μm	Sieving effect	Separation of suspended materials
Ultra-filtration	Hydrostatic pressure difference 100-1000 kPa	Symmetrical porous membrane with a pore radius of 1 to 20 nm	Sieving effect	Concentration, fractionation and cleaning of macromolecular solutions
Reverse osmosis	Hydrostatic pressure difference 1000-10000 kPa	Asymmetric membrane from different homogeneous polymers	Solubility and diffusion in the homogeneous polymer matrix	Concentration of components with low molecular weight
Dialysis	Concentration difference	Symmetrical porous membrane	Diffusion in a convection-free layer	Separation of components with low molecular weight from macromolecular solutions and suspensions
Electro-dialysis	Difference in electrical potential	Ion exchange membrane	Different charges of the components in solution	Desalting and de-acidifying of solutions containing neutral components
Gas separation	Hydrostatic pressure difference 1000-15000 kPa	Asymmetrical membrane from a homogeneous polymer	Solution and diffusion in the homogeneous polymer matrix	Separation of gases and vapors
Pervaporation	Partial pressure difference 0 to 100 kPa	Asymmetrical solubility membrane from a homogeneous polymer	Solution and diffusion in the homogeneous polymer matrix	Separation of solvents and azeotropic mixtures
Membrane distillation	Partial pressure difference 0 to 100 kPa	Symmetrical hydrophobic, microporous membrane	Differences in vapor pressure	Desalting of water, concentration of solutions

TABLE III   Membrane modules, their properties and applications

Type of module	Membrane area per volume (m^2/m^3)	Price	Control of concentration polarization	Application
Tube 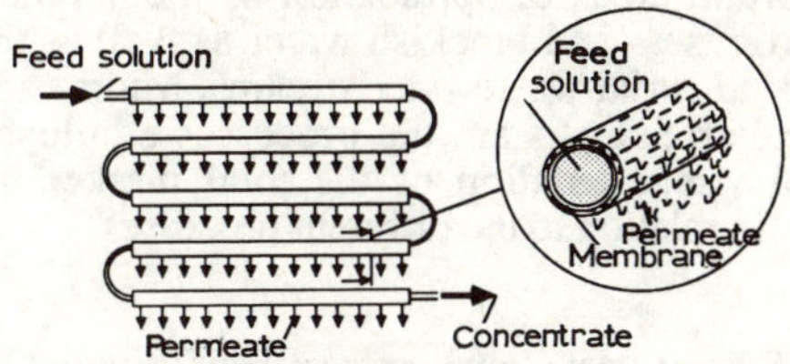	20-30	Very high	Very good	Cross-flow filtration of solutions with high solids content
Plate-and-frame 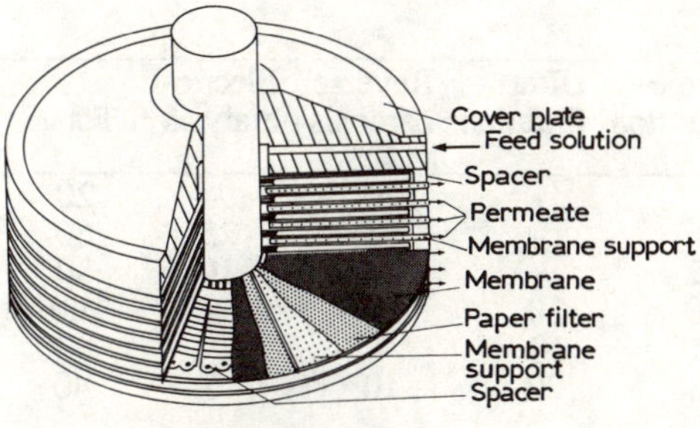	400-600	High	Fair	Filtration, pervaporation, gas separation and reverse osmosis
Spiral-wound 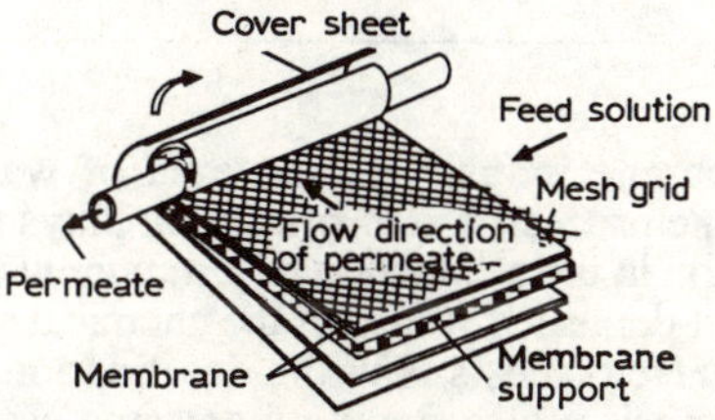	800-1000	Low	Poor	Ultrafiltration, reverse osmosis, pervaporation and gas separation
Capillary tube 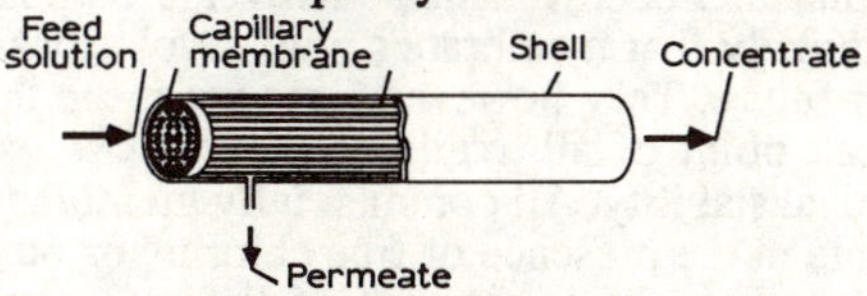	600-1200	Low	Good	Ultrafiltration, pervaporation liquid membranes
Hollow fiber 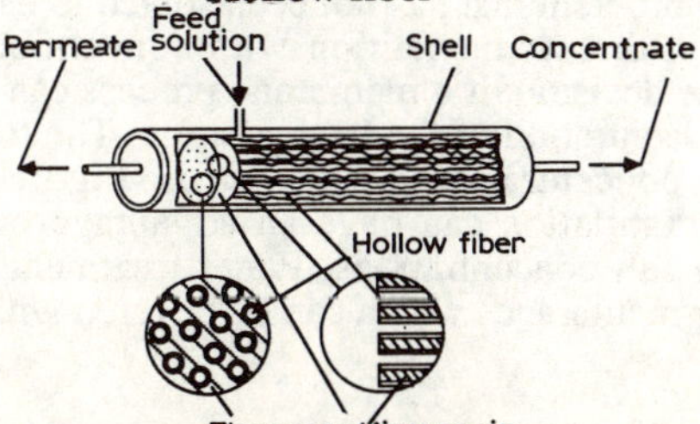		Very low	Very bad	Reverse osmosis, gas separation

3. TECHNICALLY RELEVANT MEMBRANE APPLICATIONS AND THEIR ECONOMICAL SIGNIFICANCE

Membranes are used today in numerous applications from medicine to wastewater treatment. The applications show great differences in their technical and economical significance.

3.1 Membrane Processes in Water Treatment

Water treatment is one of the most important areas of application of membranes. This includes the production of potable water from sea- and brackish water as well as the production of process water and the cleaning of industrial wastewater streams. Micro- and ultrafiltration as well as reverse osmosis and electrodialysis are the processes employed mainly in water treatment. Table IV shows the segmentation of the total market for membranes and membrane modules, respectively, in relation to the different processes.

TABLE IV   Sales of the membrane industry (US $ per year) in water treatment, divided by processes and applications

Applications	Micro-filtration	Ultra-filtration	Reverse osmosis	Electro-dialysis	Total
Desalting of seawater	-	-	20	-	20
Desalting of brackish water	-	-	35	25	60
Pretreatment of boiler and feed water	5	-	15	10	30
Ultrapure water	120	30	10	-	160
Sterile and lowpyrogen water	80	10	-	-	90
Industrial wastewater	5	20	10	5	40
Total	210	60	90	40	400

This table shows, that total sales of the membrane industry in the area of water treatment is 400 million US $ per year. Of all membrane processes, microfiltration plays the most important economical role because of its application in the production of ultrapure water in the semiconductor industry and of sterile and pyrogen-depleated water for the chemical and pharmaceutical industry. The main application of reverse osmosis is in the desalination of sea- and brackish water, but more recently reverse osmosis is also finding increasing use in the production of ultrapure water. The level that has been reached in reverse osmosis membrane performance today, makes it quite unlikely that membranes with revolutionary improved properties will be available in the near future. This, however, does not mean that further development is not necessary. The weak point of all available reverse osmosis membranes today is their poor thermal and chemical stability. All commercially membranes are either sensitive against hydrolysis or not stable in the presence of free chlorine or other strong oxidizing agents. This makes a rather extensive pretreatment of the raw water necessary. In ultrafiltration membrane fouling is a problem that has not been solved to date. In water desalination membrane processes are in direct competition with conventional processes such as distillation or ion exchange. The decision, if a membrane process can be used economically depends mainly on the salt concentration of the feed solution. The total costs for desalted water as a function of salt concentration are shown in Figure 3. Conventional processes like ion exchange and distillation can have an advantage over membrane processes for very high and very low salt concentrations. Water treatment is, however, in any case a very interesting market for membranes which can be covered with a relatively small number of products.

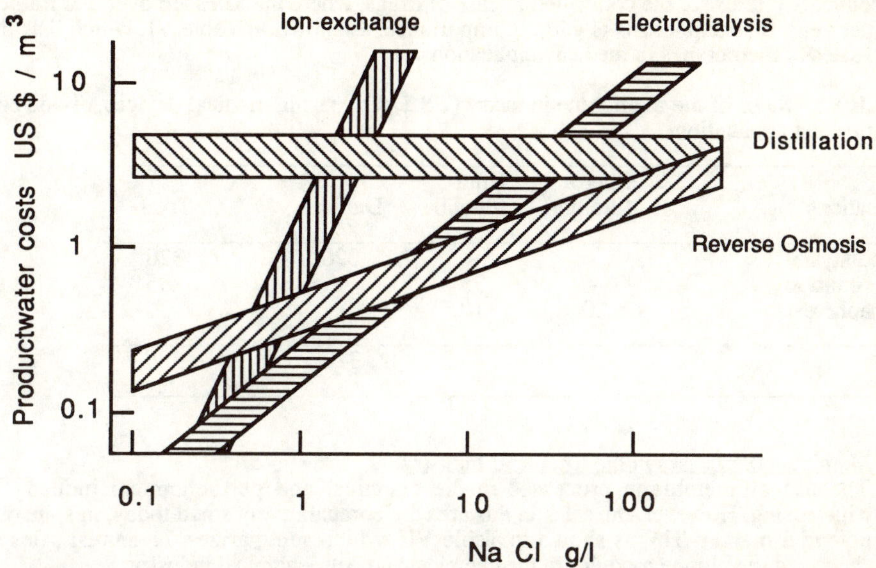

FIGURE 3   Product water costs as a function of the feed solution salt concentration for different desalination processes

3.2   Membrane Processes in the Food Industry

Another interesting area of application for membranes is the food industry as indicated in Table V, which shows the total annual sales of membranes and membrane modules in the food industry.

TABLE V   Sales of the membrane industry (US $ per year) in the food industry, divided by processes and applications

Applications	Micro-filtration	Ultra-filtration	Reverse osmosis	Electro-dialysis	Total
Dairy industry	-	35	-	10	45
Beverage industry	75	10	5	5	95
Meat processing	-	5	5	-	10
Starch industry	-	5	5	-	10
Total	75	55	15	15	160

Micro- and ultrafiltration, reverse osmosis and electrodialysis are mainly used in this application. Microfiltration dominates in the beverage industry for sterile filtration and clarification, while ultrafiltration and electrodialysis are used mainly in the dairy industry. The application of new membrane processes like pervaporation is just beginning. These processes will soon be used on an industrial scale. The membrane market in this area amounts to about 160 million US $ per year.

3.3   Membrane Processes in Medical Devices

Membranes are used in medical applications for more than 25 years. The artificial kidney is the by far most interesting membrane application [1] besides membrane-controlled therapeutical systems for the controlled release of drugs, where the sales are over 200 million US $ per year. Plasmapheresis is gaining importance, as shown in Table VI, which lists the annual sales of membranes in medical applications.

TABLE VI   Sales of the membrane industry (US $ per year) in medical devices, divided by processes and applications

Applications	Micro-filtration	Ultra-filtration	Dialysis	Total
Hemodialysis	-	-	320	320
Hemofiltration	-	35	-	35
Plasmapheresis	20	10	-	30
Total	20	45	320	385

3.4 Membrane Processes in the Chemical Industry

The use of membrane processes in the chemical and petrochemical industry is rapidly increasing. However, the sales in this area are comparatively small today, in spite of a large potential market. This is shown in Table VII, which summarizes the annual sales of membranes and membrane modules in the chemical and petrochemical industry.

TABLE VII   Sales of the membrane industry (US $ per year) in the chemical industry divided by processes and applications

Processes / Applications	Micro-filtra-tion	Ultra-filtra-tion	Reverse osmosis	Gas sepa-ration	Per-vapor-ation	Electro-dia-lysis	Electro-lysis	Total
Process water pretreatment	25	-	5	-	-	-	-	30
Fractionation of molecular mixtures	10	15	5	-	-	-	-	30
H$_2$-recycling	-	-	-	20	-	-	-	20
N$_2$-production	-	-	-	5	-	-	-	5
CO$_2$/CH$_4$ sepa-ration	-	-	-	5	-	-	-	5
O$_2$-enrichment	-	-	-	5	-	-	-	5
Separation of azeo-tropic mixtures	-	-	-	-	5	-	-	5
Desalting of pro-cess solutions, salt production	-	-	-	-	-	20	-	20
Chlorine alkaline electrolysis	-	-	-	-	-	-	70	70
Total	35	15	10	35	5	20	70	190

An interesting relatively new application is separation of gases, such as the recovery of hydrogen from the ammonia synthesis and the separation of carbon dioxide from methane in enhanced oil recovery. The separation of oxygen and nitrogen by membranes is of interest for the production of oxygen-enriched air or inert gas. But membranes have to be improved in terms of selectivity and permeability before oxygen-nitrogen separation on a large scale becomes economically feasible.

Pervaporation and perstraction are new membrane processes which will probably soon find their way into the chemical and petrochemical industry. The heterogeneous character of the applications is slowing down a widespread introduction of membrane processes.

3.5 Membrane Processes in Biotechnology

Biotechnology is an industry where membranes can offer great advantages. They are especially suitable for the separation of sensitive biological substances because the separation by membranes is a physical procedure which can be carried out at room temperature. All membrane processes including electrodialysis and pervaporation are of interest in biotechnology [2]. Presently, however, sales of membranes for biotechnology applications are still rather low and in the order of 10 to 20 million US $. A reason, for the relatively slow introduction of membrane processes in biotechnology is that the modern industrial biotechnology is still very young, and membranes and membrane processes have not yet been sufficiently adapted to the specific requirements of biotechnology. The list of industrial applications of membranes which is given here is by far not complete. There are many other applications, especially in the laboratory, which contribute substantially to the total sales of the membrane industry.

4. REGIONAL DISTRIBUTION OF THE MEMBRANE INDUSTRY AND ITS MARKETS

A look at the regional distribution of the membrane industry and its markets provides some interesting information. Almost all companies operate internationally and market their products worldwide. Worldwide more than 100 companies are involved in one way or another in the membrane technology [3]. However, only 60 companies are at the same time manufacturers of membranes or modules on a commercial basis. The other companies are mainly involved in process design and plant engineering using membranes as a component.The USA has a large part of the membrane-based industry with more than 35 companies offering membranes on a commercial basis. However, six of these companies account for 80 % of the sales of all the USA-based membrane industry. An analysis of the regional distribution of the sales in the membrane industry in relation with the location of the companies shows that companies based in the USA account for 63 % of the total worldwide sales of about 1,2 billion US $ per year. Japanese companies account for 17 %, companies based in the Federal Republic of Germany for 13 %, and the rest of the European countries for 6 % of the worldwide membrane sales. This distribution is shown in Figure 4.

Membranes in therapeutical systems and the sales of the membrane-producing companies in Eastern Europe, especially in the USSR and also in the People´s Republic of China are not taken into account in these considerations, because of the difficulty to obtain reliable data.The regional distribution of the membrane market is not identical with the regional distribution of the membrane industry as shown in Figure 5, which illustrates the worldwide distribution of the membrane market.

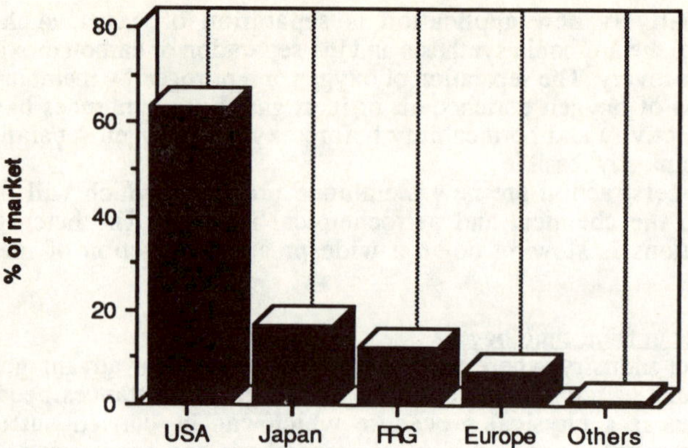

FIGURE 4   Regional distribution of the membrane industry (total sales approximately 1.2 billion US $ per year)

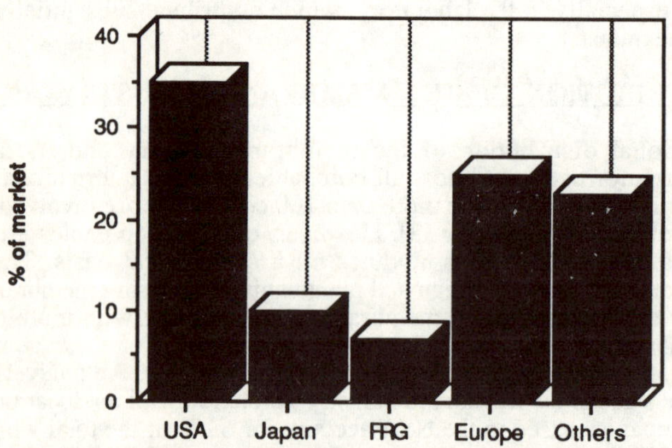

FIGURE 5   Regional division of the membrane market (total market about 1.2 billion US $ per year)

About 35 % of the membrane market is located in the USA, 10 % in Japan, 30 % in Europe, and 25 % in other countries. This is, however, a very global analysis. The distribution of the market can be totally different for the different membrane processes or membrane applications. For example, the market share of the Federal Republic of Germany in the area of the artificial kidney is 10 %, and that for electrodialysis only 1 % of the entire world market.

It is also obvious that the existence of a regional market is of great significance for the build-up of the related industry. Therefore, there is no manufacturer of ion exchange membranes in the Federal Republic of Germany, but almost 80 % of all hemodialysis membranes are manufactured here [1].

5. MEMBRANE-RELATED RESEARCH AND DEVELOPMENT

The R&D work in membrane science and technology is carried out at different levels and with different goals. A complete evaluation of the R&D activities is extremely difficult. All data concerning this topic are therefore questionable and can only be considered as an analysis of current trends. It is often difficult and arbitrary to decide which activities are research and development, "product development", or quality control.

The research and development activities carried out in the industry often differ significantly from such activities carried out at universities in terms of their level of sophistication and goals. The research in industry is in general dictated by economical considerations and concentrated on the development and optimization of products and processes. The universities, on the other hand, are involved more in basic research which is at least to some extent independent of economical considerations.Since the various membrane processes show great differences in their level of technical development, research work is concentrated on different topics in the different processes. Table VIII shows some of the subjects on which R&D efforts are concentrated in membrane processes available today. Processes which are state-of-the-art and used industrially include microfiltration, ultrafiltration, reverse osmosis, dialysis, electrodialysis, and with some restrictions also pervaporation and gas separation.

TABLE VIII   State of development of membrane processes with industrial use and present R&D efforts to improve membranes and processes in new applications

Process	Availability	Problems	Goals of R&D-Work		
			Membranes	Process	Applications
Microfiltration	State-of-the-Art	Membrane life	Asymmetric, inorganic	Cross-flow filtration	Biotechnology
Ultrafiltration	State-of-the-Art	Membrane fouling	Surface modifications	Membrane cleaning	Biotechnology, chemical industry
Reverse Osmosis	State-of-the-Art	Chlorine stability, biofouling	Chlorine resistance	Water pre-treatment	Biotechnology, wastewater
Dialysis	State-of-the-Art	Biocompatibility	Improved permeability	Process optimization	Biotechnology, food technology
Electrodialysis	State-of-the-Art	Alkaline stability	Ionselective and bipolar membranes	Multilple cell systems	Chemical industry, wastewater
Gas separation	Pilot plants	Membrane selectivity and flux	Membranes for specific gases	Process optimization	Air/solvent separation
Pervaporation	Laboratory loops	Membrane selectivity, process development	Membranes for specific materials	Process development and optimization	Wastewater treatment, biotechnology

These processes have the greatest economical relevance today. The R&D work in this area is concentrated on product improvement as well as process optimization and different aspects of application. Gas separation and pervaporation are of minor economical importance today, and the membrane sales for these processes are hampered by lack of process experiences. There is a number of membrane processes which exist today only as basic research or as concepts. This includes membranes with active transport properties, membranes in energy generation or energy conversion or in communications. It seems too early to speculate about the future economical significance of these membranes. However, there are some new developments related to membranes and membrane processes which are quite interesting. Short-term economical and technical use of these processes seems possible. Among others, there is the production of microporous structures with the aid of x-ray lithography [4]. This method is suitable for the production of microfiltration membranes with high porosity and extremely uniform pores. Another method is the use of certain parts of the cell walls of some microorganisms with an extremely sharp molecular cut-off [5]. Significant progress has also been made recently in gas separation membranes in using selective carrier substances. Membranes with immobilized biocatalyst are developed and used in micro-biological production processes and diagnostic systems.

6. R&D EXPENDITURES IN MEMBRANE TECHNOLOGY

There are significant differences in the funding of membrane-related research and development work in different countries. The total membrane-related R&D expenditures are estimated to be ca. 110 million US $, 67 % of this amount is spent in the membrane-based industry and 33 % is spent at universities and research institutions.
The regional distribution of the estimated R&D expenditures by industry as well as by universities is shown in Table IX.

TABLE IX   Estimated R&D expenses in the area of membrane technology given as percent of the membrane sales and divided regionally

Basis of companies	Membrane sales	Percentage of the estimated membrane sales	R&D-Expenditures				
			Industry		Institutions	Total	
	Mio. US $ p.a.	%	Mio. US $	% of sales	Mio. US $	Mio. US $	% of sales
USA	750	62	41	5.5	12	53	7
Japan	200	17	19	9.5	4	23	11.5
Fed. Rep. Germany	160	13	11	7	4	13	8
Europe (except F.R.G.)	80	7	7	9	13	20	25
Others	10	1	1	10	4	11	100
Total	1200		79	6.6	37	120	10

With about 53 million US $ the largest amount of money is spent in the USA on membrane-related R&d work, with about 41 million US $ spent in industry, and 12 million US $ spent in universities and research institutes. The R&D expenditures are about 7 % of the membrane sales by US-based companies. The R&D expenditures in Japan are, compared to the sales, significantly higher than in the USA or, e.g., in the Federal Republic of Germany, where the R&D expenditures are about 8 % of the sales. The amount of R&D spendings in other European countries are comparatively high in relation to the sales. The research is carried out here mainly in government-supported institutions.

It should be pointed out again that an estimation of the membrane-related R&D expenditures is extremely difficult. The numbers in Table IX are only valid for industrialized Western countries with a comparable cost structure. There is little information on the R&D activities in the Eastern European countries. The numbers in Table IX only give a rough idea about the worldwide membrane-related research work.

7. <u>REFERENCES</u>

[1]   Chmiel, H., Strathmann, H.: Chem. Ing. Tech. <u>55</u>, 282 (1983)
[2]   Strathmann, H., Chmiel, H.: Chem. Ing. Tech. <u>57</u>, 581 (1985)
[3]   4th Membrane Technology Planning Conference, Nov. 5-7, New York (1986)
[4]   Ehrfeld, W., Einhaus, R., Münchmeyer, D., Strathmann, H.: Microfabrication of membranes with extreme porosity and uniform pore size; Proceedings 5th International Symposium on Synthetic Membranes in Science and Industry, Tübingen, 1986, J. Membrane Sci. <u>36</u>, 67-77 (1988)
[5]   Sára, M., Sleytr, U.B.:Isoporous ultrafiltration membranes from bacterial cell envelope layers; Proceedings 5th International Symposium on Synthetic Membranes in Science and Industry, Tübingen, 1986, J. Membrane Sci. <u>36</u>,179-186 (1988)

STUDIES ON TRANSPORT PHENOMENA IN HOLLOW FIBRE MEMBRANES
BY MEANS OF DIMENSIONLESS GROUPS

Hans Gorissen

Akzo Engineering bv P.O. Box 9300

6800 SB ARNHEM The Netherlands

Summary

The performance of hollow fibre gas separation membranes is not determined by process conditions and membrane polymer properties alone. The fibre dimensions and the gas viscosity are important factors as well. In order to quantify their influence a mathematical model has been developed which describes the transport phenomena in gas separation modules in which the flow of the gas mixture is perpendicular to the hollow fibre bundle (see figure 1).
In the transport of the permeate gas two resistances in series can be distinguished:
- a membrane resistance, which is selective
- a viscous flow resistance, which is non-selective
The largest resistance will dominate the permeate flow. Therefore the separation will only be sufficiently selective if the viscous flow resistance is smaller than the membrane resistance. This can be achieved by selecting suitable fibre dimensions.
It turns out that the influence of the viscous flow resistance on the performance of a module is mainly determined by a single dimesionless parameter, the transport modulus, Φ_T, which in fact stands for the ratio between the above mentioned two resistances for the key component. At small values of Φ_T (< 0.1) the flow resistance inside the fibres is negligable and hence optimum selectivity is obtained. At large values of Φ_T (> 10) the flow resistance inside the fibres is dominating, and selectivity is strongly reduced.
Simplification of the equations enables short cut analytical approximations with sufficient accuracy for practical applications.
This leads to simple expressions for the capacity and selectivity of membrane modules as a function of physical parameters and fibre dimensions.

1. <u>INTRODUCTION</u>

Calculations of membrane separation processes are in general complicated. This is not obvious because the basic equations for mass transfer through membranes are fairly simple. But because membrane separation processes are kinetically controlled it is necessary to fit all transport steps into the calculations. This leads to a complex set of differential equations, which in general only can be solved numerically.

The aim of this paper is to improve the understanding of the transport phenomena involved in gas separations with hollow fibre membranes. This can be achieved by making the transport equations dimensionless and by simplifying them. From the presented analysis follows a dimesionless group which provides a useful criterion for the selection of fibre dimensions.

On the basis of the obtained insight a short-cut calculation method is developed, which provides a fast and easy tool for design calculations of membrane gas separation units.

The study is limited to cross-flow modules (see fig. 1)

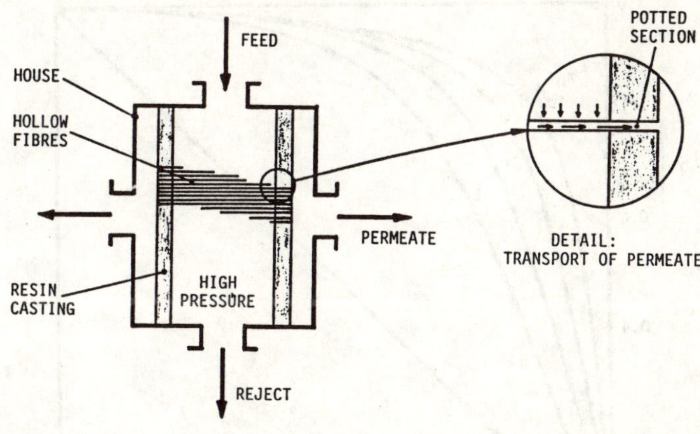

Fig. 1, Schematic representation of a cross-flow hollow fibre module

Before studying the influence of the fibre dimensions on the separation it is important to look at the influence of the pressure ratio (feed/permeate) on the selectivity of the separation. Therefore the separation of a two-component gas mixture in a single stage permeator with perfect mixing on both sides of the membrane [1] is studied. In such a permeator the permeate composition can be expressed as:

$$y = \frac{C - \sqrt{C^2 - 4\,\pi_H\,x\,\alpha^*\,(\alpha^*-1)}}{2\,(\alpha^*-1)} \tag{1}$$

with : $\quad C = \pi_H = (\alpha^*-1)\,(\pi_H\,x + 1)$

$$\pi_H = P_H/P_L \qquad\qquad \text{(pressure ratio)} \tag{2}$$

$$\alpha^* = Q_1/Q_2 \qquad\qquad \text{(ideal separation factor)} \tag{3}$$

Here x and y are the respective mole fractions of the fast gas on the feed and the permeate side of the membrane.

Figure 2 shows an example of the permeate composition as a function of the feed-side composition for a membrane with ideal separation factor 30 for different values of the pressure ratio.

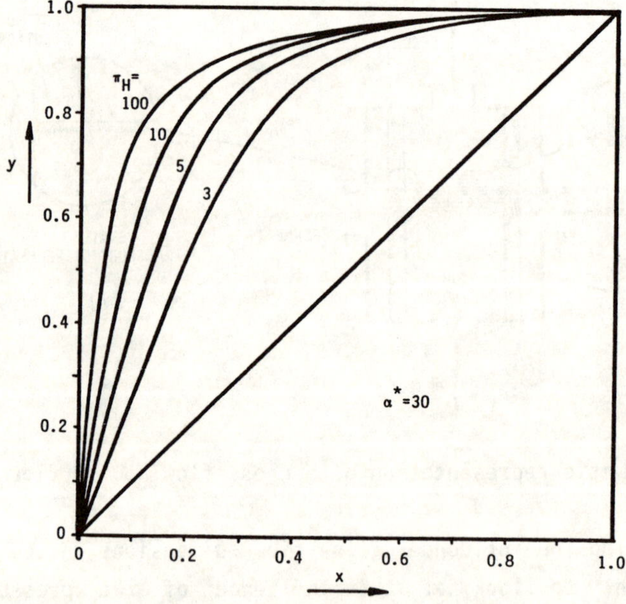

Fig. 2, Influence of the pressure ratio on the selectivity

It can be seen that especially at lower values of x the selectivity of the separation is strongly influenced by the magnitude of π_H.

2. <u>CALCULATION OF MASS TRANSFER IN HOLLOW FIBRE MEMBRANES</u>

In the transport of the permeate gas two resistances in series can be distinguished (see also fig. 1 and 3):
- a membrane resistance, which is selective
- a non-selective viscous flow resistance, which is due to the pressure drop required for the flow of the permeate gas towards the fibre ends.

The largest resistance will dominate the permeate flow. Therefore the separation will only be sufficiently selective if the viscous flow resistance is smaller than the membrane resistance. This can be achieved by selecting suitable fibre dimensions. Such fibre dimensions can be found by solving the transport equations.

A mathematical description of the influence of both resistances on the permeate transport is based on:
- equations for the permeate fluxes of all components
- a differential pressure drop equation for the flow of the permeate gas towards the fibre ends (Hagen-Poisseuille)
- differential material balances of all components.

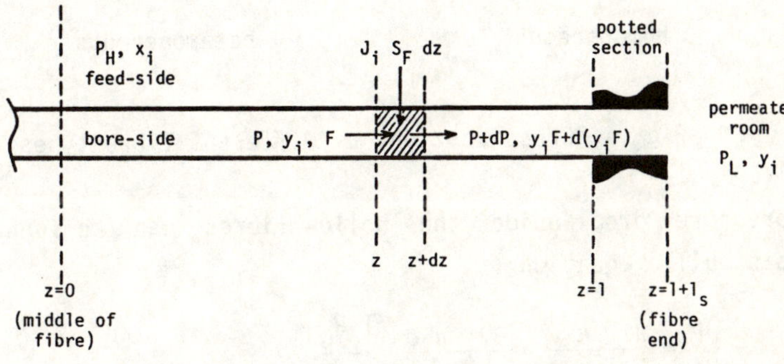

Fig. 3, Permeate transport in a hollow fibre

The permeate flux of component i can be found from:

$$J_i = \frac{Q_i}{d_w} (x_i P_H - y_i^\circ P) \qquad \text{for } 0 \le z \le 1 \qquad (4a)$$

$$J_i = 0 \qquad \text{for } 1 < z < 1+1_s \qquad (4b)$$

Because of the cylindrical shape of the fibres the wall thickness, d_w, should be calculated as follows:

$$d_w = \tfrac{1}{2}\, d_i \ln \frac{d_u}{d_i} \qquad \text{for homogeneous (melt spun) membranes} \qquad (5a)$$

$$d_w = d_w' \frac{d_i}{d_u} \qquad \text{for heterogeneous (wet spun) membranes} \qquad (5b)$$

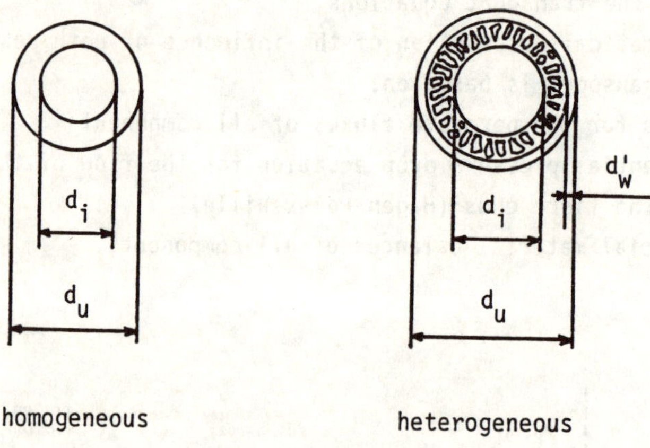

Fig. 4, Cross-sections of different fibre types

The pressure drop inside the hollow fibres can be found from the Hagen-Poisseuille equation:

$$- \frac{dP}{dz} = \frac{32\, \mu\, v}{d_i^2} = \frac{32\, \mu\, \phi}{d_i^2\, A_F} \frac{T\, P_0}{T_0\, P} \qquad (6)$$

The material balance of component i in the shaded area of fig. 3 is as follows:

$$d(y_i\, \phi) = J_i\, S_F\, dz \qquad (7)$$

with:

$$S_F = \frac{4\, A_F}{d_i} \qquad (8)$$

In equation 4a y_i° can be calculated from:

$$y_i^\circ = y_i \qquad \text{for homogeneous fibres} \qquad (9a)$$

$$y_i^\circ = \frac{J_i}{J} \qquad \text{for heterogeneous fibres} \qquad (9b)$$

with: $J = \sum_i J_i$ $\qquad\qquad\qquad\qquad\qquad\qquad (10)$

The boundary conditions for this set of equations are:

$$\phi = 0 \quad \text{and} \quad y_i = \frac{J_i}{J} \quad \text{at} \quad z = 0 \qquad (11a)$$

and: $P = P_L$ $\qquad\qquad$ at $\quad z = 1+1_s$ $\qquad\qquad (11b)$

The following assumptions are made:
- the system is isothermal
- the permeate is a perfect gas
- the viscosity of the permeate is constant
- P_H and x_i are constant along the fibre
- laminar flow without back-mixing inside the fibres
- the permeabilities of all components are constant
- the fibres have perfect cylindrical shapes

For the set of differential equations no analytical solutions have been found. Due to a boundary value problem numerical solution is rather difficult.

For a n-component system 2+n independent variables (z, ϕ, P and $y_1 \ldots y_{n-1}$) and 8+2n parameters can be distinguished. The large number of parameters hinders a neat presentation of results. However, the number of parameters can be reduced by rendering the equations dimensionless. The following dimensionless variables and parameters are therefore introduced:

pressure: $\pi = \dfrac{P}{P_L}$ $\qquad\qquad\qquad\qquad\qquad\qquad (12)$

place: $\xi = \dfrac{z}{l}$ $\qquad\qquad\qquad\qquad\qquad\qquad\qquad (13)$

flow:

$$\lambda = \frac{32 \; \mu \, l}{d_i^2 \, A_F} \frac{P_o \; T}{P_L^2 \; T_o} \phi$$

$$= \frac{40.74 \; \mu \, l}{d_i^4} \frac{P_o \; T}{P_L^2 \; T_o} \phi \tag{14}$$

pressure ratio:

$$\pi_H = \frac{P_H}{P_L} \tag{2}$$

transport modulus:

$$\Phi_T = \frac{128 \; Q_f \, \mu \, l^2}{d_i^3 \, d_w} \frac{P_o \; T}{P_L \; T_o} \tag{15}$$

potted section:

$$s = \frac{l_s}{l} \tag{16}$$

selectivity factors:

$$\alpha_i^* = \frac{Q_f}{Q_i} \tag{3}$$

In these equations a component (f) is selected as the fast key component. The dimensionless differential equations are as follows:

pressure drop:

$$-\frac{d\pi}{d\xi} = \frac{\lambda}{\pi} \tag{17}$$

material balance:

$$\frac{d(y_i \, \lambda)}{d\xi} = \frac{\Phi_T}{\alpha_i^*} (x_i \, \pi_H - y_i^\circ \, \pi) \qquad \text{for } 0 \le \xi \le 1 \tag{18a}$$

$$\frac{d(y_i \cdot \lambda)}{d\xi} = 0 \qquad \text{for } 1 \le \xi \le 1+s \tag{18b}$$

With boundary conditions:

$$\lambda=0 \text{ and } y_i = \frac{J_i}{J} \qquad \text{at } \xi = 0 \tag{19a}$$

$$\text{and: } \pi=1 \qquad \text{at } \xi = 1+s \tag{19b}$$

Thus the number of dimensionless parameters is now 1+2n. For a two-component system the number of independent parameters is then reduced from 12 to 5.

It is striking that one of these parameters, the transport modulus, Φ_T, contains μ, Q_f and all fibre dimensions. The meaning of Φ_T will be explained in paragraph 4.

3. <u>SIMPLIFIED CALCULATIONS</u>

Simplification of the calculations leads to reduced numbers of independent parameters. In one case an analytical solution can be found.

<u>Non-compressible fluid, one-component</u>

In the case of a non-compressible pure fluid the compressibility-therms (P_oT/P_LT_o) in the definitions of λ and Φ_T may be left out. The dimensionless differential equations are then as follows:

$$- \frac{d\pi}{d\xi} = \lambda \tag{20}$$

$$\frac{d\lambda}{d\xi} = \Phi_T (\pi_H - \pi) \qquad \text{for } 0 \leq \xi \leq 1 \tag{21a}$$

$$\frac{d\lambda}{d\xi} = 0 \qquad \text{for } 1 \leq \xi \leq 1+s \tag{21b}$$

For this set of equations the following analytical solution is found:

$$\pi = \pi_H - (\pi_H - 1) \frac{\cosh(\sqrt{\Phi_T}\, \xi)}{\cosh\sqrt{\Phi_T} + s\sqrt{\Phi_T}\, \sinh\sqrt{\Phi_T}} \qquad \text{for } 0 \leq \xi \leq 1 \tag{22a}$$

$$\pi = 1 + \lambda_1 (1 + s - \xi) \qquad \text{for } 1 \leq \xi \leq 1+s \tag{22b}$$

$$\lambda = (\pi_H - 1) \frac{\sqrt{\Phi_T}\, \sinh(\sqrt{\Phi_t}\, \xi)}{\cosh\sqrt{\Phi_T} + s\sqrt{\Phi_T}\, \sinh\sqrt{\Phi_T}} \qquad \text{for } 0 \leq \xi \leq 1 \tag{23a}$$

$$\lambda = \lambda_1 \qquad \text{for } 1 \leq \xi \leq 1+s \tag{23b}$$

$$\text{with:} \qquad \lambda_1 = \lambda_{(\xi=1)} = (\pi_H - 1) \frac{\sqrt{\Phi_T}\, \tanh\sqrt{\Phi_T}}{1 + s\sqrt{\Phi_T}\, \tanh\sqrt{\Phi_T}} \tag{24}$$

If the permeate viscosity is equal to zero (no pressure drop inside the fibres) then $\pi=1$ for $0 \leq \xi \leq 1+s$, giving maximum λ_1. Then equation 21a becomes:

$$\frac{d\lambda}{d\xi} = \Phi_T (\pi_H - 1) \qquad \text{for } 0 \leq \xi \leq 1 \tag{25}$$

$$\text{Giving:} \qquad \lambda_{1,NPD} = \Phi_T (\pi_H - 1) \tag{26}$$

The influence of the viscous flow resistance on the permeate flow can be seen from the transport efficiency, η_T, which is defined as:

$$\eta_T \equiv \frac{\lambda_1}{\lambda_{1,NPD}}$$

$$= \frac{\eta_{T,s=0}}{1 + s\ \Phi_T\ \eta_{T,s=0}} \tag{27}$$

with: $\qquad \eta_{T,s=0} = \frac{\tanh \sqrt{\Phi_T}}{\sqrt{\Phi_T}} \tag{28}$

If $\eta_T \to 1$ then the pressure drop is very low. The transport of the permeate is then dominated by the permeation resistance. If $\eta_T \to 0$ the transport is dominated by the viscous flow resistance. The transport efficiency appears to be independent of π_H. Figure 5 gives η_T versus the transport modulus, Φ_T, for several values of s.

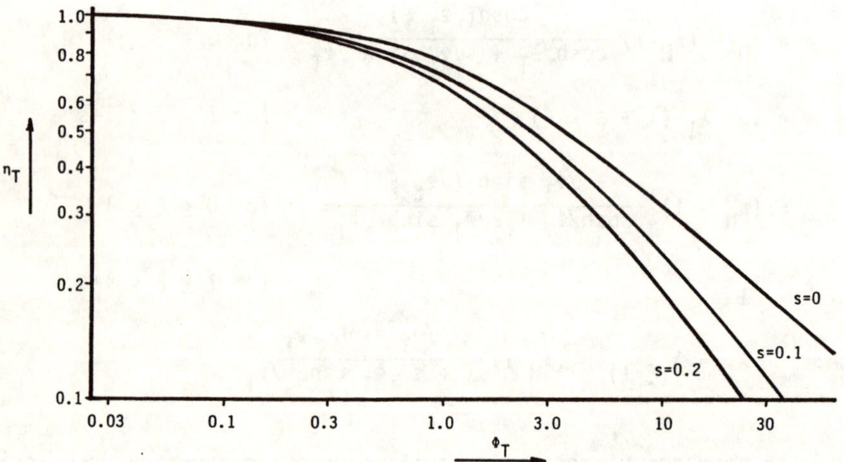

Fig. 5, Transport efficiency versus the transport modulus
(non compressible one-component system)

Figure 5 shows that Φ_T is an important measure for the influence of the viscous flow resistance. For $\Phi_T < 0.1$ this resistance is not important, for $\Phi_T > 10$ the permeate flow is completely dominated by it.

<u>Pure gas, without a potted section</u>

In this case the set of dimensionless differential equations exists of equations 17 and 21. No analytical solution has been found. After numerical solution the transport efficiency appears to depend on Φ_T and π_H, as is shown in figure 6.

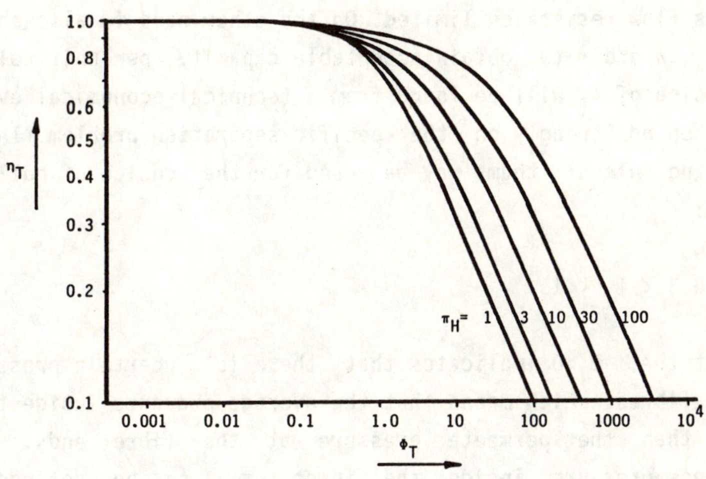

Figure 6, Transport efficiency as a function of the transport modulus
(pure gas, no potted section)

Also in this case the viscous flow resistance plays no role for $\Phi_T < 0.10$ and it is dominating if $\Phi_T \gg 10$.

4. <u>THE MEANING OF THE TRANSPORT MODULUS</u>

The preceding paragraphs show that the transport modulus, Φ_T, is an important measure for the influence of the viscous flow resistance on the permeate transport. In fact Φ_T is proportional to the ratio between the viscous flow resistance inside the hollow fibres and the permeation resistance in the fibre walls.

For fixed values of Q_f, μ, T and P_L the transport modulus can only be influenced by changing the fibre-dimensions. In order to limit the influence of the viscous flow resistance it is necessary to use short, thick and thick-walled fibres. This is conflicting with the demand for thin and thin-walled fibres in order to get sufficient capacity per unit separator volume. The choice of 1 is limited, because the fibre length is related to the external dimensions of the separator.

On one hand Φ_T should not be too high in order to keep the influence of the viscous flow resistance limited. On the other hand Φ_T also should not be too low in order to obtain acceptable capacity per unit volume. The optimum choice of Φ_T will be found from a technical-economical evaluation. This will depend strongly on the specific separation problem. In general the following rule of thumb may be used for the choice of hollow fibre dimensions:

$$0.3 \leq \Phi_T \leq 1 \tag{29}$$

The fact that $\Phi_T > 0$ implicates that there is a certain pressure drop inside the fibres, which means that the average pressure inside the fibres is higher than the permeate pressure at the fibre ends. The mean dimensionless pressure inside the fibres, π_F, can be defined by the following equation (one-component system):

$$\lambda_1 = \Phi_T \ (\pi_H - \pi_F) \tag{30}$$

From equations 26 and 27 the following equation for λ_1 can be derived:

$$\lambda_1 = \eta_T \ \Phi_T \ (\pi_H - 1) \tag{31}$$

The average pressure can be derived from equations 30 and 31:

$$\pi_F = 1 + (1 - \eta_T) \ (\pi_H - 1) \tag{32}$$

Paragraph 1 shows that π_H has a major influence on the selectivity of the separation. If $\pi_F > 1$ then the actual pressure ratio will be smaller than π_H, resulting in decreasing selectivity in the case of a multi-component separation.

The resulting effective pressure ratio, $\pi_{H,eff.}$ is defined by:

$$\pi_{H,eff.} \equiv \frac{\pi_H}{\pi_F} \tag{33}$$

Figure 7 shows $\pi_{H,eff.}$ as a function of Φ_T for several values of π_H.

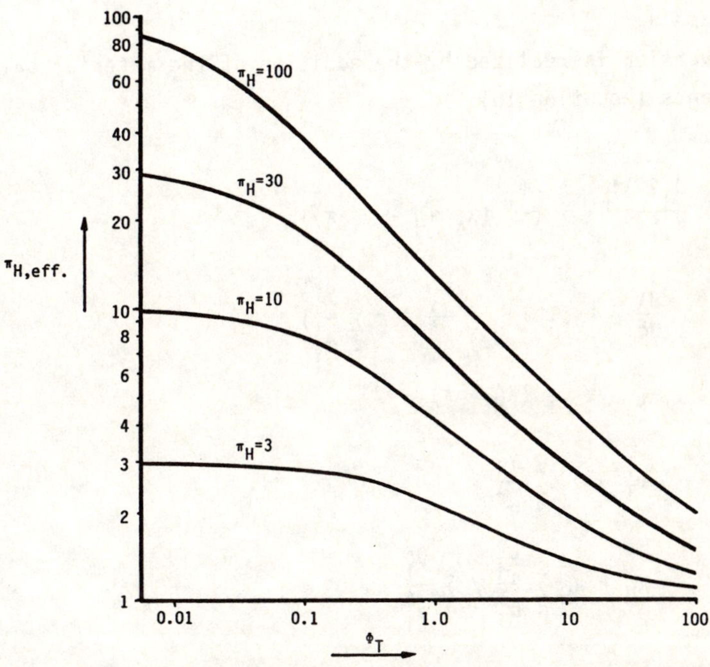

Figure 7, The effective pressure ratio as a function of the transport modulus (pure gas, s=0)

5. SHORT-CUT CALCULATIONS, THE π_F-METHOD

So far calculations were limited to one-component systems. This paragraph provides a method to estimate permeate flow and composition for multi-component systems on the basis of the transport modulus. Next this calculation method will be used for a complete unit calculation.

Permeate flow and composition

The calculation method is based on estimating the mean permeate pressure inside the fibres. If a good estimate for π_F is available, the calculation is equal to the method presented in paragraph 1. In paragraph 4 a method is described for the calculation of π_F in a one-component system. Estimation of π_F in a multi-component system requires conversion of the transport equations of this system to the equations of a pure gas with the same π_F.

The conversion is realized by the addition of the material balances for all components (equation 18):

$$\sum_i \frac{d(y_i\lambda)}{d\xi} = \sum_i \left(\frac{\Phi_T}{\alpha_i^*} (x_i\,\pi_H - y_i^\circ\,\pi)\right)$$

$$\rightarrow \qquad \frac{d\lambda}{d\xi} = \Phi_T \left(\pi_H \sum_i \frac{x_i}{\alpha_i^*} - \pi \sum_i \frac{y_i^\circ}{\alpha_i^*}\right)$$

$$= \Phi_T' \,(\pi_H' - \pi) \tag{34a}$$

with:

$$\Phi_T' = \Phi_T \sum_i \frac{y_i^\circ}{\alpha_i^*} \tag{34b}$$

and:

$$\pi_H' = \pi_H \sum_i \frac{x_i}{\alpha_i^*} \Big/ \sum_i \frac{y_i^\circ}{\alpha_i^*} \tag{34c}$$

Equation 34a is similar to equation 21a, except for the fact that Φ_T' and π_H' depend on y_i° and therefore vary with ξ. However for sufficiently low values of Φ_T y_i° will be almost constant and equal to y_i at $\xi=1$ (low pressure drop inside the fibres). In that case equations 34a and 21a are equal.

For the calculation of π_H' and Φ_T' an estimate of the permeate composition (y_i) is required. This estimate can be found by assuming $\pi_F \approx 1$ and following the procedure of paragraph 1.

For the calculation of π_F from π_H', Φ_T' and s an empyrical correlation is required. π_F can be calculated from equation 32, using the following correlation for η_T:

$$\eta_T = \frac{\tanh \sqrt{\Phi_c}}{\sqrt{\Phi_c}} \tag{35a}$$

with:
$$\Phi_c = \Phi_T' \left(1 - \frac{\pi_H'-1}{\pi_H'+2} C_1\right) (1+3 \cdot s) \tag{35b}$$

$$C_1 = 1 - \exp \left(C_3 - \sqrt[m]{C_2 \, \Phi_T' + C_3^m}\right) \tag{35c}$$

$$C_2 = (\pi_H'+20)^{2.356}/2590 \tag{35d}$$

$$C_3 = 0.119 \, (\pi_H')^{0.346} \tag{35e}$$

$$m = 0.95 \, \ln(\pi_H'+3) \tag{35f}$$

Finally the permeate composition is calculated following the procedure of paragraph 1 and by using $\pi_{H,eff}$. The permeate flux can be calculated from equation 4a and 9b, using π_F:

$$J = \frac{Q_i}{d_w} \left(\frac{x_i}{y_i} \, P_H - \pi_F \, P_L\right) \tag{36}$$

The short-cut calculation method has a reasonable accuracy. The absolute error in y_i is in general less than 0.005 if $\Phi_T < 1$. The deviation in the permeate flux is less than 2%.

Unit calculation

So far separate fibres have been studied. In a cross-flow module different fibres with varying x_i and therefore also different π_F are present. Calculations for such modules, assuming $\pi_F=1$, are described in the literature [2]. Figure 8 gives a schematic representation of the cross-flow module.

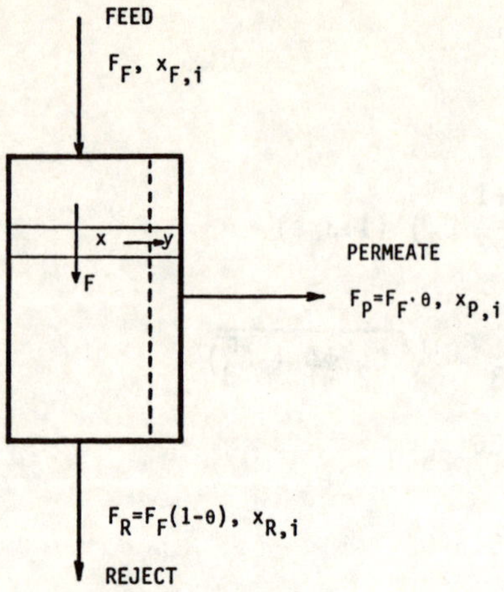

Fig. 8 Cross-flow module

For such a module the mean permeate composition can be calculated from the overall material balance:

$$\bar{y}_i = \frac{x_{i,F} - x_{i,R}(1-\theta)}{\theta} \tag{37}$$

The stage-cut, θ, can be calculated from the differential material balances:

$$- d((1-\theta)x_i) = y_i^\circ \, d\theta$$

$$\rightarrow \qquad d(\ln(1-\theta)) = \frac{dx_i}{y_i^\circ - x_i} \tag{38}$$

The relation between y_i° and x_i depends on membrane selectivity, operating conditions and module arrangement. For the separation of a binary mixture in a cross-flow module with $\pi_F = 1$ the following equation for θ is found:

$$\theta = 1 - \left(\frac{y_R}{y_F}\right)^{A1} \left(\frac{1-y_F}{1-y_R}\right)^{A2} \left(\frac{A3-y_R}{A3-y_F}\right) \tag{39a}$$

with:
$$A1 = \frac{\alpha^*}{\alpha^*-1} \frac{\pi_H}{\pi_H-1} - 1 \tag{39b}$$

$$A2 = \frac{1}{\alpha^*-1} \frac{\pi_H}{\pi_H-1} + 1 \tag{39c}$$

$$A3 = \frac{\alpha^*}{\alpha^*-1} \tag{39d}$$

Here y_F and y_R are the respective local mole fractions of the fast gas in the permeate near to the feed and the reject.

The required membrane surface area can be calculated from:

$$J_i \frac{d_i}{d_u} \, dA = F_F \, y_i \, d\theta$$

$$\rightarrow \quad A = \frac{F_F \, d_w \, d_u}{Q_1 \, P_L \, d_i} \frac{\theta \, (1 + (1-\bar{y}) \, (\alpha^*-1))}{\pi_H-1} \tag{40}$$

In this equation A is the required membrane surface area (outside of the hollow fibres).

If the pressure drop in all fibres is the same the influence of this pressure drop on $\bar{y}$, θ and A can be calculated very easily by replacing π_H and P_L by $\pi_{H,eff.}$ and $\pi_F \, P_L$ in equations 39 and 40.

In general π_F will vary with x, which means that π_F is not equal for all fibres (the modified equations 39 and 40 are not valid). However, for estimative purposes a mean π_F can be used in the modified equations 39 and 40, which is calculated for the mean feed-side composition:

$$\bar{x} = \frac{x_F + x_R}{2} \tag{41}$$

In general the error in the calculated membrane surface area from this approximation is less than 10%.

6. <u>CONCLUSIONS</u>

The transport modulus, Φ_T, is an important measure for the influence of the viscous flow resistance inside the hollow fibres on the performance of a membrane separation. Φ_T can be used as a criterion for the selection of suitable hollow fibre dimensions.

Based on Φ_T a short-cut calculation method for cross-flow gas separation modules is developed. Whereas in general complicated computer models are necessary for the calculation of gas separation modules it is now possible to get reasonable estimates from a pocket calculator.

The method will also be applicable to other membrane gas separation modules, like co- and counter current arrangements. Bruining [3] made a more or less similar study for liquid systems with the feed-side inside the fibres. This example indicates a rather universal applicability of the transport modulus in studies of hollow fibre membranes.

<u>ACKNOWLEDGEMENTS</u>

This paper is based upon work supported by the EEC under project no. RI 1B-0098-NL. The author wishes to thank Mr. Den Decker for carrying out the numerical calculations.

<u>REFERENCES</u>

1)   Weller and Steiner, Separation of gases by fractional permeation through membranes, J.Appl.Phys., 21, 279 (1950).

2)   Pan, Gas separation by high-flux, asymmetric hollow-fibre membrane, AIChe Journal, Vol. 32 No. 12, 2020 (1986).

3)   Bruining, A general description of flows and pressures in hollow fiber membrane modules, accepted for publication in Chem.Eng.Sci.

NOMENCLATURE

A	membrane surface area (outside fibres)	m^2
A_F	cross section of a fibre bore	m^2
d_i	internal fibre diameter	m
d_u	external fibre diameter	m
d_w	effective wall thickness	m
F_F	feed flow	m_o^3/s
J	permeate flux	m_o^3/s
l	half fibre length	m
l_s	length of the potted section of a fibre	m
P_H	feed-side pressure	Pa
P_L	outlet pressure of the permeate gas	Pa
P	local permeate pressure inside the fibre bore	Pa
Q	permeability	$m_o^3/m\ Pa\ s$
s	potting ratio	-
S_F	circumference of a fibre bore	m
T	temperature	K
v	local gas velocity	m/s
x	feed-side mole fraction (of the fast gas)	-
y	permeate mole fraction (of the fast gas)	-
z	distance from the middle of the fibre	m
α^*	ideal separation factor	-
η_T	transport efficiency	-
θ	stage cut (permeate/feed ratio)	-
λ	dimensionless permeate flow	-
λ_1	λ at fibre end	-
μ	permeate gas viscosity	$Pa\ S$
ξ	dimensionless z	-
π	dimensionless permeate pressure	-
π_H	pressure ratio	-
π_F	mean dimensionless permeate pressure	-
ϕ	local permeate flow inside the fibre bore	m_o^3/s
Φ_T	transport modulus	-

Subscipts

f	fast key component
i	component i
o	standard conditions

INDUSTRIAL APPLICATIONS

Recent developments in testing new membrane systems at Aere Harwell for nuclear applications

Current applications for drying acetic and formic acid and commercialised techniques

Studies on supported liquid membranes performed by scientist of the Casaccia Research Centre

Use of liquid membranes for treatment of nuclear wastes

Characteristics and performance of new types of ultrafiltration membranes with chemically modified surfaces

Membrane separation of CO_2 and H_2S from mixtures with gaseous hydrocarbons

Development of gaseous permeation membranes adapted to the purification of hydrocarbons

Optimization of an UF pilot plant for the treatment of radioactive waste

Use of membrane technology for processing waste water from olive-oil plants with recovery of useful components

A study on membrane technology for the textile industry : membrane preparation and characterization

Application of electro-osmosis to radioactive waste processing

Application of reverse osmosis to the treatment of liquid effluents produced by nuclear power plants

Fouling analysis and control

Industrial applications of microfiltration

RECENT DEVELOPMENTS IN TESTING NEW MEMBRANE SYSTEMS AT
AERE HARWELL FOR NUCLEAR APPLICATIONS

I W CUMMING AND A D TURNER
CHEMICAL ENGINEERING AND MATERIALS DEVELOPMENT DIVISION
HARWELL LABORATORY, UKAEA, ENGLAND

Summary

The development of a crossflow filtration process for the treatment
of radioactive liquid waste has been carried out at the Harwell
Laboratory. A number of different membranes in the ultrafiltration
and microfiltration range have been tested on simulated and real
radioactive effluents under both normal crossflow and electrically
cleaned conditions. The membrane fluxes and alpha rejection at
different operating conditions have been determined for the Harwell
site low level waste. It has also been shown that the flux can be
restored by chemical cleaning. For normal crossflow the membrane
flux corrected to 25°C has been found to decline to a steady state
value of 1.5m/d for a Tech Sep M4, 2m/d for a Tech Sep M6, 4m/d for
a Ceraflo 0.2μm filter and 2.8m/d for a Membralox 0.2μm filter. The
alpha removal using all these crossflow filters was to below
1.5mBq/mℓ when operating at pH5.
   The periodic direct electrical cleaning of conductive membranes
permits the use of lower crossflow velocities and pressures while
maintaining or even enhancing membrane flux. A Tech-Sep M4 gave a
permeation rate (again corrected to 25°C) of 2.3m/d for the Harwell
low level waste at 2 bar and 1 m/s crossflow with 1 second, 50mA/cm^2
cleaning pulses. The MA1 sintered stainless steel fibre
microfiltration membrane cleaned by 5 second 200mA/cm^2 pulses gave
corresponding fluxes of 30-8m/d at 1-2 bar and 1m/s crossflow for
50ppm Fe(OH)$_3$ treated low level waste over the concentration range
150ppm - 5% solids. Electrical cleaning has been found not to
compromise membrane life.

1.   INTRODUCTION
   Crossflow filtration has been developed at Harwell for a range of
different membranes for the separation of precipitated active solids from
radioactive liquids. This paper describes the testing of these systems
for the treatment of liquid active waste at the Harwell Laboratory both
under conventional filtration conditions and also with the assistance of
a novel electrical defouling process.
   The solids to be separated arise from either precipitated active
particulates or from finely divided inorganic ion-exchangers which have
been added to the waste to absorb soluble radionuclides. The
precipitated active material may be produced by the adjustment of the
waste pH and this can sometimes be enhanced by the use of a
co-precipitant such as ferric hydroxide. In some cases it is possible to
tailor the precipitation conditions of the dispersed phase to improve the
subsequent solid/liquid separation.
   The technique of crossflow filtration has been chosen in preference
to other solid/liquid separation processes for reasons of process
flexibility and simplicity. As a separation technique it can operate
well over a wide range of feed solids concentration without affecting the
efficiency of solids removal. The operating conditions of the process

such as pH can be changed or a different absorber can be dosed into the waste feed without modifying the plant. The technique can also produce a sludge with a high solids content and, due to the efficiency of separation, the activity decontamination factor is maximised. The slurry produced by the process is also pumpable which makes its subsequent handling more convenient. A further advantage of the process is that as only small quantities of absorbers are used, a better volume reduction factor can be achieved than for many other solid/liquid separation techniques – such as conventional floc processes.

For normal pressure–driven filtration a filter pore size less than the particles being filtered is required in order to give a high solids retention and hence a good radioactive decontamination factor. However, in time, a fouling layer is deposited on the filter surface. As this layer has a permeability significantly lower than the filter itself, the filtration rate will decline. In crossflow filtration the thickness of this layer is controlled at a significantly reduced level by the establishment of a hydrodynamic boundary layer parallel to the membrane surface. This can result in the maintenance of good membrane fluxes even when filtering slurries with a high solids content. Although the crossflow does maintain high fluxes over appreciable operating periods, there will still be a long term flux decline – possibly to some equilibrium value. Although this can sometimes be restored by backflushing of the membrane, this does not work for all effluents and can in some cases reduce the membrane life. Eventually, after an extended period of operation, the membrane flux is likely to decline to a level where a periodic chemical cleaning of the membrane is necessary to return the filter to near its initial performance.

For electrically conductive membranes, there is an alternative method of controlling surface fouling which has advantages over backwashing the membrane and is also considerably more effective. This involves the in situ generation of microscopic gas bubbles at the membrane surface (Direct Membrane Cleaning – DMC) by the passage of an electrolytic current supplied from a counter electrode through the waste being treated.

"In-to-out" filtration at tubular membranes has been selected as the most suitable for scale-up in nuclear applications. This is partly the result of the desire to minimise the length of seal between the concentrate and permeate side of the filter for a given membrane area. However, it is also the simplest means of preserving the uniformity of flow across the filter surface. This enables the best use to be made of the available membrane area by minimising fouling and also reduces wear caused by turbulence arising from changes in slurry flow direction.

In considering the specification for membrane selection and filter module design for both conventional and electrically enhanced filtration, radiation and chemical resistance and the prevention of tube blockage are important factors. Large bore tubular membranes have been chosen rather than capillary membranes, as the former have been shown to be less susceptible to tube blockage [1] and can therefore operate at higher solids concentrations. Inorganic rather than organic membranes have been selected for a number of reasons. They are generally more chemically resistant than organic membranes and cleaning can therefore be carried out using a wider range of reagents. As a consequence, a greater range of fouling materials can be removed from the membrane. A further advantage of inorganic membranes is that they have a significantly higher radiation stability.

2. <u>PERFORMANCE OF MEMBRANES FILTERING A RADIOACTIVE WASTE UNDER NORMAL CROSSFLOW</u>

Four different inorganic membranes have been assessed whilst processing a real radioactive liquid waste. The effluent used was the Harwell Site Low Level Waste which arises from the low active drain system on the Harwell site. This waste contains alpha, beta and gamma activity – all at levels of less than 0.4 Bq/ml. The solids content of the waste is variable between 50–500 mg/l, though normally at ~ 100mg/l. The tests were carried out in a crossflow filtration pilot plant, a simple schematic of which is shown in Figure 1. This is described in more detail in a later paper. This plant has been operated at a circuit solids concentration of between 1–2 wt% solids. Before processing the waste through the membrane, it was prefiltered through a coarse 0.5mm screen prior to pH adjustment to ~ 5, which had been found to give optimum alpha removal. Throughout the tests the membrane crossflow velocity was maintained at between 4–4.5 m/s and the temperature of the waste in the filter circuit at between 30–40°C. During plant operation the concentrated slurry in the recirculation loop was discharged for a few seconds approximately twice an hour. The effect of this was to depressurize the circuit during the discharge. All of the data quoted in section 2 of this paper have been normalized to a standard temperature of 25°C by correcting for the viscosity change from the operating temperature to that at 25°C.

2.1 Tech Sep M4 membrane

The M4 Tech Sep membranes were supplied in a stainless steel cased module containing an array of 37 tubes. They consisted of a zirconia layer supported on the inside surfaces of porous carbon tubes. The tube length is 1.2m with an internal diameter of 6mm and a wall thickness of 2mm. These membranes are quoted as being operable in the pH range between 0 and 14. This ultrafilter was the only one tested having a nominal molecular weight cut-off of 20,000 (approximately equivalent to a surface pore size of ~ 2nm).

During most of the tests on this membrane the permeate flux was held constant at ~ 1.2m/d. The transmembrane pressure therefore rose so as to maintain this flux as surface fouling occurred. The membrane was operated without being cleaned for a period of 1500 hours. At the start of the experiment the average transmembrane pressure was ~ 2.1 bar and this had risen by the end of the test to ~ 4 bar. The membrane declined in performance from an initial permeability of 0.8m/d/bar to ~ 0.4m/d/bar very rapidly, before reaching a steady value – suggesting that the membrane had reached an equilibrium level of fouling (Figure 2).

A second test using this membrane was conducted at an average transmembrane pressure of 4.7 bar. The membrane permeability declined from 0.6m/d/bar to 0.3m/d/bar over 580 hours of operation, as shown in Figure 3. This corresponds to a decline in flux from 3.1m/d to 1.5m/d. Raising the transmembrane pressure, therefore, gave some improvement in membrane throughput but also caused significantly increased fouling of the membrane. Again it would appear that in time the level of fouling approached an equilibrium value. Throughout these tests the permeate alpha content was normally at or below 1.5mBq/ml – which was the limit of detection of the counting equipment used. The membrane was cleaned after each trial by recirculating 1M sodium hydroxide solution around the filtration circuit for 1 hour at 50°C. This was followed by recirculating 1M nitric acid also for 1 hour at 50°C. This cleaning procedure gave a very good recovery of permeability back to near its

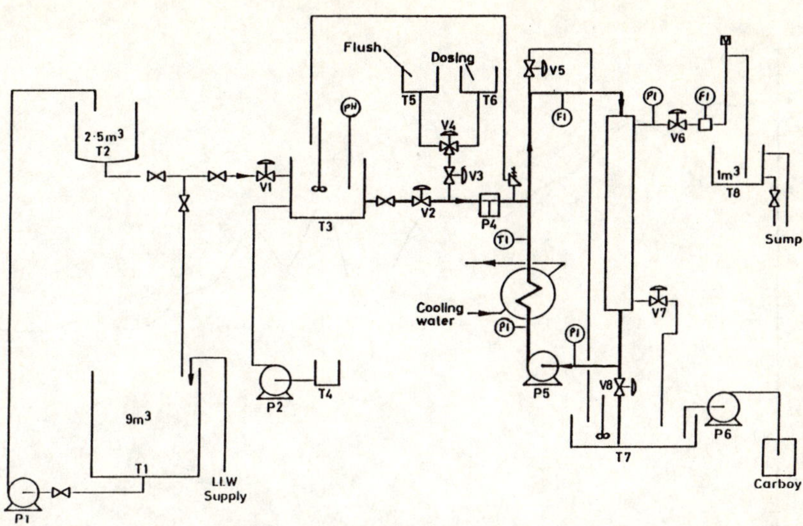

Figure 1 Schematic diagram of Harwell LLW pilot plant

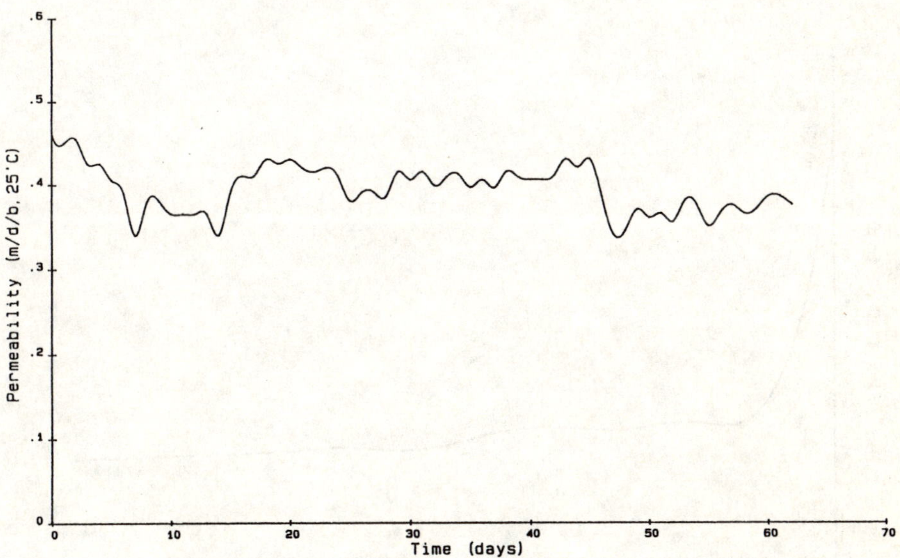

Figure 2 Permeability of Tech Sep M4 membrane when processing
Harwell LLW at a throughput of 1.2 m/d

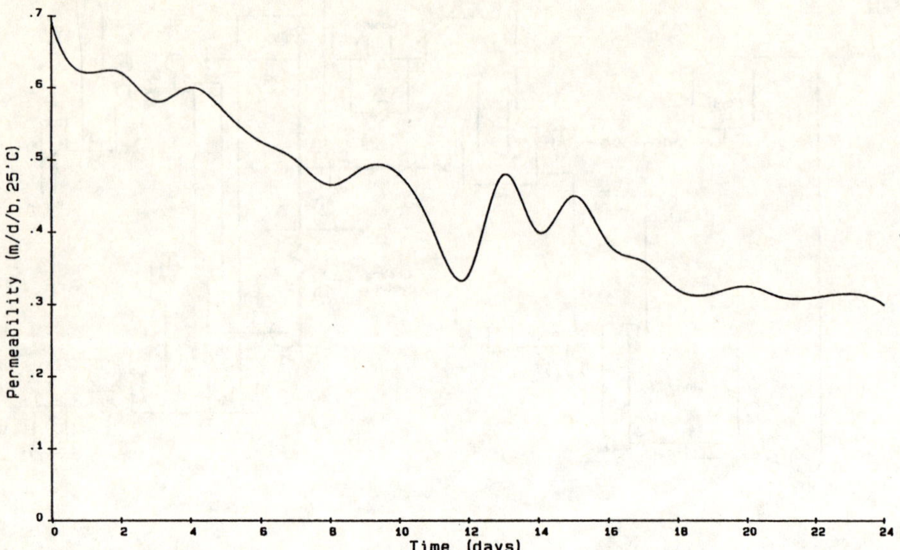

Figure 3 Permeability of M4 Tech Sep membrane processing
Harwell LLW at a transmembrane pressure of 4.7 bar

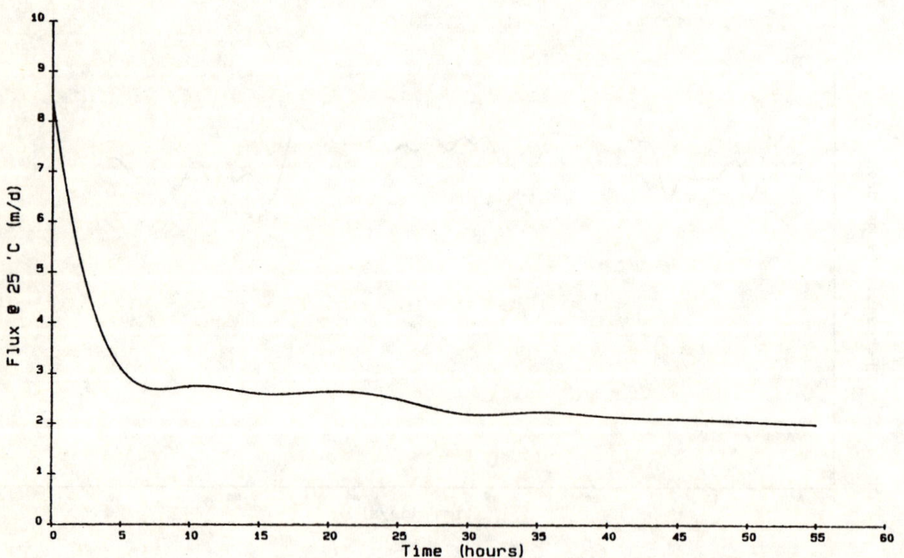

Figure 4 Flux profile of a Tech Sep M6 membrane processing
Harwell LLW at a transmembrane pressure of 4.2 bar

initial value.

2.2 Tech Sep M6 membrane

The materials and dimensions of the M6 Tech Sep membrane are similar
to the M4 membrane previously described. The module also contains 37
tubes. This filtration medium, however, was different in having a much
more open structure with a nominal molecular weight cut-off of ~ 10^6
(equivalent to a surface pore size of approximately 0.08μm). The clean
water permeability of the M6 membrane was about twice that of the M4.

The test was carried out over a 50 hour period using a transmembrane
pressure of between 4-4.5 bar. During this time, the filtration flux
rapidly declined from 8.5 m/d to 2.0 m/d as shown in Figure 4. After ~
24 hours, the flux was ~ 2.3 m/d (a permeability of 0.55 m/d/bar) - which
was no higher than that achieved using the finer pore size M4 membrane.
The alpha activity detected in the permeate was again at or below 1.5
mBq/mℓ. However, although the alpha rejection was still good, there did
not appear to be any advantage in using this membrane to treat this
particular effluent stream. The membrane was again cleaned using the 1M
sodium hydroxide and 1M nitric acid washing procedure to give a flux
recovery back to near its original value.

2.3 'Ceraflo' membrane

The 'Ceraflo' membrane is manufactured by the Norton Corporation of
America and is a microfilter with a nominal pore size of 0.2μm. The
membrane tubes are constructed from alumina and have an internal diameter
of 2.8 mm with a length of 0.45m. The stainless steel cased module
contained 56 of these tubes with a total membrane area of 0.205m². As
the alumina material making up the membrane was believed to have a lower
tolerance to acid and alkali than the Tech Sep membranes, a limit for
cleaning reagent concentration was set at 0.5M.

This membrane was operated at a membrane crossflow velocity of ~ 4.2
m/s with an average transmembrane pressure which varied between 3-3.6
bar. The test was continued for 55 hours. The flux declined initially
very rapidly to ~ 7 m/d and then more slowly to ~ 4.5 m/d, as shown in
Figure 5. A repeat experiment over 102 hours indicated that the flux
remained stable at ~ 4 m/d. Tests on this membrane showed that it had
about twice the flux rate of the Tech Sep membranes when processing
Harwell low level waste, while still giving a permeate with an alpha
content consistently at or below 1.5 mBq/mℓ. It would appear that the
more open pore microfilters were able to give as good a rejection of the
alpha content of this waste as the ultrafilter. The membrane was cleaned
by recirculating 0.5 M sodium hydroxide at 50°C for one hour followed by
recirculating 0.5M nitric acid at 50°C for a further hour. This cleaning
process again gave a good flux recovery. Examination of the module after
cleaning showed that although the membrane tubes were not showing any
signs of blockage, the tube plate at the module inlet was covered with
fibrous material from the effluent. This would appear to be a problem
associated with the smaller tube size and use of this particular membrane
would require better feed prefiltration to remove fibrous material from
the waste.

2.4 Membralox membrane

The Membralox membranes are manufactured by SCT from alumina. A
pore size of approximately 0.2μm was selected for testing. The module
contained 29 tubes which had a 7mm inside diameter and a wall thickness
of 1.5mm. The tube length was 0.75m with an active filtration area of

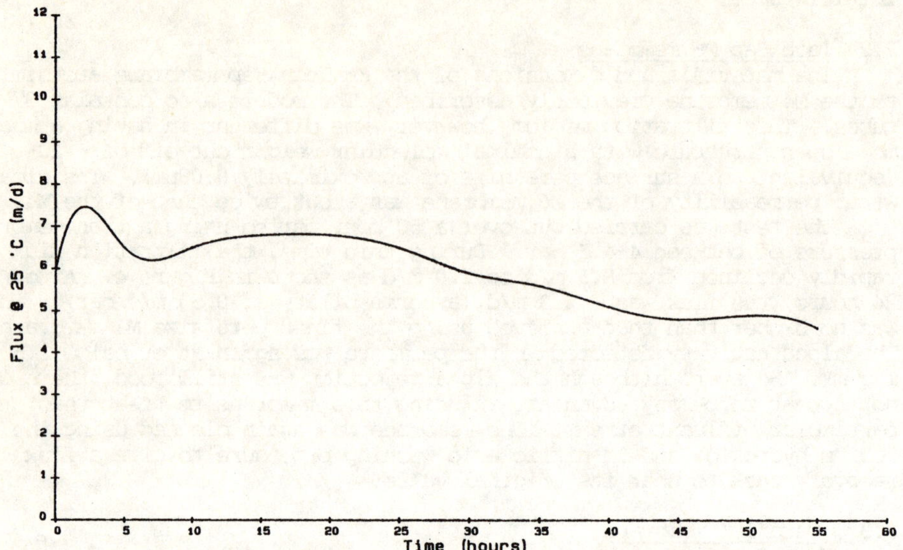

Figure 5 Flux profile of a Ceraflo membrane processing
Harwell LLW at a transmembrane pressure of 3.6 bar.

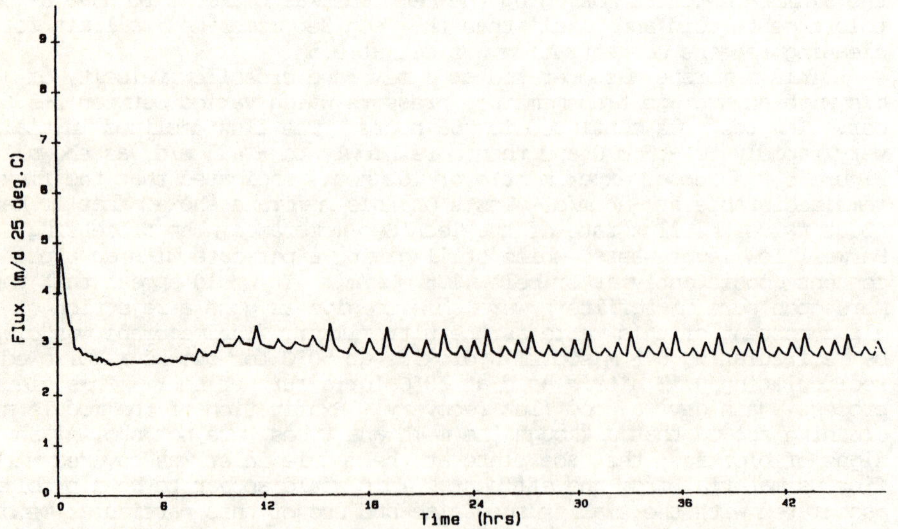

Figure 6 Flux profile for a Membralox module processing Harwell
LLW at a transmembrane pressure of 2 bar.

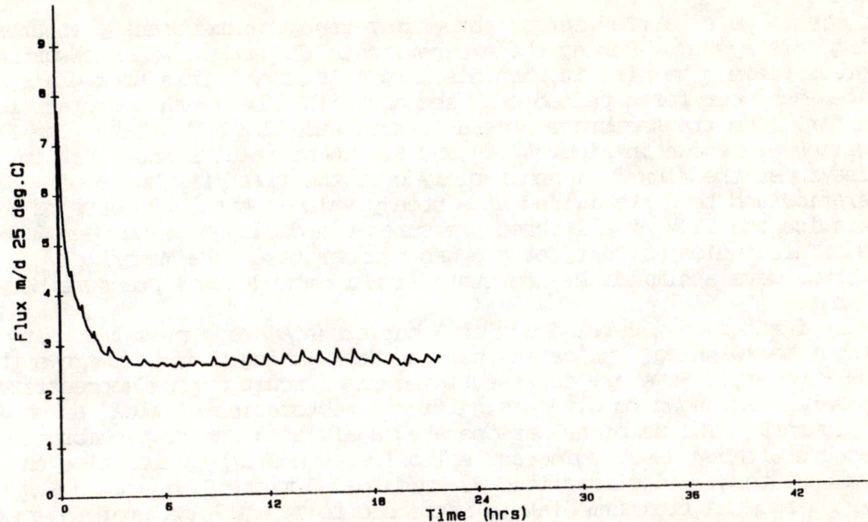

Figure 7   Flux profile for a Membralox module processing
Harwell LLW at a transmembrane pressure of 3 bar.

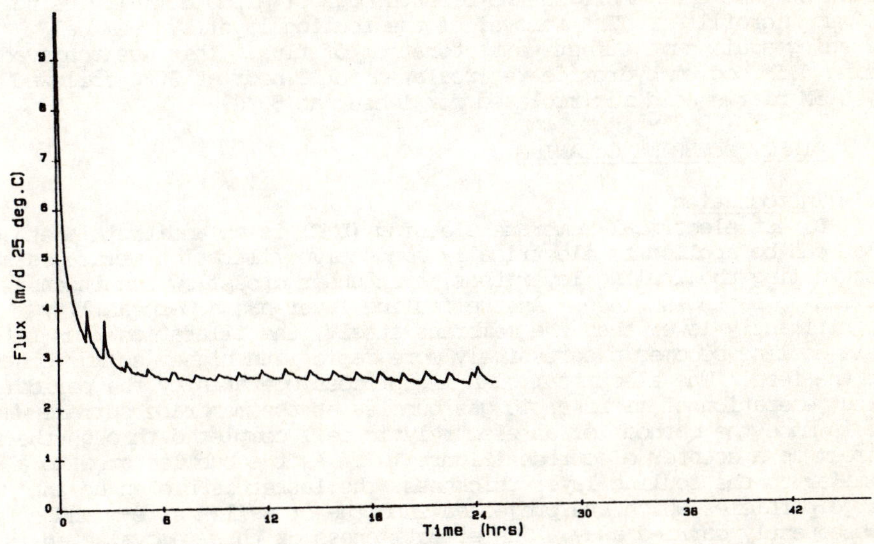

Figure 8   Flux profile for a Membralox module processing
Harwell LLW at a transmembrane pressure of 4 bar.

0.48m².

Tests were carried out at three different transmembrane pressures using this system. During these experiments the module was backwashed on depressurizing the circuit when discharging slurry. This took place twice each hour for a period of 4 seconds with a backwash overpressure of ~ 1 bar. The transmembrane pressures used were 2,3 and 4 bar. The flux profiles are shown in Figures 6,7 and 8. These results show that in all three runs, the flux declined rapidly over the first few hours of operation and then stabilised at a steady value. After 20 hours of operation the flux at all three pressures had declined to between 2.5-2.8 m/d. This indicates that for the waste under test, the Membralox membrane gave a similar steady state flux independent of pressure between 2-4 bar.

A further test carried out at 3 bar transmembrane pressure, with and without backwashing, indicated that the backwashing did not improve the flux recovery. However, depressurising the circuit during concentrate discharge was shown to give a significant restoration of flux, as shown in Figure 9. The membrane was operated again at 3 bar transmembrane pressure without discharging any solid for 48 hours by which time the flux had steadily declined from 2.6 m/d to 0.8 m/d. Discharging was then started and the flux immediately increased to 1.8 m/d and over the next 48 hours of operation recovered to 2.1 m/d.

During all these tests the alpha content in the permeate was consistently below 1.5 mBq/mℓ. A sample taken 5 minutes after plant start-up gave no higher measurement in permeate alpha activity, which would suggest that there was no breakthrough of alpha activity using a 0.2 μm microfilter. The removal of the fouling layer by chemical cleaning again gave effective restoration of flux. This was achieved using 0.5M sodium hydroxide recirculated for 1 hour at 50°C followed by 0.5M nitric acid recirculated for 1 hour at 50°C.

3   DIRECT MEMBRANE CLEANING

3.1 Introduction

Direct electrical Membrane Cleaning (DMC) is an additional method that can be applied to electrically conductive filtration membranes for controlling the fouling layer that, even under crossflow conditions, builds up in the surface. As the fouling layer has a permeability significantly lower than the membrane itself, the filtration flux falls. This decline becomes progressively more rapid with higher solids contents in the feed. The electrical cleaning process operates by the periodic in situ generation of microscopic gas bubbles at the membrane surface when it is made the cathode of an electrolytic cell completed through the feed stream to a counter electrode (Figure 10). As the bubbles grow to a size similar to the fouling layer thickness, the latter is broken up into large particles which are projected into the crossflow stream and subsequently carried away. The effectiveness of this removal step is enhanced by a temporary reduction in transmembrane pressure during current application, as this not only increases the rate of bubble growth for a given current through the restriction of external pressure, but also by increasing the net velocity of fouling layer removal by limiting permeate flow in the opposite direction.

DMC not only increases the permeability of membranes by keeping their surfaces clean, but also reduces the magnitude of crossflow necessary to achieve a high solids content in the product, as it provides an alternative way of controlling membrane fouling. This has important implications in reducing plant wear and pumping energy, as well as pump

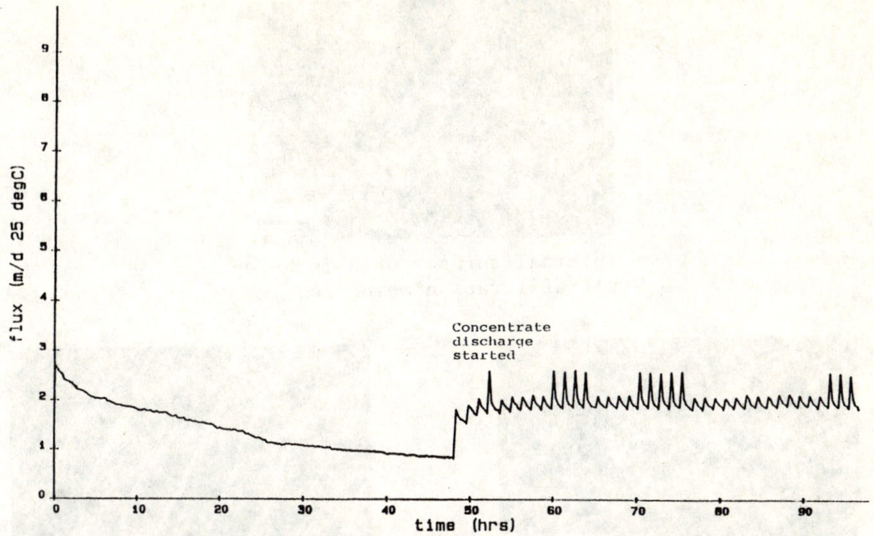

Figure 9 Flux profile for a Membralox module processing Harwell LLW at a transmembrane pressure of 3 bar with and without concentrate discharge.

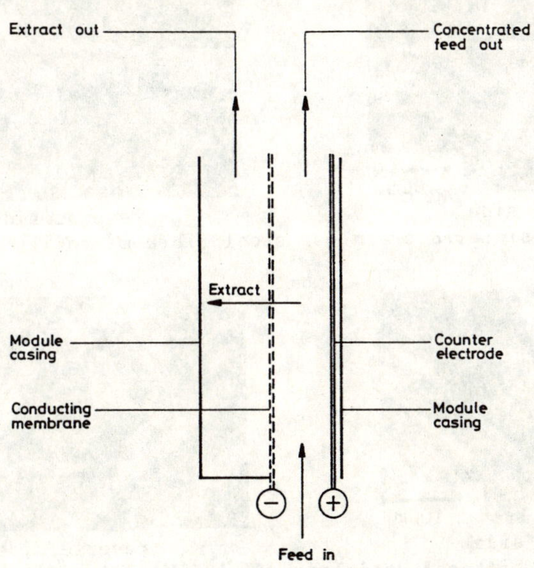

Figure 10 Schematic Direct Membrane Cleaning Filtration Cell.

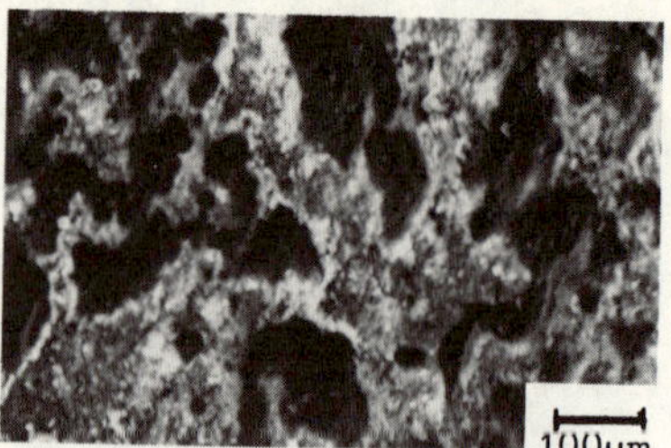

Internal surface of a Tech-Sep
ultrafiltration membrane.

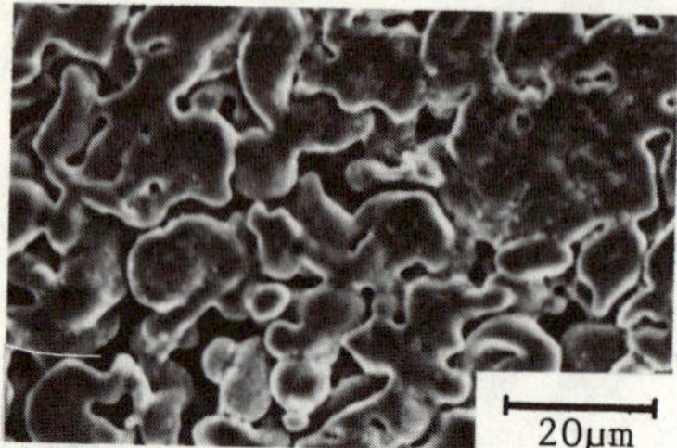

5μm sintered powder stainless
steel microfilter

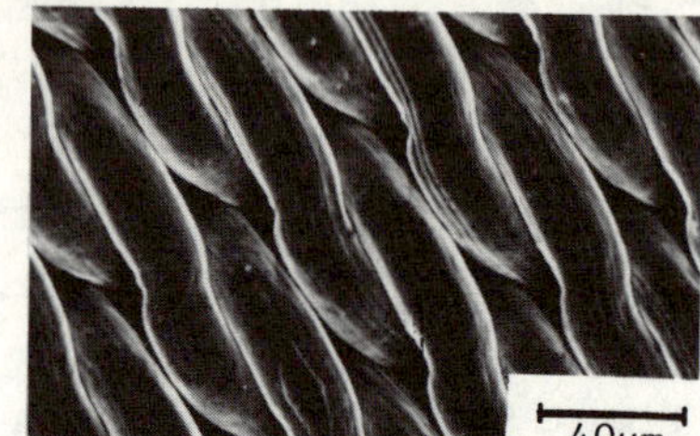

7μm Twilled Dutch Weave stainless steel filtration cloth

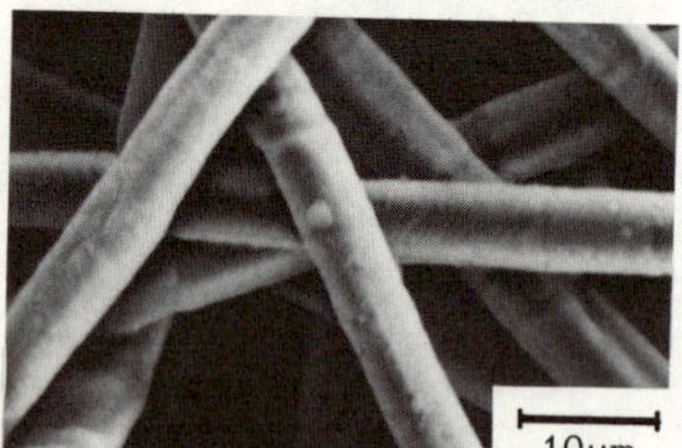

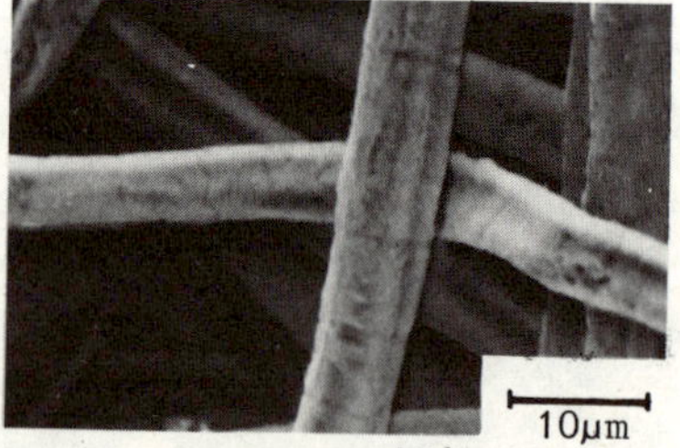

normal side   reverse side
3μm MA1 sintered stainless steel fibre microfilter

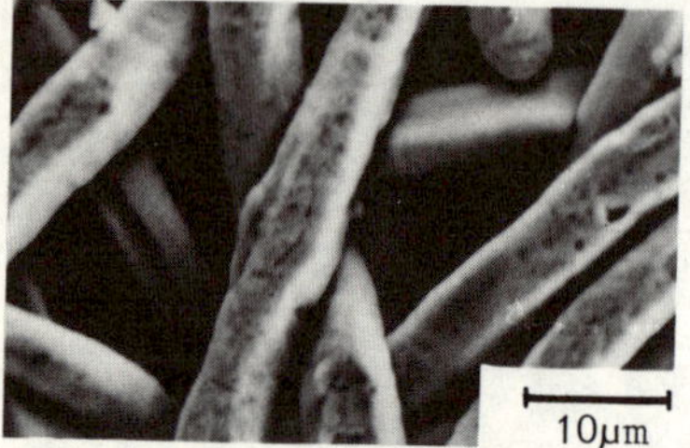

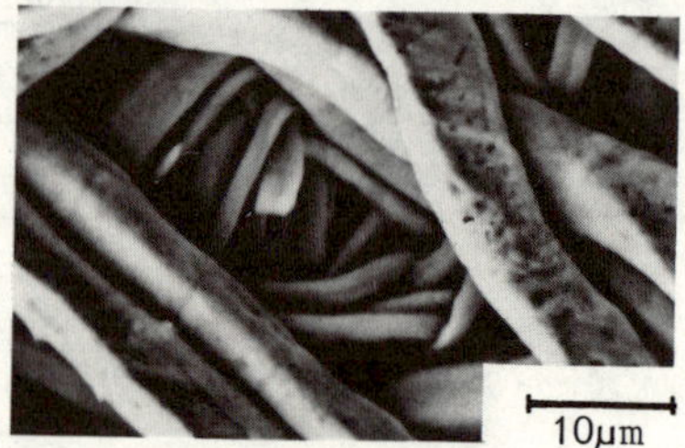

normal side   reverse side
2μm FM5 sintered stainless steel fibre microfilter

Figure 11   SEM micrographs of a range of conductive ultrafilter
and microfilter membrane surfaces.

and filter sizes. In addition, due to the reductions in shear forces on
the solids being separated, particulate degradation should also be
minimized. As the cleaning is conducted rapidly and in situ, high plant
availablity is achieved, since there is no need to interrupt the
filtration process - as would be the case for backflushing or chemical
washing.

3.2 Membrane Examination

DMC has been demonstrated at a range of filtration membranes -
including microporous stainless steel and ultrafine porous ZrO_2 supported
on microporous graphite (Tech-Sep M4). The stainless steel media
examined included woven fibres (Twilled Dutch Weave, 7µm), sintered
powder (5µm) and sintered fibres (7,3,2µm). The scanning electron
microscope has proved an invaluable tool in the examination of the
structure of these membranes, as shown in Figure 11. Tech-Sep M4
membranes comprise a microporous graphite tube (5µm pore size) of 10mm
outside diameter, 6mm internal diameter - with the bore coated in a fine
layer of ultrafine porous zirconia (1-10nm). It is this layer which
gives the membrane its ultrafiltration properties. However, as can be
seen in Figure 11, this filtration layer is not homogeneous in its
structure, but rather consists of a graphite surface covered with islands
of zirconia situated over the entrances to the micropores.

Figure 11 also shows woven fibre and sintered powder filters.
However, in comparison with sintered fibre filters, also shown in Figure
11, they have a relatively low effective filtration area due to their
small void volume. Sintered fibre membranes, on the other hand, have a
pore volume of ~ 85% which gives them a higher permeability. Figure 11
shows that they can have an asymmetric structure with a wider pore
"normal" side and a narrower pore "reverse" side produced by a finer
fibre in-filling a few microns below the surface. This gives a filter
capable of rejecting very fine particles as seen in Figure 12.

Not only has the physical form and structure of membranes been
examined, but their filtration behaviour and enhancement produced by DMC
have been evaluated in small test cells. As Tech-Sep M4 membranes are in
tubular form, small sections were mounted in a cylindrical cell (Figure
13) with an axial platinum wire anode. As the sintered stainless steel
powder membrane was available as a commercial cylindrical cartridge, this
was modified to an annular geometry (Figure 14a) for evaluation under
DMC. On the other hand, the remaining stainless steel media were in the
form of flat sheets which could be simply mounted in a planar cell of
$100cm^2$ membrane area (Figure 14b).

3.3 Filtration Testing

Filtration performance was evaluated using a range of test feeds of
relevance to the nuclear industry - including simulants based on $Fe(OH)_3$
floc and finely divided TiO_2 ion-exchanger. In addition, genuine wastes
have been tested - comprising untreated low level waste and a low level
$Fe(OH)_3$ floc containing significant other suspended solids - including
soot. Using the test loop in Figure 14c, permeation rates were
determined as a function of transmembrane pressure, crossflow velocity,
feed solids content and preparation route/history. By maintaining the
recycled feed at a constant temperature of 20°C through the heat
exchanger, comparisons could be made more easily between experiments.
The effectiveness of DMC in enhancing filtration could then be examined
as a function of current density, pulse duration, repetition rate and
transmembrane pressure during pulse application as well as the other
variables listed above, by comparing permeation rates in its presence

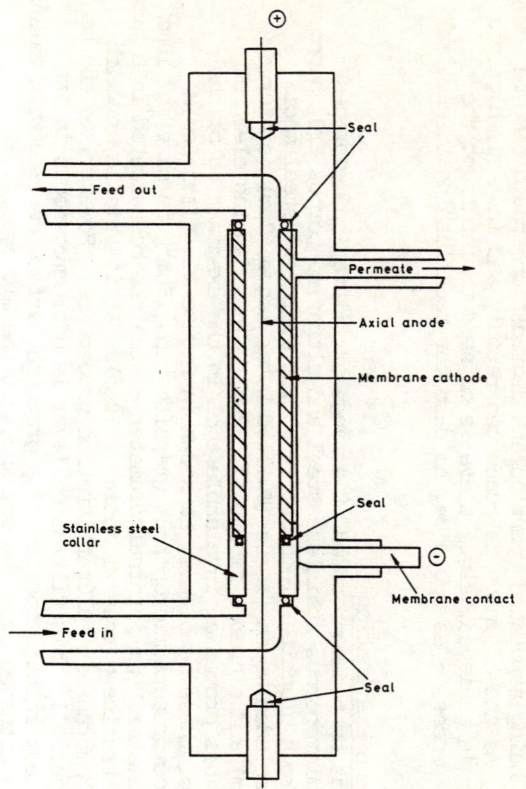

Figure 12 Filtration efficiency as a function of particle size for sintered stainless steel fibre microfilters.

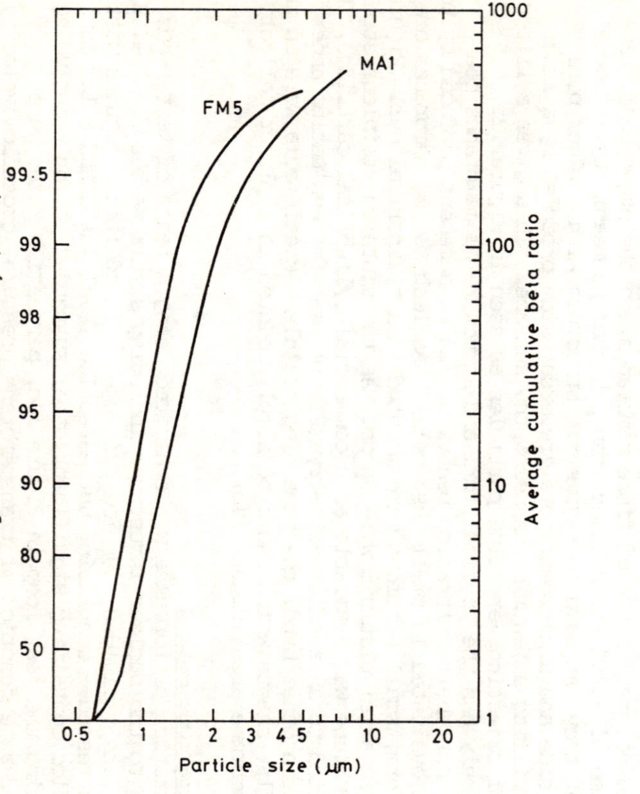

Figure 13 Small scale cell for the evaluation of DMC at Tech-Sep M4 ultrafiltration membranes.

(a) 5µm sintered stainless steel powder cartridge modified for crossflow DMC microfiltration.

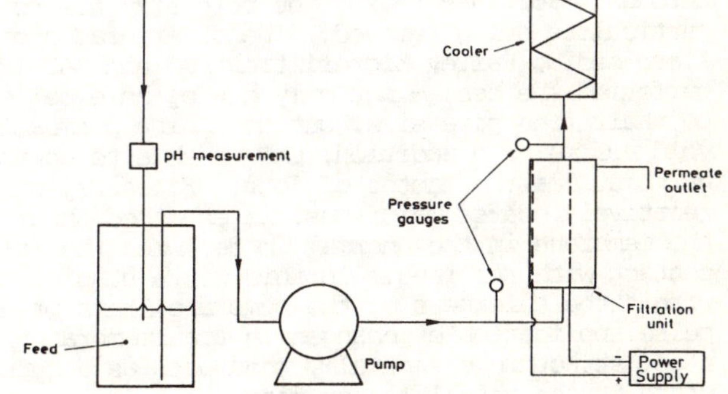

(b) 100 cm^2 planar crossflow DMC molule.

(c) Crossflow DMC test loop

Figure 14 DMC microfiltration test equipment.

and absence. A particularly useful form of experiment involved the return of the permeate to the feed after flow measurement in order to maintain a constant solids content. Typically, DMC is able to enhance permeation rates by a factor of between 2-4 at crossflow velocities of only 1/5 those normally used for crossflow filtration (Figure 15).

In selecting a membrane for a particular duty, the pore size should not be significantly larger than the particles being filtered, in order to give a high solids retention and hence a high decontamination factor. This is of particular relevance after membrane cleaning when fine particles might otherwise be able to pass through the membrane before partial fouling reduces the effective pore size of the filter again. The ultrafiltration membrane acts essentially as a surface filter due to the fineness of the pores through the ZrO_2 layer. In this way any fouling is restricted to the surface of the membrane. On application of the current pulse these solids are effectively removed by the 'scrubbing' action of electrolytic gas bubbles liberated from the exposed graphite surface. No solids breakthrough was observed into the permeate due to the fineness of the pores – rejecting even colloidal size particles. This has been confirmed for the treatment of an active waste, where the colloidal α activity in the permeate has been maintained at <1.5mBq/mℓ. From optimization tests 1 second duration current pulses of only 50mA/cm^2 are effective in cleaning the membrane (Figure 16) when the transmembrane pressure is transiently reduced during pulse application. Not only does the small current density enable a larger membrane area to be cleaned for a given size of power supply, but together with the short pulse length, the minimization of electrical charge passed reduces the gas production to an insignificant level. The frequency of cleaning depends on the solids content of the feed and the crossflow velocity, varying between 1 and 12 pulses/hour. As can be seen from Figure 16, at 1-2% solids four pulses/hour are adequate to maintain a permeability of 1m/bar/day – approaching that for water – with a crossflow velocity of only 1m/s instead of the more usual 4.5m/s.

With the coarser stainless steel microfiltration media the situation is slightly more complex. Not only do the larger pore sizes give a higher permeation rate, but they also increase the risk of fines passing through. For this reason the pore size has to be matched to the size of particulate being removed. The finer grades of sintered stainless steel fibre media, Fairey Microfiltrex FM5 and MA1 (Figure 12), have therefore performed the best – not only giving an excellent solids separation due to their fine pore size, but at a high permeation rate due to their high void volume. In addition, DMC is able to enhance their permeability by up to at least a factor of four. Fines rejection, even at these relatively coarse membranes, is enhanced under DMC conditions by using the membrane in the 'normal' mode, when the larger pore side is in contact with the feed. In this way a stable, thin pre-coat is formed within the thickness of the membrane, thus preventing breakthrough on DMC pulse application or changes in transmembrane pressure. The stainless steel structure essentially then acts as a support for the filter cake which is the actual filter medium.

Due to the rougher filter surface, higher electrical charges are required to restore the filtration flux – up to 200mA/cm^2 for 5s duration. However, as with the Tech-Sep membrane, typically only 4 cleaning pulses/h are required even with 10% solids in the feed, as shown in Figure 17. For systems containing flocs or dispersed ion-exchangers, virtually quantitative activity retention is achieved at microfiltration membranes with permeabilities of 0.6-1.1 m/h/bar at least an order of

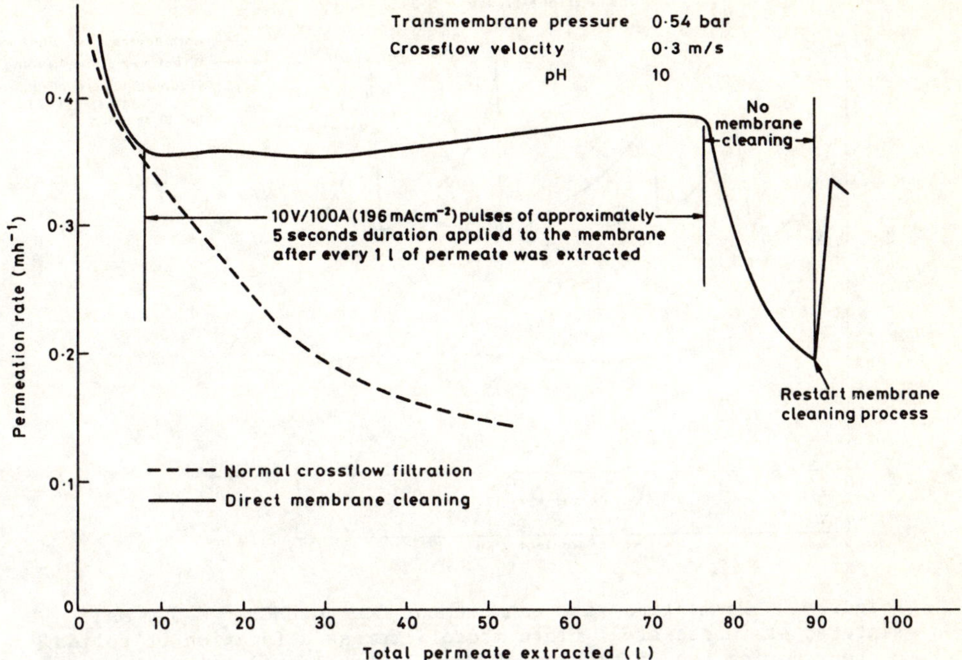

Figure 15   Permeation rate enhancement for the dewatering of a 0.5% $Fe(OH)_3$ suspension by DMC at an annular 5μm sintered stainless steel powder crossflow microfilter.

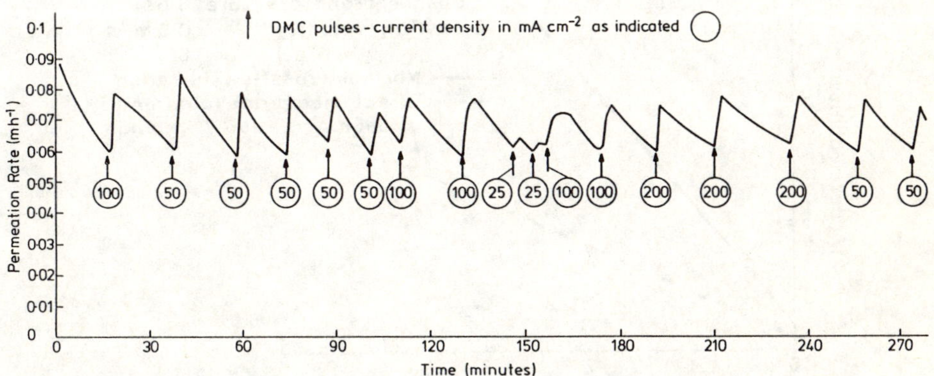

Figure 16   Permeation rate as a function of time for the filtration of a 1% solids Harwell low level waste of constant concentration at a Tech-Sep M4 ultrafiltration membrane with a transmembrane pressure of 2 bar and crossflow velocity of 1m/s.   One second cleaning pulses were applied at intervals of 10-20 minutes.

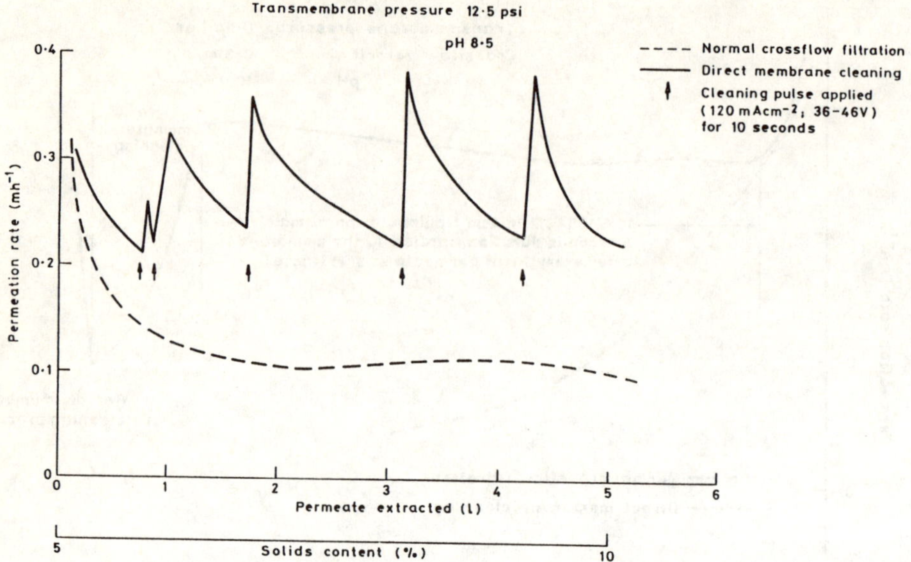

Figure 17   Permeation rate enhancement due to DMC at a 3μm MA1 sintered stainless steel fibre microfilter as a function of solids content during the filtration of Harwell low level active slurry.

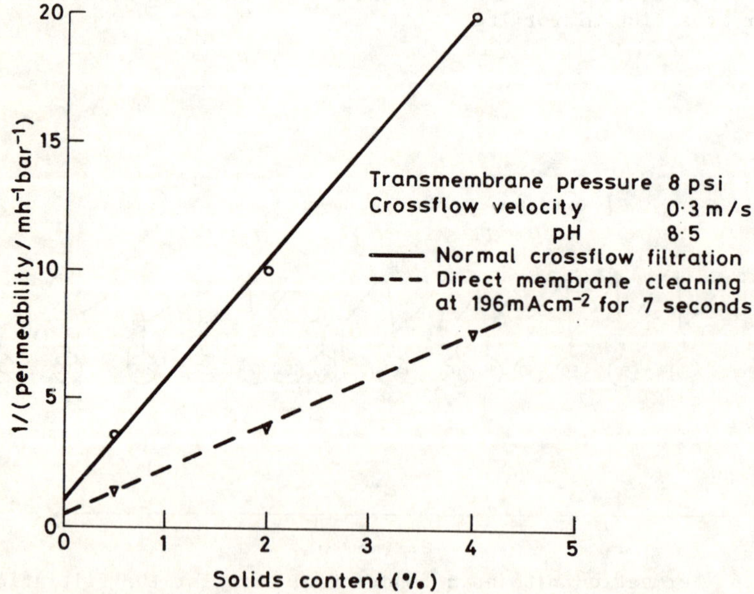

Figure 18   Inverse permeability of a 5μm sintered stainless steel powder membrane as a function of Fe(OH)$_3$ solids content at pH 8.5 during crossflow dewatering – with and without DMC.

magnitude greater than for the equivalent ultrafiltration system.

As with any pressure driven filtration process, the permeation rate falls with increasing solids content. This occurs even with DMC, albeit with a much higher filtration flux, as shown in Figure 18. This is due to the more substantial deposition of solids on the membrane surface. As a result, high processing rates are maintained up to 5-10% solids for gelatinous feeds (e.g Fe(OH)$_3$) or > 20% for more granular dispersions (e.g TiO$_2$). However, precipitation conditions able to enhance the degree of crystallinity of flocculant materials should also increase the maximum solids content possible. DMC cleaned filtration systems can also be used for precipitate washing to reduce their salt content - either in preparation for immobilization or further dewatering by electro-osmosis. The minimum volume of rinse water is required when this is added to the concentrated slurry at a rate equivalent to the permeation rate. An order of magnitude reduction in salt content can be achieved by the addition of only 2.3 times the concentrated waste volume.

3.4 <u>Membrane Life</u>

While data are available from the industrial use of commercially available membranes such as Tech-Sep M4 and MA1 sintered stainless steel fibre under conventional filtration, no such information is available for DMC. In order to demonstrate that DMC has the potential to become a practical industrial process, tests were carried out to increase confidence that it would not adversely affect useful membrane life. However, as, a filter may be in service typically for at least a year, accelerated tests were required to give an early estimate of long term performance. These were carried out on both Tech Sep M4 and Fairey Microfiltrex MA1/S membranes under both continuous and pulsed current in 0.1M NaNO$_3$ at pH10 to simulate a medium level waste after floc treatment, and 0.01M NaNO$_3$ at pH5 to simulate an untreated low level waste. Weight losses and water permeabilities were monitored for the Tech Sep M4. After 30 hours of continuous 50mA/cm^2 they were virtually unchanged with the ZrO$_2$ layer still intact. Pulsed current (1 second on, 5 seconds off) for the same charge passed showed identical performance. For a 200 day operating year this 30 hour accelerated test is equivalent to 5.6 years of elapsed time for the 4 pulses/h required to maintain the permeation flux for a 2% suspension. It is predicted that lives considerably longer than this are possible, as due to the reduction in crossflow velocity enabled by the use of DMC, any abrasive removal of ZrO$_2$ from the filter by the feed flow will be significantly reduced.

Similar weight loss experiments on the MA1/S membrane conducted in the same solutions but at 150mA/cm^2 under DC and pulsed current, together with subsequent microscopic examination, indicated that lifetimes of at least five years are expected.

During the remainder of this programme developments in new conductive ceramic and metallic filtration media will continue to be evaluated. It is hoped that we may be able to test on radioactive waste streams membranes developed under a new BRITE programme. The goal of that programme is to engineer fine pore size, stable materials and structures able to support high fluxes without sacrificing solids retention. As they will be fabricated from conductive ceramics, they should be simply and effectively cleaned by DMC. However, even with the materials examined to date, high performance effluent processing has been demonstrated - yielding high decontamination factors from compact plants.

4.   <u>CONCLUSIONS</u>

Operating at normal crossflow, membrane fluxes were found to be very similar for a range of filtration media. However, the alumina microfilters tested did have approximately twice the flux of the Tech Sep M4 membrane whilst still achieving a good alpha removal from Harwell low level waste. The flux achievable using an alumina microfilter was also found to be independent of transmembrane pressures from 2-4 bar. In addition, it has been shown that periodic depressurization of the crossflow filtration circuit causes the membrane flux to come to a steady state value rather than continuing to decline. For all the membranes tested a good recovery of flux could be achieved using sodium hydroxide and nitric acid sequentially as cleaning reagents. Advanced electrical cleaning techniques hold the promise of further reducing plant size and hence cost by providing an alternative to high crossflow velocities as a method of controlling membrane fouling and enhancing permeation rates. Depending on the nature of the feed, products of between 5-25% solids may be produced at only 1m/s - thus significantly reducing pumping energy and wear within the plant. This technique is currently being evaluated at a pilot plant scale, although additional experience is required in order to make a fair comparison with the current "state of the art" ultrafiltration system.

<u>Reference</u>

[1] Gutman, R.G. et al, Active liquid treatment by a combination of precipitation and membrane processes. Final report to the CEC EUR-10822EN, October 1980-June 1985.

<u>Acknowledgement</u>

This work has been part funded by the Commission of the European Communities in the framework of its research programme on radioactive waste management. In addition, the work has also been carried out with financial support from the U.K. Department of Environment, U.K. Department of Energy, British Nuclear Fuels plc, and Ministry of Defence. In the D.o.E content, the results will be used in the formulation of Government Policy, but at this stage they do not necessarily represent Government Policy.

CURRENT APPLICATIONS FOR DRYING ACETIC AND FORMIC ACID AND COMMERCIALISED TECHNIQUES

J. B. COOPER
Chemical Engineering Manager
Research & Development Department
B P Chemicals Limited

Summary

It is unfortunate that this Brite Project has not yet started in earnest. BPCI's contribution is in evaluation and testing of the membranes we hope the project will develop. This paper cannot therefore deal with experimental results but instead describes the areas where we hope to find applications and in loose terms (for reasons of confidentiality) outlines the nature of alternative acid drying techniques with which membranes will have to compete.

1. INTRODUCTION

World acetic acid production is about 4 million tonnes per annum. Formic acid production is about 0.2 million tonnes. Most of this acid must be separated from water during manufacture. Most of the end uses for acetic acid for example cellulose acetate and vinyl acetate manufacture also involve large scale separation of acid from water.

The predominant acetic acid process (methanol carbonylation) requires over half a tonne of steam for every tonne of acid to separate recycle water from product and servicing the capital cost of this section of the plant represents an even greater cost.

A robust membrane could find wide application in the industry if flux and selectivity were adequate to make it competitive with current technology.

2. THE PROBLEM

Acetic acid does not actually form an azeotrope with water but a combination of high reflux ratios and many trays are required to obtain a near pure water stream overhead; minimum reflux ratio is 2.6:1. Obtaining a water free acid stream is considerably easier; a boil up ratio of 1:1 will remove 2% water from acetic acid.

Formic acid forms a high boiling azeotrope containing 22% water at atmospheric pressure.

3. CONVENTIONAL SOLUTIONS AND MEMBRANE APPLICATIONS

3.1 Straight Distillation

In methanol carbonylation to acetic acid there is no requirement to remove water from the system and conventional distillation to give a dry acid product stream and a dilute acid stream for recycle, is employed.

The steam requirement for this duty and the size of column required can be reduced by increasing the strength of the wet acid feed. This is an ideal area of application for a low selectivity acid resistant membrane because there is no requirement for either the water rich permeate (which would be recycled to the process) or the acid rich retentate which passes to the drying column to be pure. Low grade heat could supply the heat to vaporise the permeate in place of some of the medium pressure steam used to boil up pure acetic acid at the base of the column.

The trend in technology for methanol carbonylation, however, is towards enhancing the reactor catalyst package to allow much lower water concentrations in the reactor thus reducing or eliminating the drying column duty.

A membrane capable of removing the last few percent of water to produce a dry acid (99.9%+) could find application in such a system because the distillation cost then becomes insensitive to the quantity of water to be removed.

As the water concentration decreases, however, the driving force for pervaporation becomes smaller and much harder to create in terms of the vacuum required on the permeate side.

Formic acid is sold with residual water of up to 15% in order to avoid the difficulties of the final drying. Membranes could find application in upgrading this product to broaden its application in the market and in producing 85% acid from the atmospheric pressure azeotrope (22.6% water) thus simplifying distillation.

3.2 Azeotroping Agents

Oxygenated solvents (esters, ketones etc) have a higher affinity for formic and acetic acid than for water. They are commonly employed to extract acid from water in multistage liquid/liquid extraction giving an acid free water stream and a solvent stream containing all the acid together with some residual water. The same solvents (for the same reason) enhance the vapour pressure of water such that distillation in their presence can produce dry formic or acetic acid and an overhead stream containing acid free water.

In some applications such as ester production the product itself will act as an azeotroping agent.

Liquid extraction is fundamentally a low cost process but when the agent is not already in the system it adds considerably to the process complexity. Membranes may therefore find applications for small scale separations or debottlenecks where the relative simplicity of a membrane and its linear dependence on scale give lower investment cost.

4. TARGETS FOR THE BRITE PROJECT

The joint research programme is targeted at producing an acid resistant membrane capable of economically removing residual water from formic or acetic acid streams to provide saleable strength final product.

STUDIES ON SUPPORTED LIQUID MEMBRANES PERFORMED BY SCIENTIST OF THE CASACCIA RESEARCH CENTRE

P.R. DANESI* and G.M. GASPARINI
ENEA- CRE Casaccia, Rome, Italy

Summary

A short survey of the work carried out by ENEA'S scientist at the Casaccia Center and at ANL on Supported Liquid Membranes (SLM) is presented. Simple relationships, permitting a quantitative prediction and control of the separative processes, have been proposed and verified on several SLM's systems.Some economic considerations concerning large scale processes are reported.
The studies on radioactive liquid wastes are briefly described.

1. GENERAL CONSIDERATIONS

The use of extractants immobilized on a thin microporous inert polymeric film interposed between two aqueous solutions (feed and strip) represents a new attractive separation methodology(1).
The technique is simple and economically promising. For this reason it has been proposed as an alternative approach to other more traditional techniques for the separation, recovery and purification of metal species in solution.

Due to the small amount of extractant which is needed for the separation, the technique allows one to utilize the sophisticated and expensive selective extractants which have been synthesized and studied during the last few years.

The availability of hollow-fibers permits also to utilize very compact separation modules characterized by high membrane area densities. This should enable one to obtain energy savings, low capital cost and the possibility to reach

---

*present address:
IAEA's Laboratories at Seibersdorf.
IAEA P.O. Box 200, A-1400, Vienna, Austria

elevated separation factors in a single stage, while utilizing high feed/strip ratios(2).

It must be however emphasized that Supported Liquid Membranes (SLM) do not represent an alternative technology to traditional solvent extraction but rather a complementary method, particularly suitable for metal species present at high dilution.

One major limitation to the use of SLM is in some cases represented by the longer time needed to perform the separation.

In the field of nuclear applications problems exist in their use for the reprocessing of irradiated nuclear fuel, where radiolytic problems impose to minimize contact times. On the other hand SLMs can have suitable applications in the treatment of medium and low level radioactive liquid wastes. Here they can be utilize for the isolation of small amounts of highly radiotoxic nuclides from the large matrix of inorganic inactive salts.

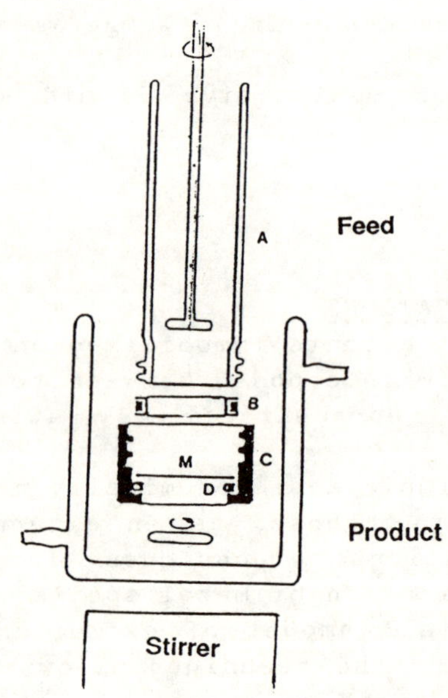

Fig.1 Experimental apparatus for horizontal flat-sheets.

2. BASIC STUDIES

Stimulated by these encouraging perspectives a number of studies have been performed by us during the recent years.As a consequence the transport properties of SLMs are now sufficiently well understood and characterized from both the experimental and the theoretical point of view.

In a series of basic studies, Danesi et al. (1) have elaborated a simple transport model through SLMs both for the case of counter-transport and co-transport of metal species. Simple relationships, holding for the two limiting cases of metal species present at low concentrations ($\leq 10^{-3}$ M) and high concentrations ($\geq 10^{-2}$ M), have been proposed and verified at the laboratory level on several SLM's systems, using a variety of metal cations and membrane carriers (extractants).

Length = 18 cm, diameter 1.8 cm
External Area = 216 cm^2
Internal Area = 145 cm^2
Number of fibers = 25

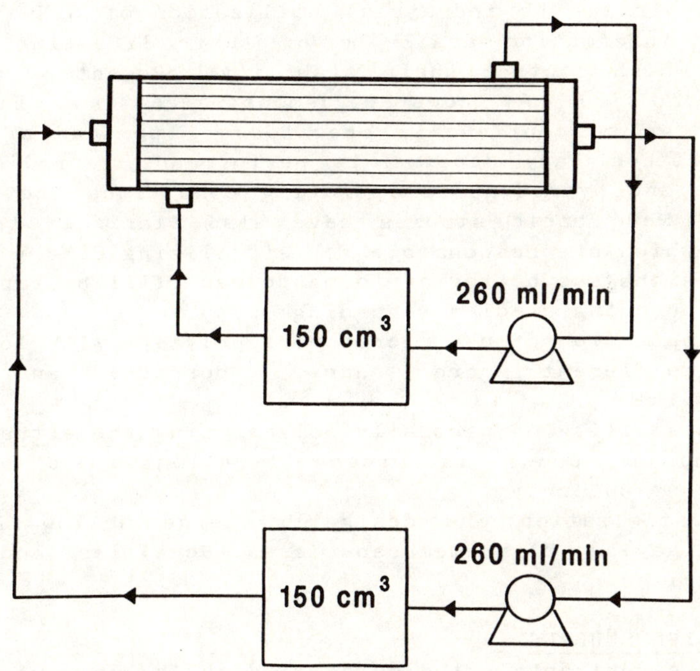

FIG.2 Schematic representation of an hollow-fiber SLM module for separation of metal species.

The equations derived permit a quantitative prediction and control of the separation process. The majority of the studies have been performed using flat-sheet SLM's or small hollow-fiber modules, containing a limited number of fibers (Fig.1,2).

In spite of their promising advantages no large scale process applications have been so far reported for SLMs in the field of metal separation, recovery and purification. The main reason for this is mainly economic. Since the metal industry is at present in a depressed state, no need for new plants seems to exist. The only possibility would then be a rebuilding of existing solvent extraction plants into SLM plants. However this would not be economical as the main advantage of SLMs is their low investement costs, being the operating costs very similar to that of a solvent extraction plant. Different considerations may nevertheless apply for the waste treatment of both nuclear and non nuclear origin.

Other reasons that till a few years ago were unfavourable to the industrial utilization of SLMs were the lack of information available on their life-time and the factors which control their stability. Recent work in this field (3) has however demonstrated that laboratory modules can be continuosly operated for periods of several months without any decline in performance, providing the chemical and hydrodynamic operating conditions are properly chosen. More basic studies have also clarified the major factors which are responsible for stabilizing SLMs (4).

Nevertheless better performance can still be expected by further pursuing studies regarding:
- the preparation of new microporous polymers with lower pore size, different pore shape, porosity and higher hydrophobicity;
- the possibility of partially polimerizing the extractant in the membrane phase in order to entangle it with the microporous support;
- the optimization and design of large hollow-fiber SLM modules having high membrane area densities and better hydrodynamics.

3. <u>APPLIED STUDIES</u>

The experimental studies carried out by us have followed two main directions. The first one, described in the previous section, has been towards the elucidation of the basic aspects of SLM permeation (trasport mechanisms, kinetics, modelling etc.) (1-3).

The second one has been towards the synthesis and the utilization of new extractants as membrane carriers and their practical applications (5-7).

In particular we have tested tri alkyl aceto hydroxamic acids (5,6):

$$\begin{matrix} R \\ R \\ R \end{matrix} \!\!\!\Big\rangle\, C(O)NHOH$$

(where R is Me,Et,Pr,Bu,Am and the total number of carbon atoms is 12 or 13) which are powerful extractants of Pu(IV),Fe(III),Mo,V,Zr,Nb, from 0.1 to 5 M HNO_3, HCl, H_2SO_4 and $HClO_4$ (Fig. 3 and 4).

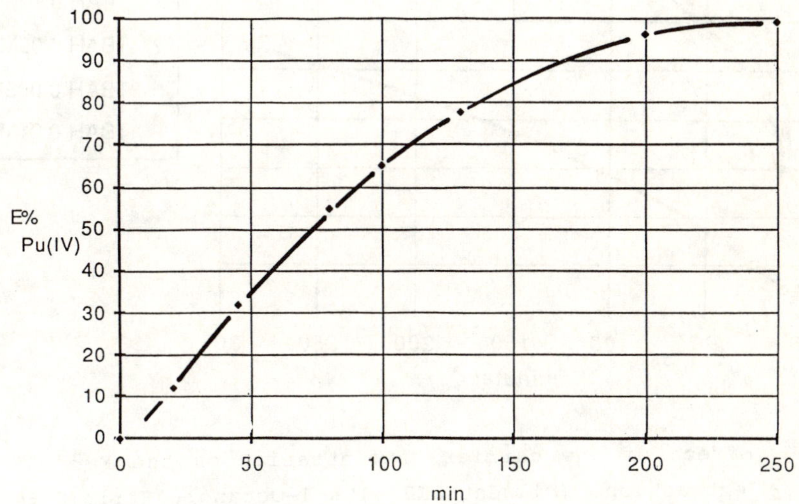

Fig.3   The Extraction of Pu^{IV} versus time.
   (Membrane:Celgard 2500, TBAH 0.1M DEB+10%1-Octanol;
   Feed:1M Na_2CO_3,0.6M $NaNO_3$; Strip:0.5M $(COOH)_2$).

The possibility of practically utilizing SLMs for the purification of Pu(IV) present in carbonate solutions deriving from the regeneration section of a reprocessing plant, has been examined using these extractants as carriers (7). Several membranes and different experimental procedures have been tested using Fe^{59} (as Pu simulant) and Pu^{239}.
The effects of different variables on the permeation rates have been experimentally analyzed.

SLM's consisting of a solution of 0.25M n-octyl (phenyl)
-N,N'-diisobutylcarbamoylmethylphosphinoxide (CMPO) and 0,75M
TBP (tri-butylphosphate) in decaline, absorbed on thin
microporous polypropylene supports, have been also studied
for their ability to perform selective separations and

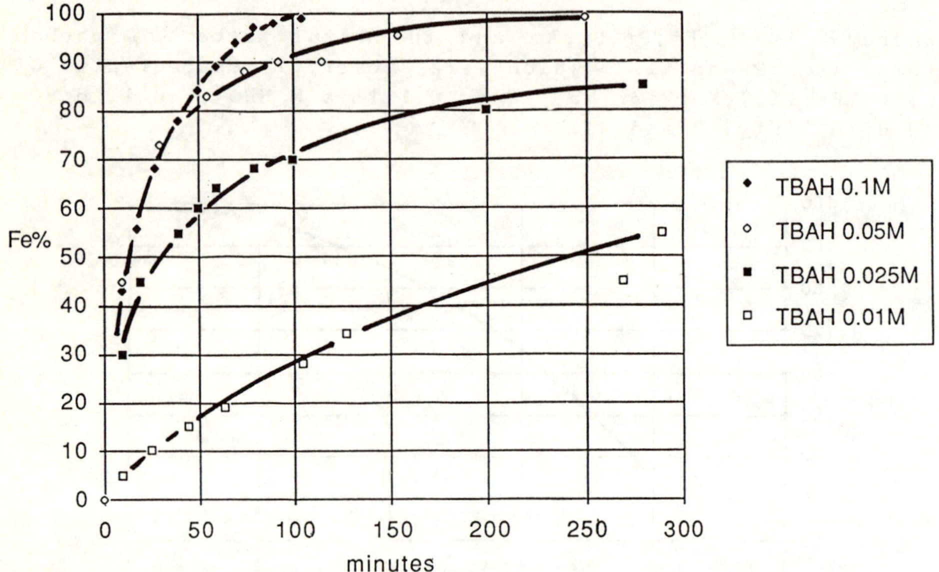

Fig.4   Effect of the carrier concentration on the Fe^{59}
extraction. (Diluent:DEB +10% 1-octanol, Strip:0.5M
$(COOH)_2$.

concentrations of actinide and lanthanide ions from synthetic
nuclear wastes derived from PUREX processes (8).   The
permeability coefficients of selected actinides (Am,Pu,U,Np)
and of some of the other major components of the wastes have
been measured using SLMs in flat-sheet and hollow-fiber
configurations.   The results have shown that the membrane
permeation process is mainly controlled by the rate of
diffusion through the aqueous boundary layer. Hollow-fibers
SLMs exhibited lower permeability coefficients and longer
life-times. The esperiments have shown that the actinides can
be   efficiently removed from   synthetic waste
solutions to the point that the resulting solutions can be
considered non transuranic wastes (less than 100 microCuries

per gram of disposed form). The work has also demonstrated
that actinides removal from PUREX synthetic wastes solutions
is a feasible chemical process at the laboratory scale level.
The process is characterized by the typical features of SLM's
processes,i.e.(a) very small quantities of extractants, (b)
high feed/strip volume ratios, resulting in a corresponding
concentration factor of the actinides, (c) simplicity of
operation and (d) negligible solvent entrainment.

<u>References</u>

(1) P.R. Danesi, Separation Science and Tech. <u>19</u>, 857
     (1984-85)
(2) P.R. Danesi, J. Membr. Science, <u>20</u>, 231 (1984)
(3) P.R. Danesi and P.G. Rickert, Solv. Extr. Ion Exch. <u>4</u>,149
     (1986)
(4) P.R. Danesi, L; Reichley-Yinger, P.G. Rickert, J. Membr.
     Science <u>31</u>, 117 (1987)
(5) G.M.Gasparini, Gazz. Chim. Ital. <u>109</u>, 357 (1979)
(6) G.M. Gasparini, Proceedings of Internat. Solvent
     Extraction Conf., Toronto, Sept. 1977 p.654
(7) G.M. Gasparini, et al. Proceedings of Internat. Solvent
     Extraction Conf., Denver, Sept 83, p.395
(8) P.R. Danesi, R. Chiarizia, P.Rickert and E.P. Horwitz
     Solv. Extr. Ion Exch. <u>3</u>,111 (1985)

USE OF LIQUID MEMBRANES FOR TREATMENT OF NUCLEAR WASTES

J.F. DOZOL
Commissariat à l'Energie Atomique
CEN-Cadarache – Saint Paul lez Durance – France

The reprocessing operations produce liquid wastes in which the main components are nitric acid and sodium nitrate. The goal of the experiments is to separate trace amounts of radioactive elements from these acidic and high sodium nitrate content solutions.

CMPO, a neutral bifunctional organophosphonus compound, and crown compounds (DC18 C6 – B21 C7) are able to extract respectively actinides strontium and cesium from these high salinity solutions.

The SLM render the use of expansive tailor-made extractant molecules like CMPO or crown ethers possible. The results obtained for the extraction of actinides and strontium are promising, but research must now be oriented towards improving the stability of the membrane.

1. INTRODUCTION

The reprocessing of spent fuels separates uranium and plutonium which are recycled from highly radioactive fission products stored in geological formation after vitrification.

These reprocessing operations produce medium active liquid wastes (MALW) in which the main components are nitric acid and sodium nitrate. At the Marcoule reprocessing plant, it is planned that the insolubilisation process by chemical precipitation be substituted by evaporation that enables the radioactivity discharged into the environment to be greatly reduced. The main disadvantage of the process is that it concentrates all inert and radioactive salts and leads to an considerable increase of stored solid wastes.

The goal of the experiments carried out is the separation of the waste into a small volume of highly radioactive waste and a large volume containing inert constituents and short-lived nuclides. The latter, after conditioning in matrices like bitumen or cement, can be stored in shallow land burial.

The selective removal of actinides, cesium and strontium for which the molar concentrations are very low (10^{-6} – 10^{-8} M) from sodium nitrate ($\sim$ 4 M) and nitric acid liquid wastes cannot be achieved by using organic or inorganic ion exchangers. Tests carried out in the laboratory show that only ammonium phosphomolybdate makes the removal of cesium possible and that an ion exchanger is unable to efficiently remove strontium or actinides from the concentrates.

2. EXTRACTANTS USED

2.1. Extractants of actinides

HORWITZ and al (1) have synthesized new selective extractants for actinides belonging to the family of the neutral bifunctional compounds and tested their ability to extract all the valences of actinides from acidic and high sodium nitrate content liquid waste. Among several compounds, the most promising is the n Octyl (phenyl) N,N diisobutylcarbamoyl methyl phosphine oxide or more simply CMPO (Fig. 1).

The reaction responsible for the extraction of americium from the acidic nitrate solution is :

$$Am^{3+} + 3\ NO_3^- + \overline{3E} \rightleftharpoons \overline{Am(NO_3)_3\ E_3}$$

E represents CMPO, the bar indicates the organic phase.

Pu^{4+} and UO_2^{2+} are also extracted according to the reaction :

$$Pu^{4+} + 4\ NO_3^- + \overline{4E} \rightleftharpoons \overline{Pu(NO_3)_4\ E_4}$$

$$UO_2^{2+} + 2\ NO_3^- + \overline{2E} \rightleftharpoons \overline{UO_2\ (NO_3)_2\ E_2}$$

TBP is added to reduce the possibility of formation of a third phase and increase the solubility of CMPO in diluent, TBP extracts nitric acid :

$$HNO_3 + TBP \rightleftharpoons TBP\ (NO_3)_3$$

In low acidity medium (pH > 2), the actinides are back-extracted.

2.2. Extractants for strontium and cesium

In 1967, PEDERSEN was the first to synthesize the crown ethers, the cyclic compounds that form strong complexes with alkaline or alkaline earth elements (2) (Fig. 2) :

$$M^+ + NO_3^- + \overline{E} \rightleftharpoons \overline{MNO_3\ E}$$

$$M^{2+} + 2NO_3^- + \overline{2E} \rightleftharpoons \overline{M(NO_3)_2\ E}$$

Crown ethers like 15 C5, 18 C6, 21 C7 which are not very lipophilic, are relatively soluble in an aqueous solution. The crown compounds with benzo or cyclohexano substituents show a more pronounced hydrophobicity that makes them utilizable in aqueous solutions (Fig. 2).

The addition of alkyl and cyclohexyl substituent groups increases the hydrophobicity of the macrocycles with minimal reduction of its complex ability. But benzo and other electron-withdrawing substituent group reduce macrocycle complexing power (3).

The benzo substituted crown ethers produce a stronger bond with alkali metal than with alkaline earth metals : the more rigid benzo substituted crown may maintain a cavity in solution thus allowing the alkali metal to profit from the fact that no rearrangement of the coordinating oxygens is necessary. But, the doubly charged alkaline earth ions require a configuration that is not easily attained by the rigid benzo crown ether (4).

As PEDERSEN has observed, complexing is weak when the crown ether is too small because the cation cannot enter the plane of oxygen atoms where the charge density is highest, or when the crown ether is too large because the cation cannot be simultaneously close to all the oxygen atoms.

<u>Table I</u>

Crown	Hole size (A)	Crystal ionic diameters	
12 C4	1.2	Li^+ : 1.20	Mg^{2+} : 1.30
14 C4	1.2 - 1.5	Na^+ : 1.90	Ca^{2+} : 1.98
15 C5	1.7 - 2.2	K^+ : 2.66	Sr^{2+} : 2.26
18 C6	2.6 - 3.2	Rb^+ : 2.96	Ba^{2+} : 2.70
21 C7	3.4 - 4.3	Cs^+ : 3.38	
24 C8	4.5		

Molecular Weight
407.58

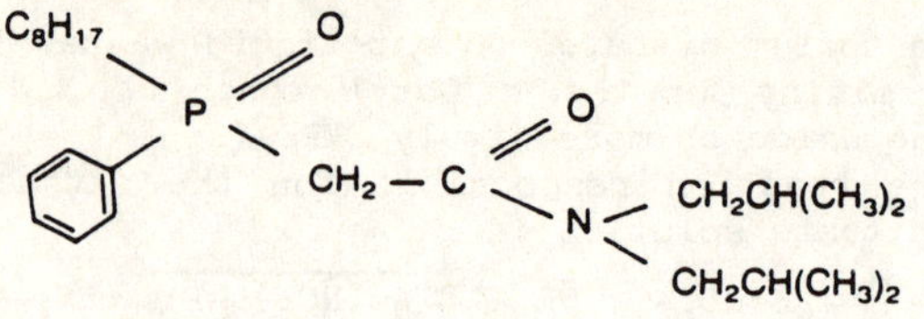

Figure 1

C.M.P.O.

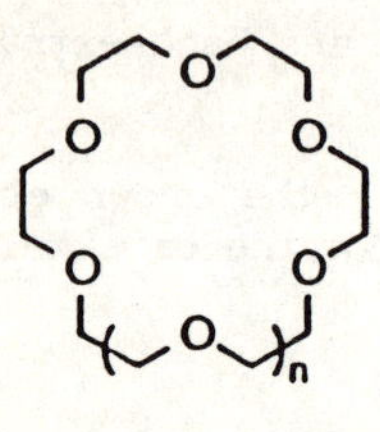

n=0 15 Crown 5
n=1 18 Crown 6
n=2 21 Crown 7
n=3 24 Crown 8

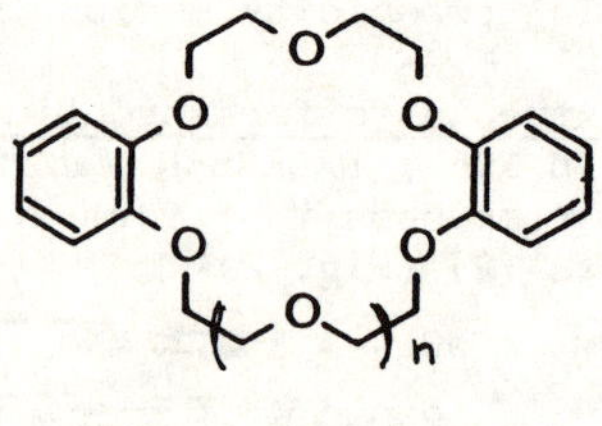

n=0 Dibenzo 15 Crown 6
n=1 Dibenzo 18 Crown 6
n=2 Dibenzo 21 Crown 7
n=3 Dibenzo 24 Crown 8

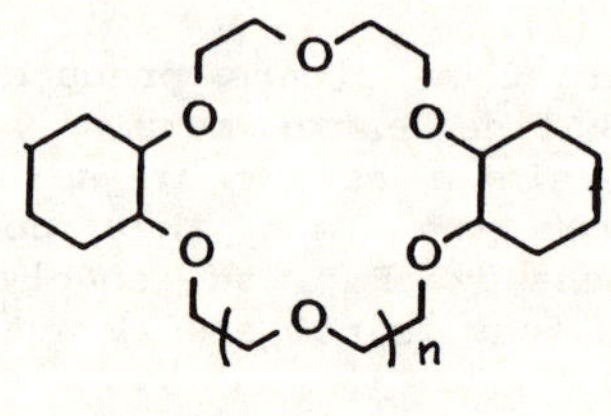

Figure 2

n=0 Dicyclohexano 15 Crown 5
n=1 Dicyclohexano 18 Crown 6
n=2 Dicyclohexano 21 Crown 7
n=3 Dicyclohexano 24 Crown 8

Experiments confirm these theoretical considerations : for strontium extraction the most efficient crown ether is DC18 C6 and for cesium extraction benzo substituent groups of 21 crown 7.

3. SUPPORTED LIQUID MEMBRANE (S.L.M.)

3.1. Principle of supported liquid membranes

A supported liquid membrane consists of an organic solution of an extracting agent (carrier), absorbed on a thin microporous support. The supported liquid membrane separates the aqueous solutions initially containing the permeating ions (feed solution) from the aqueous solution initially free of these ions (strip solution) (Fig. 3).

The transport of an ion is due to the chemical potentiel gradient existing between the two sides of the membrane. If the concentration of the permeating ions is much lower than that of the other species responsible for the driving chemical gradient, then during the permeation process the chemical gradient stays pratically constant. In the case of concentrates, the driving force is provided by the concentration gradient of nitrate through the supported liquid membrane.

The permeation of a metal ion from the feed solution to the strip solution takes place because the distribution coefficient of the metal ion into the supported liquid membrane is high. The back-extraction of the metal ion is possible if the distribution coefficient of the metal ion between the supported liquid membrane and the strip solution is low.

To favour back-extraction, some complexing organic compounds can be added to the strip solution : formic acid, sodium citrate ...

3.2. Double supported liquid membrane system

It was seen previously that nitric acid is transported by CMPO and TBP (but also by crown ethers) through the membrane ; what is more, the increase of nitrate ions in the strip solution decreases the driving force. Thus, the permeation of metal ions slows down and eventually stops.

To avoid the build-up of nitrate in the strip solution, CHIARIZIA and DANESI propose that the excess of nitrate be removed by a double liquid membrane system (5).

Nitric acid and actinides are transported by SLM 1 from the feed solution to the strip 1 solution and nitric acid only by the second membrane from strip 1 to strip 2.

To selectively remove nitric acid, CHIARIZIA and DANESI propose the long chain primary amine PRIMENE J.M.T. This amine selectively transports acid nitric to an alkaline solution where it is neutralized but does not extract the metal ions.

The PRIMENE J.M.T. liquid membrane makes it possible to control the nitrate concentration in the strip solution of actinides and to quantitatively transfer these elements.

4. EXPERIMENTAL

4.1. Reagents

CMPO is prepared by M and T Chemicals Co and must be purified by ion exchange.

DC18 C6 is an ALDRICH Chemical Co product.

B21 C7, not commercially available, was synthesized by GRAMAIN (University of Strasbourg).

All the reagents used for the synthesis were first purified :

In pyridine medium, at ice bath temperature, tosylation of hexaethy-

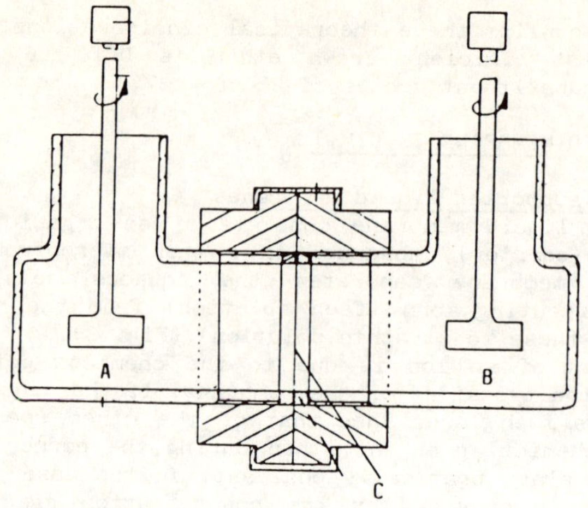

Figure 3

A : Feed solution - B : Strip solution - C : S.L.M.

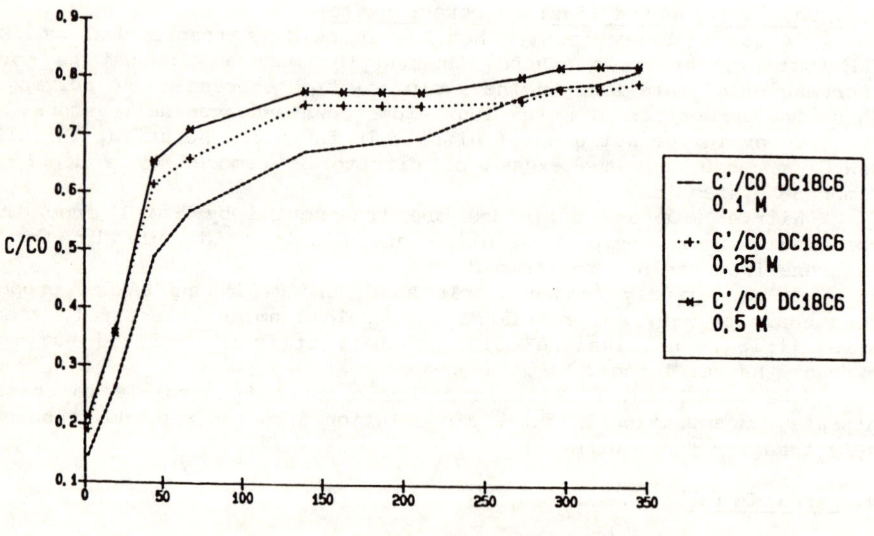

Figure 4

Influence of the concentration of crown ether

lene glycol :

$$HO\ (CH_2OCH_2)_6\ H + 2\ TsCl \rightleftharpoons TsO\ (CH_2OCH_2)_6 Ts + 2\ HCl$$

After washing, the precipitate obtained is deshydrated.

In aprotic solvent (acetonitrile), reaction of cyclization in presence of cesium (template effect) :

$$TsO\ (CH_2OCH_2)_6\ Ts + OH - C_6H_4 - OH \rightleftharpoons B21\ C7 + 2\ CsO\ Ts$$

The cristallised B21 C7 is purified in boiling heptane and the macrocycle is liberated by chromatography over a column of alumina, eluting with dichloromethane.

PRIMENE J.M.T. is obtained from ROHM and HAAS.

All reagents are analytical purity grade products from ALDRICH Chemical Co.

4.2. Analysises

Simulated wastes are prepared with ^{85}Sr and ^{137}Cs from AMERSHAM, ^{239}Pu and ^{241}Am from CEA stocks.

γ spectrometry measurements are carried out using Ge-Li detector and actinides measurements by liquid scintillation counting or α spectrometry.

All the supported liquid membrane experiments are performed at $26 \pm 1°C$. The different aqueous solutions are stirred magnetically.

The permeation of metal ions through the membrane is measured by a periodical sampling of the aqueous solutions.

4.3. Characteristics of the membranes

The supports of the membrane are microporous polypropylene films from CELANESE PLASTIC with a normal porosity of 45 % and a pore size of 0.04 µm.

CELGARD 2500 and CELGARD 2502 membranes are respectively 25 µm and 50 µm thick. For all experiments, the membrane area is 35.4 cm^2. The S.L.M. is prepared by soaking the membrane in the organic phase containing the extractant for 24 hours.

4.4. Feed solution

For some special fuels of graphite gas natural uranium reactor fuel, the clads must be chemically dissolved. The concentrate, according to the nature of the clads, results in three different compositions : MAR 400, MAR 400 (Al), MAR 400 (Mg).

Table II : Compositions of waste solutions. Molar concentrations.

	MAR 400	MAR 400 Mg	MAR 400 Al
HNO_3	1 M	1 M	1 M
$NaNO_3$	3.8 M	3.4 M	3.5 M
$Mg(NO_3)_2\ 6H_2O$	0.078 M	0.62 M	
$Al(NO_3)_2\ 9H_2O$	-	-	0.41 M
$Ca(NO_3)_2\ 4H_2O$	0.03 M	0.015 M	0.015 M
$Fe(NO_3)_3\ 9H_2O$	0.015 M	0.0032 M	0.0032 M
NaCl	0.0032 M	0.012 M	0.012 M
NaF	0.076 M	0.017 M	0.017 M
$Na_2O\ SiO_2\ 5H_2O$	0.023 M	0.001 M	0.001 M
Na_3PO_4	0.14 M	0.021 M	0.021 M
Na_2SO_4	0.165 M	0.021 M	0.021 M
$NH_4\ NO_3$		0.1 M	0.1 M
TBP		0.00056 M	0.00056 M

The presence of aluminium or magnesium does not modify the coefficients of distribution of uranium and plutonium but improves that of americium.

<u>Table III</u> : Distribution coefficients

	MAR 400	MAR 400 (AL)	MAR 400 (Mg)
$D_{UO_2^+}$	400	500	550
$D_{Pu^{4+}}$	1000	1000	1000
$D_{Am^{3+}}$	14	35	55

5. <u>EXTRACTION OF STRONTIUM</u>

The evolution of the number of moles, in the liquid phase, as a function of time is generally linear during the first hours of the experiment, an initial transfer rate is determined by the relationship :

$$V_M = (mol.h^{-1}) = \frac{C'M}{t}$$

C'M number of moles of metal ion M

The flux is defined by M TROMP (6)

$$F_M \ (mol.cm^{-2}.s^{-1}) = \frac{VM}{S}$$

The real area of contact between the organic and aqueous solutions are unknown through a lack of information above the membrane. Therefore the area of support is chosen as the exchange area.

5.1. <u>Decanol as diluent</u>

For the extraction of strontium, the distribution coefficients between the organic phase (extractant and diluent) and the feed solution are higher for crown ethers diluted in tetrachloroethane than for decanol.

Tetrachloroethane which has a relatively high water solubility (0.29 %) does not allow one to obtain a satisfactory stability for the membrane. With flat sheet supported liquid membrane, the effects of chemical degradation are magnified because of the very small inventory of extractant in the system.

Decanol is a promising diluent because alcools or phenols increase the solubilisation of the anion associated with the cation extracted by the crown ether. S.L.M. consisting of DC18 C6 at different concentrations in decanol are stable. The life-time exceeds 300 hours, but it is not possible to transfer more than 80 % of strontium from the concentrate to the deionised, water used as stripping solution (Fig. 4).

During the previous tests, a sharp decrease of pH is observed : the build up of nitrate slows down the flux of strontium. To avoid this drawback, a double liquid membrane is used, with a second membrane consisting of PRIMENE J.M.T. diluted in decaline :

Feed solution	: Concentrate
SLM 1	: Crown ether in decanol
Strip 1 solution	: Deionised water
SLM 2	: PRIMENE J.M.T. in decaline
Strip 2	: NaOH

This double membrane system is very efficient : first pH decreases

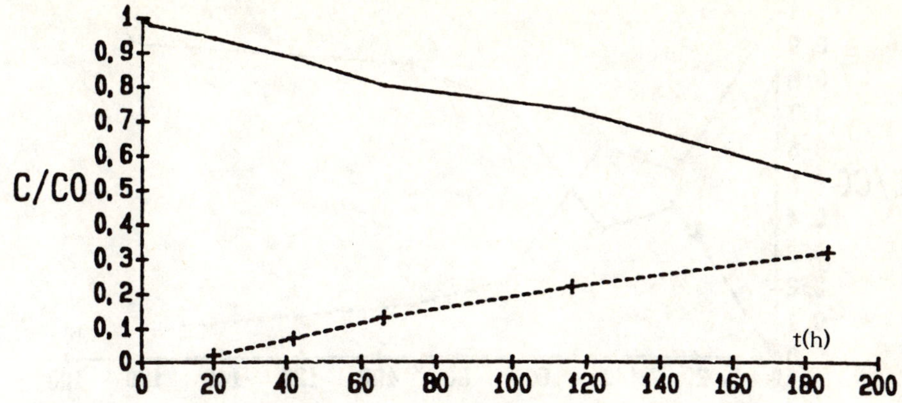

Figure 5

Extraction of Sr with a double S.L.M
DC18 C6 0.5 M   in decanol

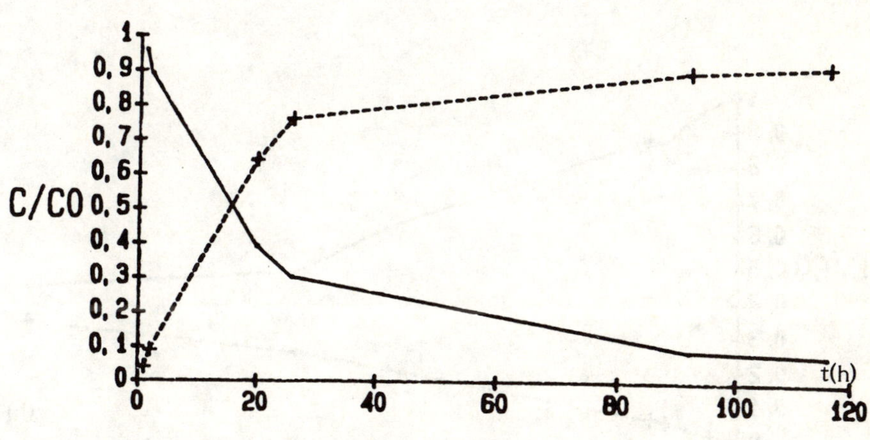

Figure 6

Extraction of Sr with one S.L.M
DC18 C6   0.5 M in hexylbenzene/decanol

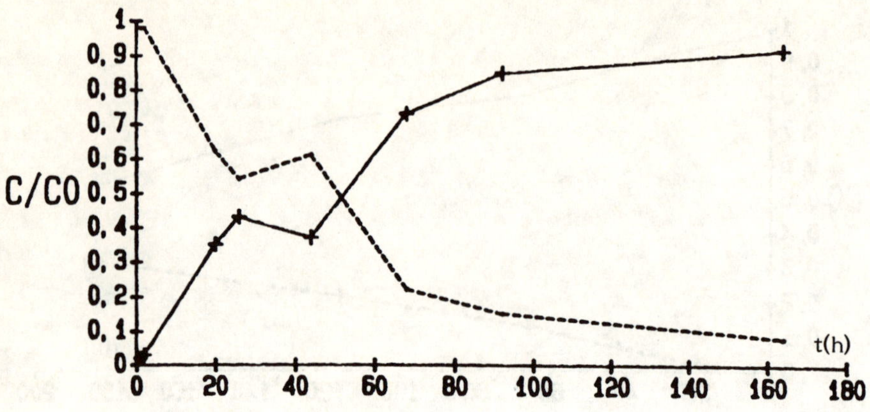

Figure 7

Extraction of Sr with one S.L.M
DC18 C6 0.1 M in Hexylbenzene/decanol

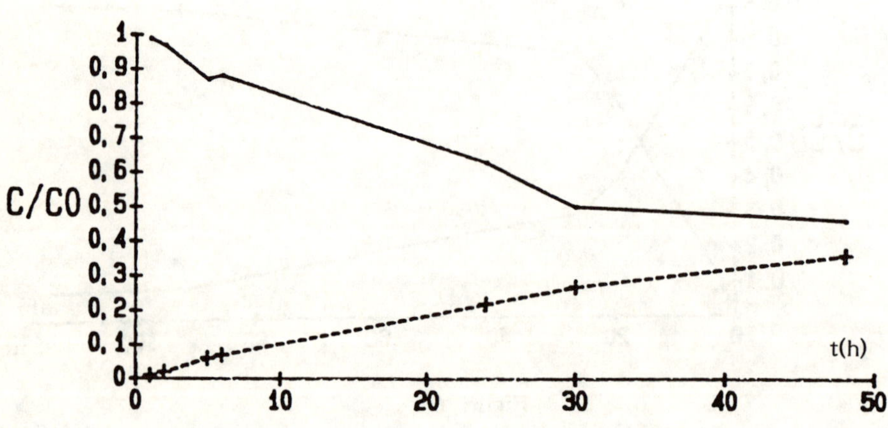

Figure 8

Extraction of Sr with a double S.L.M
DC18 C6 0.5 M in Hexylbenzene /decanol

slowly, is minimal after 60 hours, then increases and reaches a value higher than 2.6.

The flux of strontium is very slow : only 32 % transferred into the strip solution before precipitate occurs. The low rate of transfer of strontium may be due to the precipitation of salts which blind the membrane (Fig. 5).

5.2. Hexylbenzene / decanol as diluent

According to IZATT, phenylbenzene is an excellent diluent for the dicyclohexano crown ether ; its very low solubility in water and its low volatility lead to an excellent stability of the membrane. Compared to decanol, hexylbenzene presents another advantage, it does not transfer acidity. When hexylbenzene is in contact with the concentrate, a third phase occurs. The addition of decanol, at a concentration higher or equal to 0.5 M, avoids the formation of the third phase.

Using the mixture of hexylbenzene and decanol as diluent instead of pure decanol leads to better results :
- a lesser decrease in pH ;
- the flux of strontium through the membrane is increased ;
- recovery of strontium, in the second cell, is higher than 93 %.

When using lower concentrations of extractants, the rate of transfer of strontium decreases, but the same percentage of strontium is extracted (Fig. 6 and 7).

To achieve a greater DF, the double membrane system is used once more, with a less concentrated strip solution of NaOH (1 N) in order to reduce the risks of precipitation of aqueous solutions. The pH during the whole experiment exceeds 1.8 but if the flux of strontium is higher than in the previous experiment with a double SLM, it is still very low. The precipitation of salts inside the pores of the membrane seems to be responsible for the decrease of permeability of the membrane (Fig. 8).

5.3. Hexylbenzene / nonylphenol as diluent

Nonylphenol which possesses a very high viscosity may be of interest in the obtention of a more stable SLM. Like decanol, nonylphenol avoids the formation of a third phase.

The SLM constituted by the mixture phenylbenzene – nonylphenol presents a high stability. At the end of the experiment, the membrane was entirely satisfactory, the pH decreasing very slowly.

Experiments carried out with the three different concentrates confirm the excellent stability of the membrane and the high selectivity of crown ethers : The bivalent magnesium and the trivalent aluminium do not modify the results obtained with the concentrate containing mainly soldium : For the three concentrates the flux of strontium is almost the same.

6. EXTRACTION OF CESIUM

E. BLASIUS (7) showed that DB21 C7 diluted in nitrobenzene is the most suitable crown compounds for the extraction of cesium from acidic high sodium nitrate liquid waste.

P. GRAMAIN who synthesized several crown ethers at the University of Strasbourg and who supervized the synthesis of the crown ether, chose the B21 C7 whose extractive properties are comparable to DB21 C7 but whose synthesis is simpler.

The extractant power of synthesized B21 C7 is checked on a acidic nitric solution of cesium (HNO_3 0.1 M) and on the concentrate, the distribution coefficients are respectively 5.8 and 1.25.

Nitrobenzene, with a low viscosity is not suitable for obtaining a stable S.L.M. It would be advantageous to remove cesium and strontium form the concentrate with only one S.L.M. and to introduce B21 C7 into the DC18 C6/nonylphenol/hexylbenzene mixture.

To spare B21 C7 only available in small amounts, we carried out a test with a membrane (CELGARD 2500) impregnated by the mixture DC18 C6 (0.25 M) DB24 C8 (0.1 M) Nonylphenol (0.6 M) diluted in hexylbenzene.

The stability of the membrane is good, the pH decreases quite rapidly ; in these conditions more than 90 % of strontium and about 10 % of cesium are extracted. However, this low percentage extracted is promising, the DB24 C8 not being the most suitable crown ether for the removal of cesium in high content sodium nitrate solutions.

7. EXTRACTION OF ACTINIDES

7.1. Extraction with a single membrane

For the extraction of actinides, the compounds and the concentrations chosen by DANESI were used : Mixture of CMPO (0.25 M) and TBP (0.75 M) diluted in decalin. Measurements of the distribution coefficients of UO_2^{2+}, Pu^{4+}, Am^{3+} were carried out between the organic phase containing the mixture CMPO - TBP diluted in decalin and other aqueous solutions.

The results are listed in table IV.

Table IV

	MAR 400	Formic acid (1 N)	Sodium citrate (0.5 M)	Tartric acid (1 N)	H_2O	Am.acetate (0.5 M)
$D_{UO_2^{2+}}$	400	1.9	10^{-2}	.38	6.2	5.10^{-2}
$D_{Pu^{4+}}$	1000	2.10^{-2}	3.10^{-3}	4.10^{-2}	0.70	1.1
$D_{Am^{3+}}$	14	$2.6\ 10^{-2}$	9.10^{-4}	0.19	0.69	$8,5.10^{-2}$

The distribution coefficients between the stripping solution and the organic phase have to be the lowest, they are high in deionised water but very low in sodium acetate. Experiments performed with SLM using the different stripping solutions confirm that sodium acetate is the suitable solution for backextracting the actinides : To quantitatively remove actinides, the concentrations of sodium citrate in the stripping solution must be higher than 0.4 M.

A test carried out by adding separately to the concentrate one of three actinides : U, Pu, Am, after 48 h 99 % of uranium or plutonium and 98 % of americium are transfered through the membrane.

It is important to notice the great discrepancy of concentrations of uranium (6.7 g.1^{-1}) plutonium (6.2 10^{-5} g.1^{-1}) and americium (4.7 10^{-7} g.1^{-1}).

The experiment was performed again by simultaneously adding the three actinides. The percentages of actinides transferred after 54 h through the membrane for U, Pu and Am are respectively : 92, 97 and 99.4 % (Fig. 9).

7.2. Extraction with double S.L.M

The stripping solution of sodium citrate makes it possible to remove nearly quantitatively Pu and Am but the relatively high sodium content of the solution limits the concentration factor of the process. The use of a double S.L.M may enable less concentrated complexing solutions to be used. To limit the secondary wastes, concentration of the stripping solution 2 must be decreased. pH measurements were performed with different complexing

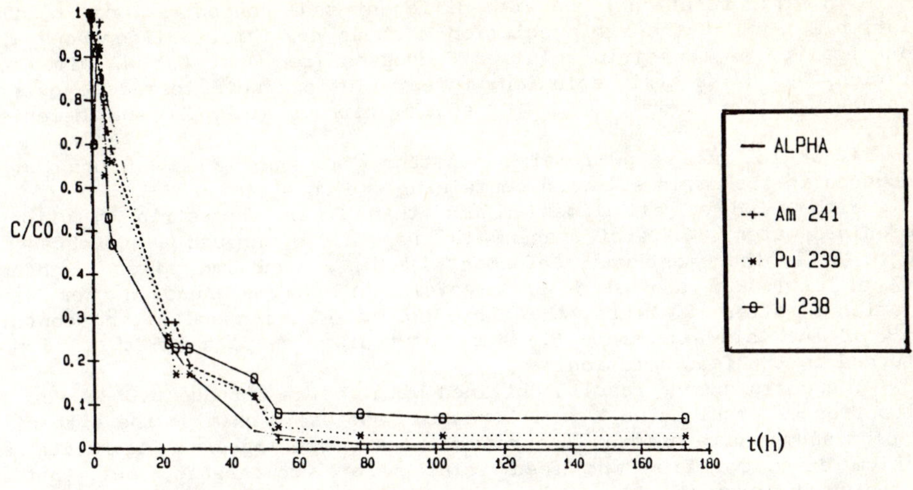

Figure 9

Extraction of actinides with one S.L.M
Strip solution : sodium citrate (0.5 M)

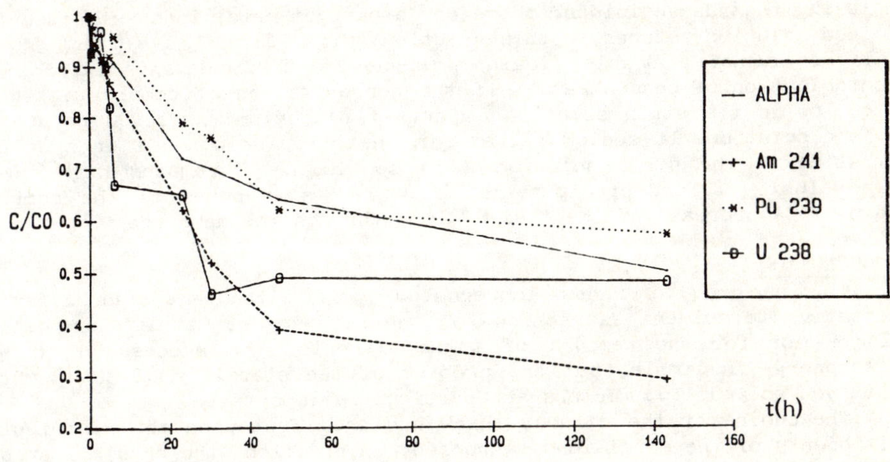

Figure 10

Extraction of actinides with double S.L.M
Strip solution 1 : sodium citrate (0.05 M)
Strip solution 2 : NaOH (0.5 M)

agents in strip solution 1 and with different soda concentrations in strip solution 2 : pH of the strip solution 1 shows negligible differences when the normality of the strip solution decreases from 4 to 0.5 N. With soda solution (0.5 N) as strip solution, three solutions were tested to extract uranium : sodium acetate (0.5 M), sodium citrate (0.05 M) and deionised water.

After 150 hours, pH remains constant, but only 25 % of uranium is recovered in the strip solution containing sodium acetate.

With a pH at all times higher than 2 in the strip solution 1 (deionised water), 80 % of uranium is transferred through the membrane in 160 h. An another test was performed with U, Pu and Am, after 55 hours, 93 % of plutonium and 81 % of americium are transferred in the strip solution 1, after 130 hours, the strip solution (deionised water) contains 90 % of americium but only 81 % of plutonium. A part of this element returned in the feed solution.

The satisfactory results obtained with sodium citrate 0.05 M for the extraction of uranium (98 % in 150 hours) lead us to examine the extraction of plutonium and americium separately. The percentages of americium and uranium extracted from the feed solution are comparable, but that of plutonium is lower (81 %).

The test is carried out again with the three actinides with the same strip solution (sodium citrate 0.05 M). The rate of transfer of plutonium is high, 92 % extracted after 53 hours, but as the pH continues to decrease (~ 0.3), a part of plutonium returns in the concentrate. The flux of americium is lower but even at low pH the americium is always transferred from the feed to the strip solution (85 % in 180 hours). The low concentration of sodium does not allow to make the strip solution sufficiently complexing towards actinides, therefore a part of metal ions extracted from the feed solution returns in this solution (Fig. 10).

The use of a double liquid membrane is difficult to use for the decontamination of concentrate : if the concentration of complexing agents is too low in the strip solution a part of plutonium extracted returns in the feed solution. If sodium citrate in the strip solution 1 or soda concentration in the strip solution 2 is too high, pH remains high in the strip solution 1, and precipitation occurs in some pores of the membrane leading to a decrease of the flux of ions through the membrane.

8. CONCLUSION

The use of membranes impregnated with extractant can offer an alternative to solvent extraction for the recovery of metals from diluted solutions or for the removal of toxic elements. The process which uses little energy is simple, but the problem of the stability of the membrane is not yet solved. Two factors are particulary important :

The choice of the diluent which has an effect upon the distribution coefficients of the metal ions extracted, in addition, the physical properties of diluent are essential so as to obtain a satisfactory stability : low solubility in the feed and strip solutions, low volatility, viscosity... The chemical nature of the diluent is also important, it can contribute to the transfer through the membrane of unwanted ions like H^+.

Most of the efforts must be carried out on the nature of the membrane or the geometry of hollow fibers (large pore dimensions so as to easily reimpregnate the fibers, a greater thickness of the membrane so as to increase the inventory of the solvent, a larger diameter of fibers so as to avoid blocking them).

Efforts must also be made to find a reliable process of reimpregnation of the membrane.

REFERENCES

(1) D.G. KALINA - E.P. HORWITZ - L. KAPLAN and A.C. MUSCATELLO, Sep. Sci. Technol., 16 1127 (1981)
(2) C.J. PEDERSEN - J.Am. Chem. Soc. 89 7017 (1967)
(3) R.M. IZATT - J.D. LAMB and R.L. BRUENING, Separation Science and Technology for energy applications. KNOXVILLE, Tennessee, Oct. 26-29, 87
(4) W.J. MAC DOWELL - B.A. MOYER - S.A. BRYAN - R.B. CHADWICK - G.N. CASE - Conf. 860921-3 - DE 86 008976
(5) R. CHIARIZIA - P.R. DANESI - Sep. Sci. Technol. 22 641 (1987)
(6) M. TROMP - These ENS de Chimie de Strasbourg (1985)
(7) E. BLASIUS - K.H. NILLES - Radiochim. Acta 35 173 (1984)

CHARACTERISTICS AND PERFORMANCE OF NEW TYPES OF ULTRAFILTRATION MEMBRANES WITH CHEMICALLY MODIFIED SURFACES

Steen Kristensen
DDS FILTRATION, Advanced Membrane Technology, Stavangervej 10,
P. O. 149, DK - 4900 Nakskov, Denmark

Summary
A special "coating" technique for making new types of surface modified ultrafiltration membranes has been developed and evaluated.
Using this method you can impart hydrophilic properties to an otherwise hydrophobic, high resistant synthetic polymeric ultrafiltration membrane. The properties of the final membrane depend on the
characteristics of the support membranes, the composition of the
coating solution and the amount applied and the reaction conditions.
The technique seems more or less universal - e.g. it is possible to
make a series of membranes with different cut-off values. Results
from measurements on different test solutions and on relevant industrial products are described. Characterization is done by
measuring hydraulic permeability and rejection of dextrans and proteins. Flux stability and fouling tendency have been examined using
different model foulants. The main advantages of the new membranes
are that higher flux values compared to a standard polyethersulfone
UF membrane are generally obtained. Some of these new membranes are
now commercially available and are supplied by DDS FILTRATION,
designated ETNA series.

1.   INTRODUCTION
Today many different types of polymeric ultrafiltration membranes
are commercially available. They are generally made by the traditional
phase inversion process using high resistant synthetic polymers like
polysulfone, fluoropolymers, polyamides, acrylics etc. (ref. 1), and they
were introduced on the market during the 1970's. Since then, these UF
membranes have established the use of ultrafiltration technology within
many industries.

In recent years, especially in the early 1980's membrane development
has mainly been concentrated on reverse osmosis membranes because of
maturation of the technique for making outstanding new types of desalination membranes, e.g. thin film composite membranes, poineered by the work
of Cadotte (refs. 2-4) and Riley (refs. 5-6). An enormous amount of literature, patents and patent applications can be found within this field.
John E. Cadotte (ref. 7) gives a short review on those types that have
survived the selection for commercial development. Different techniques
are used, which can be summarized as follows:

1 Casting of the thin barrier layer separately followed by lamination to
   a support membrane.

2 Solution coating of a support membrane by dipping in a polymer solution and and drying.

3 Solution coating in a reactive monomer solution followed by curing
   with heat or radiation.

4 Gas-phase deposition of the barrier layer from a glow discharge plasma
   or plasma treatment of the membrane surface to impart cross-linking.

5 Interfacial polymerization of reactive monomers "in situ" on the surface of a support membrane.

In principle these methods should also be applicable to ultrafiltration membranes, but until now very little has been published about similar techniques in respect to UF membranes.

In DDS FILTRATION we have looked into some of the above-mentioned
possibilities as well as other surface modification methods combined
with various chemical reactions to find a suitable technique to improve
the properties of existing UF membranes by changing their surface structure and properties. During screening studies encouraging results were
obtained, and further developments have let to the introduction of a new
series of ultrafiltration membranes now commercialized as the ETNA
series. This paper describes the preparation method and principles, basic
membrane properties and characteristics as well as laboratory, pilot, and
field tests on relevant solutions and some industrial products.

2.   BACKGROUND AND GENERAL PRINCIPLES

Our standard membranes for general ultrafiltration purposes are made
of either polysulfone (PSO), polyethersulfone (PES) or polyvinylidene-
fluoride (PVDF). Because of their chemical nature you have to take into
account the importance of adsorption phenomena and fouling. Within many
industrial applications protein fouling is a serious problem when using
these hydrophobic membranes. It is generally accepted by strong evidence
that hydrophilic membranes are less accessible to fouling (refs. 8-9).
Consequently, the basis of our studies was to modify the surface properties of an otherwise hydrophobic UF membrane to render it hydrophilic.
In the literature you can find some examples of hydrophilic/low ad-
sorbtive membranes using "surface modification techniques" as follows:

G. B. Tanny and M. Heisler (ref. 10) describe the application of a
thin film composite UF membrane which is claimed to have low surface
affinity towards adsorption of proteins. No details about preparation or
composition of the membrane are given. R. J. Petersen and co-workers,
FilmTec Corp. (refs. 11-12), have developed a novel set of thin film composite UF membranes, consisting of ultra thin barrier layers of cellulosics graft copolymers coated on microporous polysulfone supports. Graft
copolymers containing hydrophilic backbones on ionically charged side-
chains were synthesized on the basis of commercial available cellulosics
and different vinyl monomers. The grafted copolymers being highly water
soluble were fixed onto the support by rendering them insoluble by cross-
linking - thermally or chemically by heating with glyoxal under acidic
conditions. For sulfonate based copolymers cross-linking could be
achieved by incorporation of acid-catalyzed cross-linking polymers such
as N-(hydroxymethyl)acrylamid. These membranes have a smooth surface and
ultrathin barrier layers, and the described tests confirmed the assump-
tion that thin film composite membranes are far more resistant to fouling
because of smooth surface and hydrophilicity compared to microporous

polysulfone ultrafiltration membranes, and could be effectively cleaned by simple cleaning treatment.

Highly cross-linked thin film composite UF membranes are commercially available from Desalination Systems Inc. U.S.A. (refs. 13-14). They are made by an interfacial polymerization technique (polyamide reaction) and different cut-off values are offered. Filtron Corporation produces UF membranes based on polyethersulfone with modified surface chemistry (ref. 15). One series has low protein binding and another is resistant to anti-foaming agents. Fane et al (ref. 16) studied the effects of pretreatment with non-ionic surfactants on ultrafiltration flux values.

Our introductory studies included the preparation principles outlined in Table I. During a screening phase good results were obtained, especially when using the first mentioned principles, which were consequently selected for further examinations as described below.

TABLE I
Surface modification principles - a survey.

---

1. Chemical binding (grafting) of selected hydrophilic polymers and/or monomers onto the surface of a preformed UF support membrane.

2. Insolubilization (curing/cross-linking reactions) of hydrophilic polymers and/or monomers brought into contact with the surface of a support membrane.

3. Coating the surface of a support membrane by an organic solution of water indissoluble hydrophilics and curing.

4. In situ polymerization of monomers on a support membrane giving rise to thin film polymeric structures containing hydrophilic groups.

5. Ultraviolet light initiated in situ polymerizations and grafting reactions on a support membrane to impart hydrophilic properties.

---

3. THEORETICAL ASPECTS

Several reaction schemes can be imagined to get a chemical "grafting" of hydroxy or amine compounds onto a PVDF material. The procedure can be done in one or more steps, which does not mean that the reaction is as simple as indicated:

(A) "One Step Reaction"

PVDF is attacked by a strong base causing elimination of fluorine and hydrogen. Reactive sites (free radicals) are created which can combine to form double bonds, attack another chain to form cross-links, or graft to a molecule that can be attacked by free radicals.

$$\left(CF_2 - CH_2\right)_n + R - OH \xrightarrow[\text{heat}]{\text{NaOH}} \left(\underset{OR}{CF} - CH_2\right)_n$$

(B) "Two Step Reaction"

1. Elimination:

$$\left(CF_2 - CH_2\right)_n \xrightarrow[\text{heat}]{\text{NaOH}} \left(CF = CH\right)_n$$

2. Nucleophilic addition:

$$-(CF = CH)_n + ROH \xrightarrow{H+} -(CF - CH_2)_n$$
$$\qquad\qquad\qquad\qquad\qquad\quad OR$$

(C) "Three Step Reaction"
1. Elimination $\longrightarrow$ as (B) 1

2. Addition:

$$-(CF = CH)_n \xrightarrow[CH_2Cl_2]{Br_2} -(CF - CH)_n$$
$$\qquad\qquad\qquad\qquad\quad Br \quad Br$$

3. Nucleophilic substitution:

$$-(CF - CH)_n + ROH \longrightarrow -(CF - CH)_n$$
$$\quad Br \quad Br \qquad\qquad\qquad\qquad OR \quad Br$$

We decided to try the first mentioned route (A), because it is a simple procedure from a practical point of view, which is easily transformed into a continuous process using our existing membrane machinery.

In fact the reaction mechanism is not known, but nucleophilic reactions in relation to vinylidenefluoride polymers are described in the literature (refs. 17-18) as a method for curing/cross-linking of vinylidenefluoride elastomers at highly basic conditions.

4. EXPERIMENTAL
4.1. Materials and Method
Support membranes made of PVDF are prepared by the traditional phase inversion process, using knife casting directly on a non-woven polypropylene material as substrate. The properties of the support membranes are influenced mainly by casting solution composition and temperature of coagulation batch. Several types can be produced, and they are characterized by water flux and molecular weight cut-off (based upon dextran rejections) as shown in Table II.

TABLE II
Support membranes

Membrane designation	Material	Total thickness[1] μm	Water flux[2] l/m²/h	MWCO cut-off Dalton
FS20-U	PVDF	250/180	200 - 400	20,000
FS50-U	PVDF	250/180	300 - 500	50,000
FS100-U	PVDF + "filler"	250/180	>500	100,000

1) Two types of polypropylene non-woven used - membrane thickness ~ 50 μ.
2) Typical water fluxes at 2 bar, 20°C - DDS module UF 36.

In view of the good properties of membranes made of cellulosic materials, e.g. hydrophilic properties, low adsorption, good separation characteristics, such materials were adopted to a surface treatment procedure using the the reactions described earlier. Michaels et al.

(refs. 19-20) have used hydrophilic polymers, among others different non-ionic cellulosics, for surface modifications simply by exposing the up-stream surface of membranes placed in a system to a dilute solution of the polymer, followed by rinsing in pure water prior to use. Interesting results were obtained with respect to permeation rates. Based upon these facts and indications, different types of cellulosics were used alone or in combination with some (preferable) aromatic co-reactants like amines, hydroxycompounds or epoxides, containing additional (hydrophilic) functional groups.

The method used for the surface modifications is schematically indicated in Fig. 1. A solution containing the chemicals, the catalyst and possible additive(s) is brought into contact with the support membrane (dipping or pouring). Excess solution is removed by squeezing or draining. The reaction is carried out in a convection oven. Unreacted chemicals etc. are removed by a hot rinse and finally the membrane is dried.

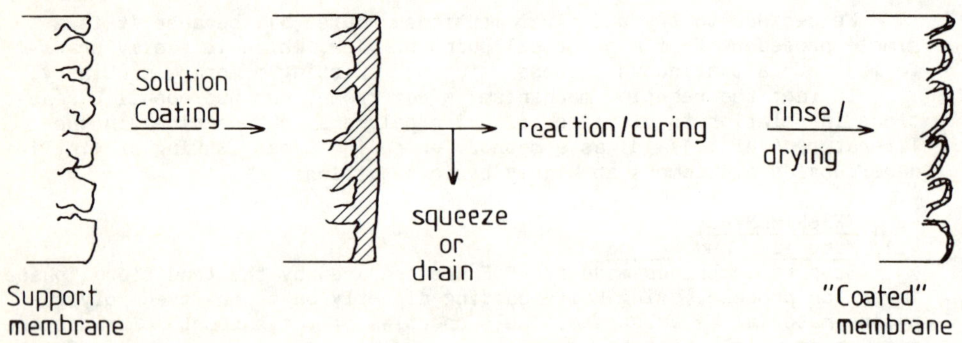

Fig. 1. Surface treatment, basic principle.

4.2. Membrane Preparation - Technical Approach

The preparation procedure can be run as a continuous process on the membrane machinery shown in the diagram, Fig. 2. The machinery is constructed with different coating/application zones, reaction/drying zones and a washing section. Typical compositions and operation parameters are as follows:

Support membrane: Asymmetric UF membrane of PVDF.
Coating solution composition:

    0.2 - 1.5 % cellulosic polymer
    0.5 - 5   % low MW co-reactant (amines/hydroxy comp.)
    1   - 3   % additives (surfactants, cross-linking agents, etc.)

Reaction/curing conditions:
Temperature: 90 - 125°C
Reaction time: 2 - 15 min.
Rinse tanks: 50 - 95°C, water containing surfactants, stabilizers, etc.
Drying temperature: 75 - 90°C

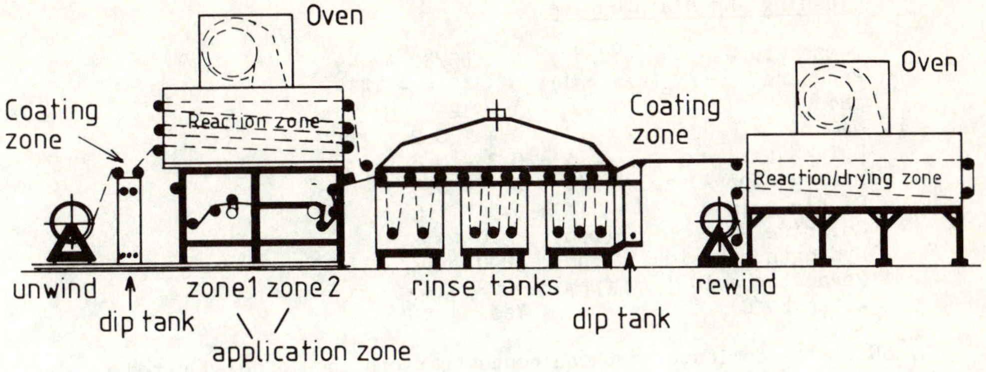

Fig. 2. Membrane machinery - general outline.

4.3. Membrane Characterization and Analysis

To establish the permeability and permselectivity our "DDS FILTRA-TION Lab. Equipment" (ref. 21) was used to get flux/permeability data on water and water solutions of sucrose, dextrans of various molecular weights and proteins.

Further tests on different industrial products were done in lab scale, pilot scale as well as in some industrial plants.

Fouling tendency was examined in UF tests on some model foulants of interest (protein, polysaccharide, and surfactant) and hydrophilic properties (wettability) were tested for by measuring aqueous contact angles.

Scanning electron microscopy was employed to examine the surface of the membranes with respect to the morphology of different coatings compared to the support membrane itself.

Chemical characterization of the coatings was investigated by infra-red spectroscopy (FTIR/ATR) and X-ray photoelectron spectroscopy (XPS/ESCA) to evaluate the possibilities to get some relevant analytical data.

5. RESULTS AND DISCUSSION
5.1. Membrane Characteristics

Until now the ETNA series consists of 3 types.

Characteristics:

Type	MW cut-off[1] Dalton	Water flux[2] $1/m^2/h$	pH	Temp. ^oC	Pressure, bar
				Operation range	
ETNA 01A	1.000	50 - 100	1 - 12	0 - 60	0 - 20
ETNA 10A	10.000	100 - 200	1 - 12	0 - 60	0 - 15
ETNA 20A	20.000	200 - 300	1 - 12	0 - 60	0 - 15

1) Based on dextran/dextrin rejections.
2) Measured at 4 bar, 20°C, DDS M36 Lab Unit.

<u>Cleaning and Disinfection</u>

```
Temperature              (°C/°F): 0 - 65/32 - 149
Pressure                 (bar/psi): 0 - 10/0 - 145
pH:                               1 - 12
NaOH3)                       (%): 0.1 - 0.2
EDTA-Na4                      (%): 0.3 - 0.5
Strong mineral acid3)        (%): 0.3 - 0.5
Citric acid                  (%): 0.5 - 1.0
Chlorine4)                 (ppm):   20
Hydrogen peroxide4)        (ppm): 1000
Peracetic acid4)           (ppm): 1000
Water 80°C:                       Yes
```

3) pH limits dominate so these concentrations should be adjusted to the
   right pH.
4) Max. temperature 25°C/77°F.

Support membrane type FS50-U is used for production of ETNA 20A and
type FS20-U for ETNA 10A and ETNA 01A.
    Some typical rejection data measured in the laboratory using our Lab
20 Module are shown in Table III.

TABLE III
Typical rejection data (%) - Lab M20: Test data[*]).

| Type | Sucrose 342 | Dextrin/dextrans (average MW) | | | | Myoglobin MW 17.500 |
		1400	4 - 6.000	9.400	17.200	
ETNA 01A	25	> 85	> 95	> 99	–	~ 100
ETNA 10A	10	–	75	> 95	–	~ 100
ETNA 20A	5	–	25 - 50	> 75	> 95	> 98
FS20-U	< 5	–	< 25	40 - 60	> 75	> 90
FS50-U	< 2	–	–	< 5	< 25	< 25

[*]) ETNA 01A at 5 bar, 20°C - other types at 2 bar, 20°C.

At the moment we are evaluating an improved characterization method,
using a mixture of dextrans and analysing feed and permeate samples
regarding concentration and molecular weight distribution by high performance liquid chromatography (HPLC). As a result retention curves are
drawn, and these curves permit a better comparison of individual membranes.

5.2. <u>Lab/Pilot and Field Tests</u>
In the laboratory tests on dairy products like whey and skimmed milk
showed flux increases for the ETNA types in the order of 20 - 50% com-
pared to their polyethersulfone UF membrane counter parts (our GR types).
These results have been verified in pilot scale for whey as shown in
Figs. 3a and 3b.

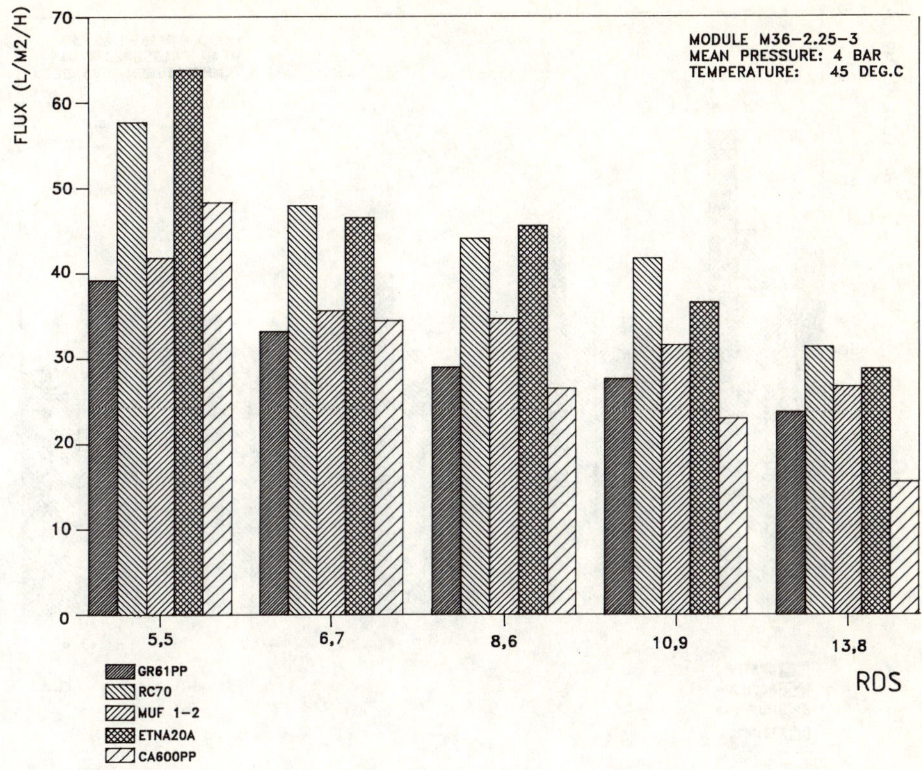

Fig. 3a. Whey, Trial No. 1

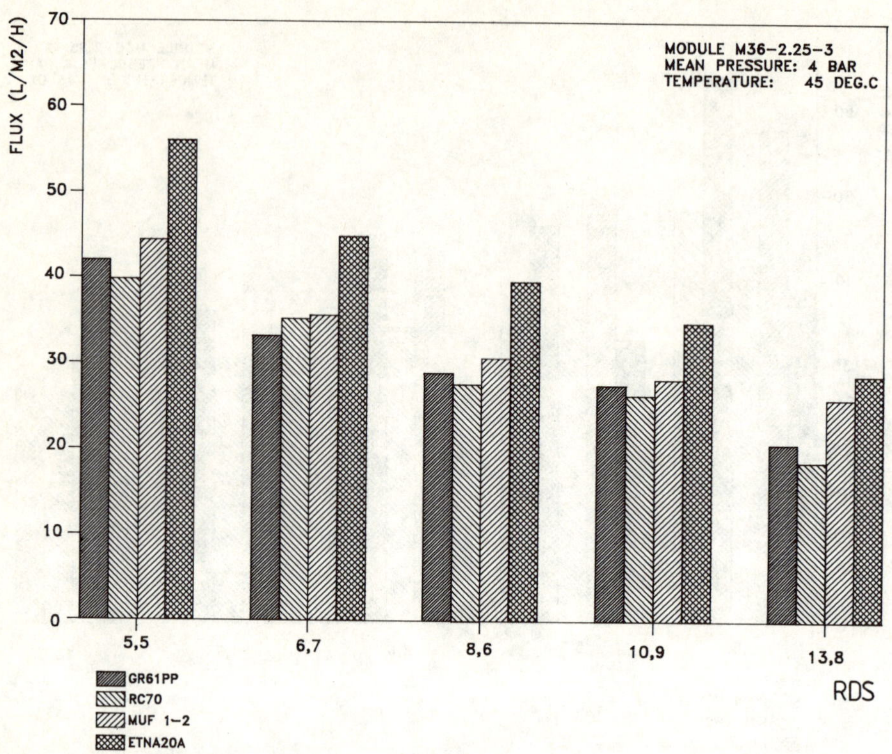

Fig. 3b. Whey, Trial No. 2

The other membranes tested are:

GR61PP = Polyethersulfone, MWCO 20.000
CA600PP = Cellulose acetate, MWCO 20.000
RC70PP = Regenerated cellulose, MWCO 10.000
MUF1-2 = Experimental/coated type based on a polyethersulfone support,
          MWCO 20.000

Flux improvements on ETNA 20A compared to GR61PP are in the order of 20-50% over the entire concentration range. The RC70PP membrane was chosen for comparison because of its hydrophilic properties and smooth surface appearance, and as can be seen this membrane also performs very well. Unfortunately, reduced film strength and high degree of swelling causing delamination put some limitations to its general use. The ETNA membranes do not suffer from marginal strength having the mechanical properties of a PVDF membrane.

At a dairy in Austria having a 6 module DDS FILTRATION UF plant for concentration of whey from approx. 5% TS to 13 - 15% , samples of ETNA 20A membranes were mounted for long term tests in a real production plant. Other tests have been done at a dairy in U.S.A. The results were identical - up to 20 - 25% higher flux values than for GR61PP. One reason for the lower flux increases compared to the pilot experiments may be attributed to the effect of concentration polarization, caused by lower fluid velocity in the production plants.

Within the biotec industry down stream processing offers interesting applications for membrane filtration, and the ETNA membranes might improve the applicability and create new processes. Drastic improvements are seen in Fig. 4 giving flux values for an unfiltered fermentation broth, where a GR61PP membrane does not work at all.

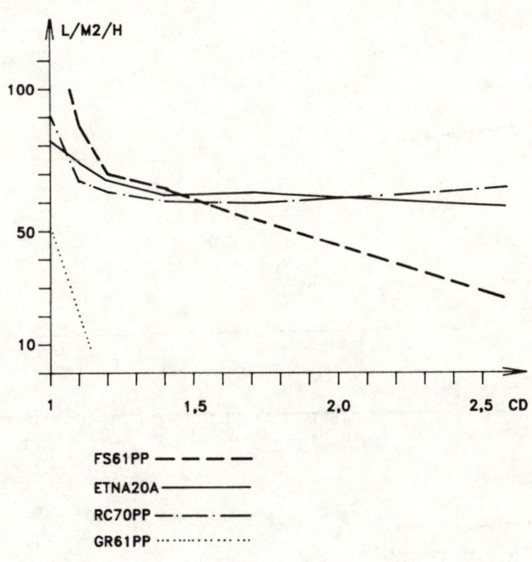

Fig. 4. Flux vs. volumetric concentration degree (CD) - Unfiltered fermentation broth

5.3. <u>Fouling and Hydrophilicity</u>

Short time tests on some model foulants have been performed as UF test, measuring flux changes for the membranes, when exposed to a solution of the foulant during a 2 hours run, Fig. 5.

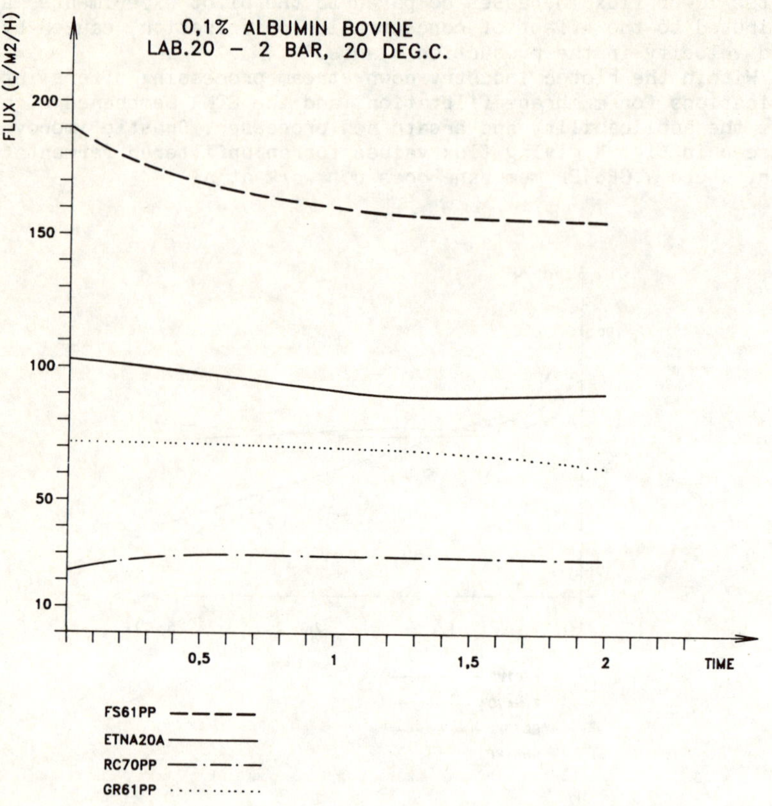

Fig. 5. Flux decline vs. time.

A protein (0.1 % albumin/bovine), a polysaccharide (50 mg/l dextran 4, average MW = 4.000 - 6.000) and a surfactant (50 mg/l Triton x 100, non-ionic) have been tested. Comparison of fouling effects can be done in different ways (ref. 22). In Table IV the relative recovery in water flux, before and after exposure to solute is calculated from above experiments. Of course you have to consider other factors (porosity, surface smoothness, actual flux values, etc.) when comparing such results, but nevertheless the almost complete flux recovery for ETNA 20A and RC7OPP is indicative for a less flux decline/fouling tendency.

TABLE IV
% water flux recovery - UF experiments.

Membrane type	Albumine bovine 0.1 %	Dextran 4 50 mg/l	Triton X-100 50 mg/l
FS61PP	66	86	75
ETNA 20A	97	100	100
GR61PP	54	88	75
RC70PP	100	100	100

Aqueous contact angle measurements can be used to indicate the differences in hydrophilicity/hydrophobicity. It is a difficult and a rather imprecise method, but the data given in Table V clearly show a consistent trend, e.g. the "coating layer" effects the wettability by making the membranes more hydrophilic.

TABLE V
Contact angle measurements[1]

Type:	GR61PP	FS50-U	ETNA 20A	ETNA 10A	ETNA 1A
Contact angle (α)	61	69	42	40	43

[1] The membranes have been throughly rinsed and dried before measurements.

5.4. SEM-Studies

Scanning electron micrographs of the cross section and the top surface of an ETNA 10A membrane are shown in Figs. 6a and 6b. The membrane exhibits a large number of macrovoids and fingers like cavities, typical for a PVDF ultrafiltration membrane. The surface structure appears inhomogeneous with cracks of different sizes. The surface morphology is distinct from the support membrane itself which seems to have more or less well defined pore openings, Fig. 7. The ETNA 01A membrane also shows surface cracks (Fig. 8), but at a lower overall porosity. The explanation and significance of the surface morphology need further investigations, but the effect of the surface treatment is obvious with respect to UF properties as described earlier.

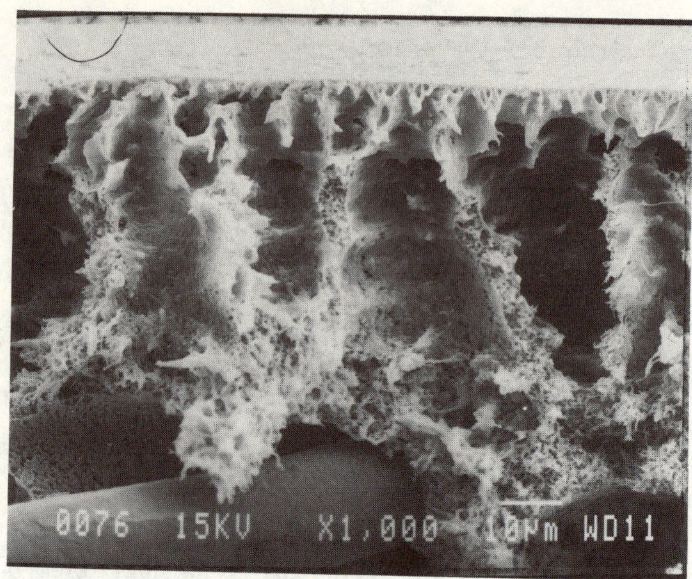

Fig. 6a. Scanning electron micrograph of cross section of ETNA 10A membrane.

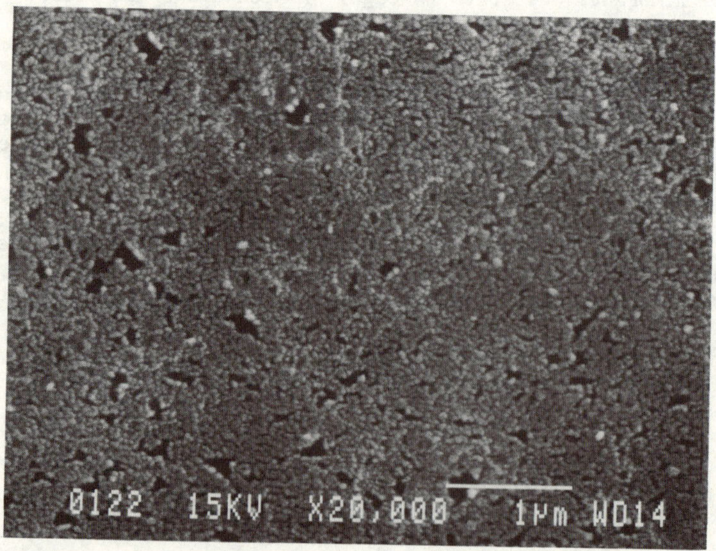

Fig. 6b. Scanning electron micrograph of top surface of ETNA 10A membrane.

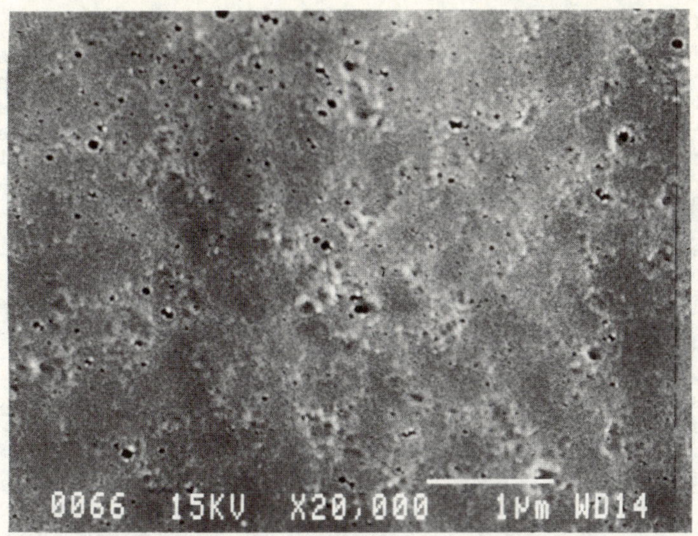

Fig. 7. Scanning electron micrograph of top surface of FS20-U support
membrane.

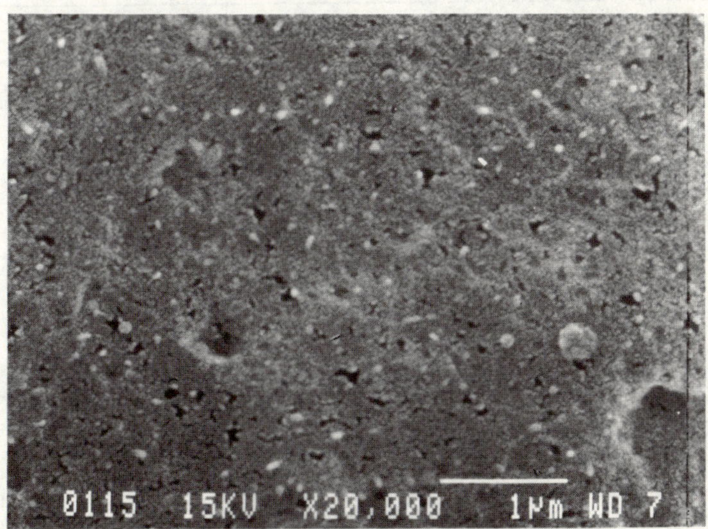

Fig. 8. Scanning electron micrograph of top surface of ETNA 01A membrane.

5.4. Surface Analysis

Infrared spectroscopy (FTIR/ATR) has previously been used successfully to characterize the separation layer on our thin film composite RO membranes. When used as a surface technique on the ETNA membranes, the IR spectra were on the whole identical showing the characteristic spectrum of the PVDF polymer. Difference spectra showed no differences either. The reason must be ascribed to the fact that very little "coating material" is present on the membrane surface.

X-ray photoelectron spectroscopy (XPS/ESCA) being a true surface analytical method, giving identification of elements, their quantitative atomic concentrations and information about chemical shifts, were tried. The Tables VI and VII show the results with respect to kinetic energy and relative atomic concentrations. The shift into a doublet for the carbon is in agreement with the nature of the polymer chains ($-CF_2-CH_2-$). Other shifts could not be identified, which might be ascribed to marginal resolutions of the peaks, and additional studies shall be done. The changes in atomic concentrations seen for the surface treated membrane are mainly caused by the elimination of fluorine during surface treatment, but the considerable increase in the concentration of oxygen must be due to the surface treatment procedure. The depth of information is max. 100 A and consequently being F and $\underline{C}$ - F signals in the spectra, the thickness of the coating layer must be less than 100 A or of inhomogeneous nature.

TABLE VI
XPS analysis of FS50-U membrane

Element	C	N	O	F
Kinetic energy (eV)	956.7; 961.1	848.1	710.7	557.8
Atomic conc. (%)	24.9; 34.7	1.7	2.8	35.9

TABLE VII
XPS analysis of ETNA 20A

Element	C	N	O	F
Kinetic energy (eV)	956.7; 961.4	847	714.6	559.8
Atomic conc. (%)	15.7; 47.0	2.9	15.7	18.7

6. CONCLUSION

A special coating technique is utilized to make a new line of ultrafiltration membranes called ETNA-series. The ETNA membranes show much higher fluxes than traditional membranes when processing liquids containing proteins. Additionally, they provide longer periods of continuous operation and shorter cleaning cycles because of less flux decline and better cleanability. These properties can be ascribed to the composition of the coatings, which are very hydrophilic and thereby reducing protein affinity. Consequently, the most profound improvements in performance compared to membranes made of hydrophobic, synthetic polymers will generally be obtained when processing products like fermentation broths, dairy products, antibiotics, beverages based on yeast, enzymes, etc.

The preparation technique seems to be more or less "universal", e.g. it should be possible to make additional UF membranes within a broader range of molecular weight cut-off, and probably even new microfiltration membranes can be made by similar coating procedures.

It should be expected that further investigations within this field will bring about other interesting new types of "surface modified" membranes with special properties and for special applications.

REFERENCES

(1) APTEL, P., CLIFTON, M., in "Synthetic Membranes: Science, Engineering and Applications." P. M. Bungay, H. K. Lonsdale and M. N. de Pinho (Ed.), D. Reidel Publishing Company, 1986, pp. 240-305.
(2) CADOTTE, J.E., ROZELLE, L.T., In Situ Formed Condensation Polymers for Reverse Osmosis Membranes. Office of Saline Water, Research and Development Progress Report No 927, National Technical Information Service, Nov. 1972.
(3) CADOTTE, J.E., PETERSEN, R.J., LARSON R.E., ERICKSON, E.E., Desalination 32 (1980), 25.
(4) CADOTTE, J.E., KING, R.S., MAJERLE, R.J., PETERSEN R.J., Interfacial Synthesis in the Preparation of Reverse Osmosis Membranes. J. Macromol Sci.-Chem. A15(5) pp. 727-755 (1981).
(5) RILEY, R.L, HIGHTOWER, G.R., and LYONS, C.R., Applied Polymer Symposium No 22, pp. 255-267 (1973).
(6) RILEY, R.L, FOX, R.L., LYONS, C.R., MILSTEAD, C.E., SEROY, M.W., and TAGAMI, M., Desalination, 19 (1976) pp. 113-123.
(7) CADOTTE, J.E., Evolution of Composite Reverse Osmosis Membranes in: "Material Science of Synthetic Membranes". D. R. Lloyd (ED.), Chap. 12, pp. 273-294. SCS Symposium Series 269, American Chemical Society, Washington D. C. 1985.
(8) MICHAELS, A.S. and MATSON, S.L., Desalination 53 (1985), pp. 231-258.
(9) TSUDA, S., OZAWA, K., and SAKAI, K. Proceedings of the 1987 International Congress on Membranes and Membrane Processes. ICOM '87. Tokyo, Japan, June 8-12, 1987, pp. 467-68.
(10) TANNY, G.B., HEISLER, M., "The Application of Novel Ultrafiltration Membranes to the Concentration of Proteins and Oil Emulsions". Polym. Sci-Technol. (80), Vol. 13, pp. 671-83.
(11) PETERSEN, R.J., KING, R.S., CADOTTE, J.E.: "Novel Thin Film Composite Ultrafiltration Membranes for Waste Water Filtration". Proc. Water Reuse Symp. (82), pp. 1363-77, No 2.
(12) PETERSEN, R.J., KING, R.S., BENNET, C.R., CADOTTE, J.E.: "Improved Pretreatment Processes for Waste Water Reuse Using Reverse Osmosis". Office of Water Research and Technology. Washington D. C. Sept. 80, PB 81 - 197600.
(13) QUINN, R.M., Desalination 46 (1983), pp. 113-123.
(14) Die Ernährungsindustrie 4 (86), 12.
(15) Membrane & Separation Technology News, Vol. 4, No 7, p. 6, March 1986.
(16) FANE, A.G., FELL, C.J.D., and KIM, K.J., Desalination 53, (1985), pp. 37-55.
(17) PACIOREK, K.L., MITCHELL, L.C, and LENK, C.T., J. Polym. Sci. 45, 405 (1960).
(18) SCHMIEGEL, W.W., Die Angewandte Makromolekulare Chemie 76/77 (1979), pp. 39-65 (No 1122).
(19) MICHAELS, A.S., ROBERTSON, C.R., and REIHANIAN, H. Proceedings of the International Membrane Technology Conference (IMTEC' 83), Sydney, Australia, Nov. 8-10, 1983, pp. 59-63.

(20) MICHAELS, A.S., ROBERTSON, C.R., and REIHANIAN, H. Proceedings of
     the 1987 International Congress on Membranes and Membrane Processes
     (ICOM' 87), Tokyo, Japan, June 8-12, 1987, pp. 17-19.
(21) DDS FILTRATION, Brochure: Lab Equipment 1624 GB.
(22) FANE, A.G., KIM, K.J., FELL, C.J.D., and SUKI, A.B. "Character-
     ization of Ultrafiltration Membrane: Flux and Surface Properties",
     in: "Characterization of Ultrafiltration Membranes", Proceedings pp.
     39-80, (Ed. Gun Trägårdh) Örenäs Slot, Sweden, Sept. 9-11, 1987.

BRITE GAS SEPARATION NR.: I1B-0098-NL (GDF)

MEMBRANE SEPARATION OF CO2 AND H2S
FROM MIXTURES WITH GASEOUS HYDROCARBONS

Henk Stam
Akzo International Research
P.O. Box 9300, 6800 SB Arnhem

Partners: - The European Economic Committee
          - Société Nationale Elf Aquitaine
          - Twente University of Technology
          - AKZO International Research B.V.

Duration: March 1986 - December 1988

Summary

The Brite-project was aimed to develop a process to separate
CO_2 and H_2S from natural gases by means of membranes. A target
with respect to permeability and selectivity of the membranes
has been set.
To reach the target properties commercial polymers, like PSO,
PES and PPO were chemically and physically modified. Although
much effort was put in these modifications, all types of
modification showed that selectivities could be improved only
at the expense of permeability, or that the permeability could
be enhanced, while reducing selectivity.
Research on spinning techniques has shown that it is possible
to produce fibres resistant to the high pressure of natural
gas by melt spinning, but not (yet) by wet spinning.
A membrane module resistant to the high natural gas pressures
has been developed.
To perform techno-economical evaluations the membrane systems
are studied closely with different models. For the hollow
fibre modelling the dimensionless number developed by Mr.
Gorissen is of major importance.
The techno-economical evaluations have shown that improvement
of the permeability has more economical advantages than
improvement of the selectivity.
Surveys on the costs of the hollow fibre membranes, based on
price/performance, have demonstrated, that wet spun hollow
fibres are to be preferred instead of melt spun hollow fibres,
provided that pin holes can be avoided, and the outer diameter
can be reduced.
Analyses of the final module costs show that the costs of the
housing are the main part and that e.g. polymer prices are of
minor importance. Therefore other, more expensive, polymers
than the original types were synthesised and tested with
promising results.
However, market studies have shown that real market
opportunities for this type of gas separation are limited.

1. <u>INTRODUCTION</u>

The Brite-project aims to develop a process to separate CO_2 and H_2S from natural gases by means of membranes. The technical opportunities as well as the economical perspectives in the sense of competitive processing compared to the existing processes should be considered.

2. <u>OBJECTIVES OF RESEARCH</u>

2.1. <u>General</u>

The objective of research is to develop <u>membrane modules</u> with improved selectivity and permeability compared to the existing polymers and resistant to the high pressure of natural gas.

The target for the membranes shows a CO_2 permeability of 1.0 E-16 (m3.m/m2.Pa.s) and a selectivity of 50 for CO_2/CH_4.

The existing polymers like PSO, PES and PPO should therefore be modified either chemically or physically.

The two possibilities of making membranes either homogeneous via melt spin process or asymmetric via wet spin process were to be considered.

These membranes or modules are being tested on a pilot plant scale.

Evaluations of the feasibility are made based upon market studies and technical-economical studies for which system modelling of the membrane process is of importance.

2.2. <u>More specified objectives</u>

The task of the project is the development of a membrane technology to be able to produce alone or in combination with conventional techniques pipeline gas quality out of sour natural gas containing CO_2 and H_2S as acid components.

The compositions of the feed gas and the pipeline gas are:

	<u>feed gas</u>	<u>pipeline gas</u>
CH_4	60 – 90 %	98 %
H_2O	to 100 % RH	
H_2S	0 – 25 %	4 ppm
CO_2	10 – 40 %	2 %

In general these requirements of the pipeline gas will only be obtained by a combination of membrane separation and the conventional, existing separation technologies.

2.3. <u>Targets for hollow fibres</u>

To achieve the required natural gas quality by membrane
technology or a membrane assisted technology on an economically
attractive basis, two membrane developments will be pursued with
the following minimum requirements to their properties:

- the first one consists of a homogeneous, melt spun hollow
  fibre with very small diameters and small wall thicknesses

- the second one consists of an asymmetric wet spun hollow fibre
  with an improved selective skin layer.

Specific Parameters	Homogeneous Hollow Fibre	Asymmetric Hollow Fibre
Selectivity CO_2-CH_4	50	50
H_2O-CH_4	100	100
H_2S-CH_4	50	50
Permeability of CO_2 (*)	1.0×10^{-16}	1.0×10^{-16}
Outside diameter (μm)	30	300
Selective wall thickness (μm)	5	0,1 - 0,5

* Permeability: 1.0×10^{-16} $m^3.m/m^2.Pa.s$ equals 13.3 Barrer.

Both hollow fibres should be resistant to a pressure of 70 bar
on the outer side of the fibres.

2.4. <u>Participants and their main activities in this project</u>

<u>Elf Aquitaine</u>
(SNEA (P), Lacq, France)
- Pilot plant testing
- Techno-economical studies

<u>University of Twente</u>
(department of chemical
 technology, Enschede)
- Chemical modification of polymers
- Preparation of asymmetric hollow
  fibre membranes

<u>AKZO</u>
- project coordination

Akzo Research Laboratories
Arnhem
- techno-economical studies
- system modelling

Akzo Research Laboratories
Obernburg
- polymer modification
- spinning of homogenous hollow
  fibres
- module development

3. PROGRESS OF MEMBRANE DEVELOPMENT

The goals on selectivity and permeability of the polymers should be reached by modifying existing polymers like polysulphone (PSO), polyethersulphone (PES), and others such as PPO.

The properties of these original polymer types are given in figure 1 and compared to the target.

Modification of these polymers has been done in three different ways:

- by chemical modification
- by physical modification
- by blending of polymers.

3.1. Chemical modification

Chemical modification means modification on-chain by substitution of aromatic protons, i.e. by means of nitration, losylation, bromination of alkyl groups and acylation.

Quite a lot of modifications have been carried out but all show the same tendency as is shown in figure 1, where the effect of nitration is shown as a main example.

Even modifications with more complex chemicals have not shown the desired result.

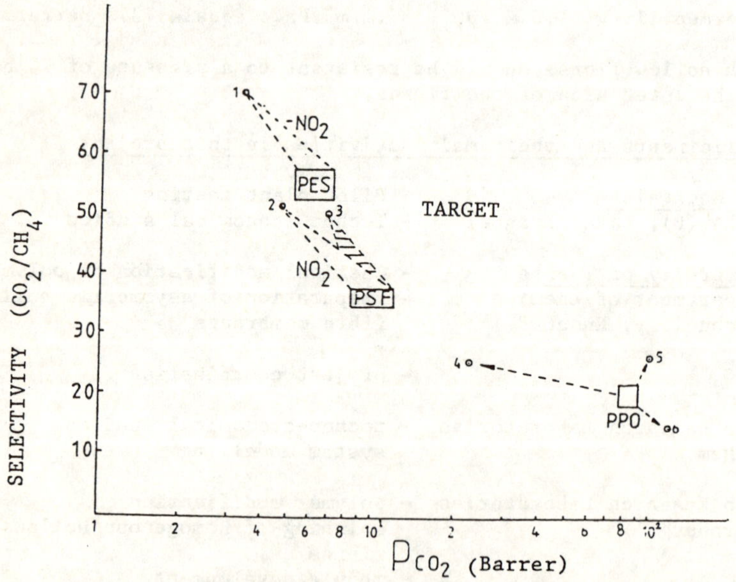

Fig. 1, Influence of chemical modification

3.2. Physical modifications

With physical modifications is meant to put additives to the
polymer material. A wide variety of inorganic components has
been added to the polymer materials.
The most successful was the addition of molecular sieves (Na-Y).
Addition of 1% b.w. of molecular sieve to the polymer membrane
has shown a relative improvement of 2% on the permeability
without effecting the selectivity. At contents of 15% molecular
sieve maximum permeability has been reached. At higher contents
selectivity was reducing again, where as permeability was not
improving any further.

3.3. Blending

Much effort was put in obtaining the desired results by blending
of polymers. These attempts were not limited to the original
polymers but also to the chemically modified types. Most of
these efforts did not meet the desired result as no homogeneous
polymer mixtures could be obtained, or mechanically seen such
doubtful structures, weak or brittle, were produced, that it was
not to be expected that a resistant hollow fibre could be
produced from this. The successful blends did not show
significant improvement on permeability and selectivity.

3.4. Conclusions

Although much effort was put in finding polymers by these
methods to meet the target, the obtained results were not
satisfactory enough. It can be concluded that:

- selectivity could be improved at cost of reduced permeability
- permeability could be improved at cost of reduced selectivity

There was no success in improving the selectivity and
permeability together sufficiently. Research on chemical and
physical modification of existing polymers has been stopped.

Further work is done on fluorinated new polymer types (see
chapter 8).

4. <u>TESTING AND MEASUREMENTS</u>

To evaluate the progress of the research activities measurements
are made at materials in different shapes:

- permeametry measurements on flat sheet membranes
- permeametry and pressure measurements on hollow fibres
- permeametry and pressure measurements on modules

All the partners have enhanced and improved their permeametry
measurement equipment on flat sheet membranes in order to
evaluate quickly their polymer development. At Elf Aquitaine
the facilities are available to test also at high pressures of
natural gases (70/80 bars). Therefore the important developments
are finally tested in the equipment of Elf Aquitaine.

In measuring membranes etc. the following aspects should be
taken into consideration carefully:

- measurement conditions
- aging and initial effects
- feed composition and temperature

4.1. <u>Measurement conditions</u>

Considerable and large differences in mainly selectivity has
been shown in relation to the measurement conditions where the
pressure on the permeate side was of utmost importance.
Differences of about 40% in selectivity were only originated
from the difference in pressure on the permeate side.

In view of this the final application measurements have been
done with a pressure above normal atmospheric pressure on the
permeate side.

4.2. <u>Aging and initial effects</u>

In the first few hours of testing the measured permeability
values are extremely high, but should be ignored for evaluation
of the polymer material. After about 24 hours of measuring
adequate figures can be obtained (see figure 2, next page)

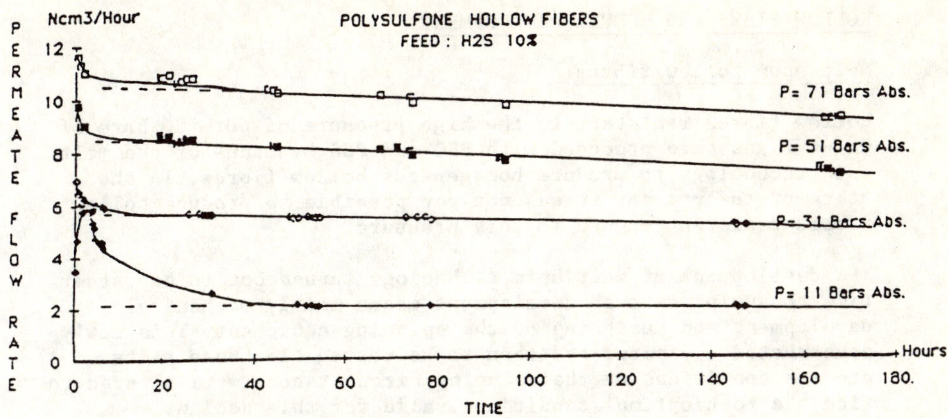

Fig. 2, Initial Effects and Aging on the permeability

On the other hand there is an aging effect on the polymer
material which reduces the permeability gradually. For
definitive conclusions about the appropriateness of the polymer
material this aging effect should be taken into account.

4.3. Feed composition and temperatures

Figure 4 shows clearly the effect of temperature and pressure
of CO_2 on the separation factor of melt spun PES-membranes.

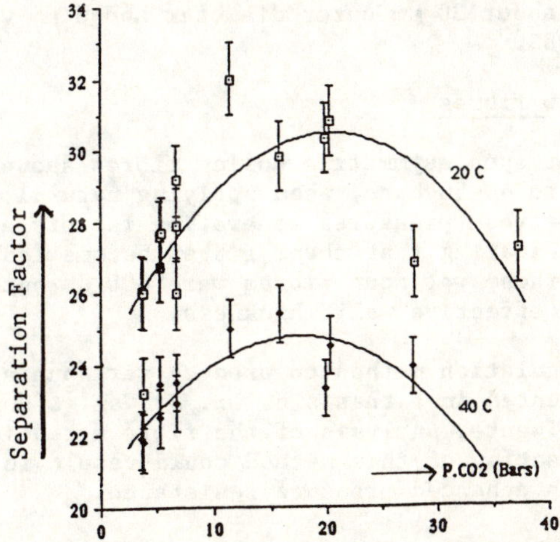

Fig. 3, Example of effect of gas composition and temperature on
the selectivity of the membranes.

5. HOLLOW FIBRE AND MODULE DEVELOPMENT

5.1. Melt spun hollow fibres

Hollow fibres resistant to the high pressure of 60 - 80 bars of
natural gas were produced with PSO and PES by means of the melt
spin technology to produce homogeneous hollow fibres. In the
start of the project it was not yet possible to produce hollow
fibres resistant enough to this pressure.

The development of melt spin technology turned out to be rather
time consuming as such development means mainly mechanical
development and designing of the spinning equipment. This newly
constructed equipment had then to be tested. In these tests
process conditions of the spinning circumstances were altered to
find the most optimal conditions valid for this design.

The pressure resistant hollow fibres could be produced by taking
into account the following aspects of the spinning procedure:

- reduction of residence time of the polymer to prevent polymer
  degradation
- maintaining a very homogeneous temperature over the spinneret
  to avoid undesired variations in wall thickness
- filtration of the polymer mix to such an extension that no
  particles remain
- winding of the final hollow fibre without mechanical stress

Thus it was possible to spin 12 hollow fibres in parallel with
dimensions of about 30 μm outer diameter and 6 μm wall thickness
with PSO and PES.

5.2. Wet spun hollow fibres

So far, the wet spun asymmetric hollow fibres showed a pressure
resistance up to 40-50 bars, when applying natural gas mixtures.
Above these upstream pressures separation factors appeared to
deteriorate indicating that these fibres become leaky. The
dimensions of these wet spun fibres were 600 μm outer diameter
and about 1 μm effective wall thickness.

A two-bath coagulation method to produce wet spun asymmetric
fibres is presented in a thesis of Mr. J. Van 't Hof at the
University of Twente. Analysis of the fibre morphology indicated
that an optimisation of this method could result in fibre
structures with enhanced pressure resistance.

5.3. Module development

It has been reached to produce "semi-permeable modules" with
melt spun symmetrical hollow fibres resistant to a pressure of
130 bars, when tested with air (see fig. 4, next page).

Fig. 4, Example of a compact membrane disc.

In such a disc construction the void volume is 50 to 60%.
This is acceptable with respect to the pressure losses.
A very high surface per volume can be obtained in the magnitude
of 30.000 to 50.000 m^2/m^3 .

The disc shape is chosen as it is very easy to install such a
unit between flanges of the (pressure resistant) housing and in
this way the flow of the permeate at almost atmospheric pressure
is facilitated on the outer border of the disc (see fig. 5).

Fig. 5, Part of the test unit for modules at Akzo Laboratories
        Obernburg. The installation of the compact membrane disc
        in the housing.

6. <u>SYSTEM MODELLING AND TECHNO-ECONOMICAL EVALUATIONS</u>

The system modelling of the membrane modules is the basis to
perform the techno-economical evaluations. These evaluations has
been used in two directions:

- to evaluate the feasibility
- to guide the research activities.

In the following paragraphs both aspects will be described.

6.1. <u>System modelling</u>

The system modelling of membrane separation processes is to be
divided in:

- description of the transport phenomena in the hollow fibre
- the performance of a membrane unit

For the <u>hollow fibre modelling</u> Mr. Gorissen developed a new
dimensionless number (see fig. 6).

$$\Phi_T = \frac{128\ Q_f\ \mu\ l^2}{d_i^3\ d_w}\ \frac{P_o\ T}{P_L\ T_o}$$

Fig. 6, Dimensionless number of Gorissen, the transport modulus.
      (Explanation of the parameters: see paper of Mr.
        Gorissen, also presented during this symposium)

This dimensionless number (Φ_T) distinguishes between two
essential transport phenomena of membrane separation viz.:

- the selective transport through the membrane wall
- the removal of the permeate through the fibre bore

Essential for gas separation is that the main resistance of the
gas transport should be in the transport of the gas through the
wall, which determines the selectivity of the process.
Considering this the value of the "number of Gorissen" or
transport modulus (Φ_T) should be smaller than 1.

The general practical use of this number is:

- to determine the optimal fibre dimensions (e.g. length), when
  the process parameters and the fibre characteristics are known
- to distinguish between different hollow fibre types as well as
  comparison between homogenous fibres and asymmetric fibres
- simplification of the calculations of the membrane unit.

The validity of the Φ_T number is shown in the experimental results as given in figure 7.

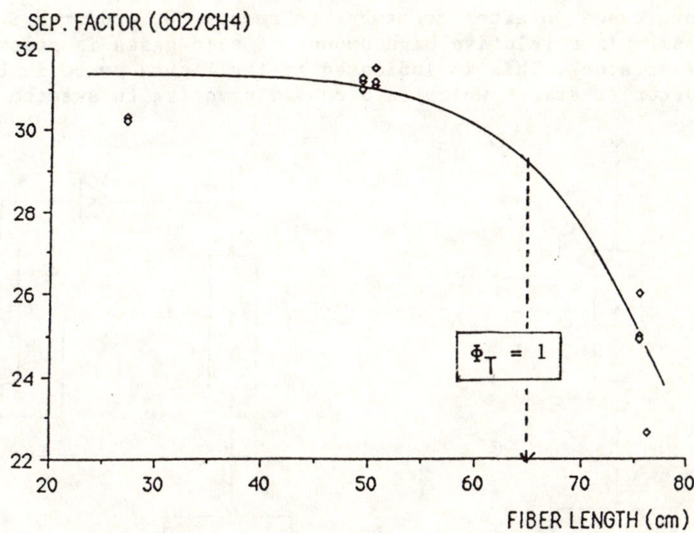

Fig. 7, Relation between Separation Factor and Fibre Length

This knowledge is an adequate tool to select the fibre dimensions of the gas separation module where in general the length of the fibre is of most importance unless mechanical constrains are putting limitations on the optimal lengths.

An accurate numerical model has been developed to calculate the behaviour and performance of the membrane unit (which basically consists of a package of crossed hollow fibres) by a lot of differential equations. The final values of the permeate and retentate flow can be calculated with the model.

For economical optimisation of membrane systems there is a practical approach by using the Φ_T-number only for a first estimate. To establish the system performance more accurately recalculations are made by using the exact calculation model for a gas separation unit.

Last but not least the required membrane area has to be corrected for non-isothermal effects of the gas separation. (See: Temperature changes involved in membrane gas separations, H. Gorissen, Chem. Eng. Process **22** (1987), p. 63–67). Because of these temperature effects the required membrane surface area might increase with more than 40% whereas the installation of gas heaters will be necessary.

6.2. Mono versus dual membrane systems

To sweeten sour natural gas two basic solutions are possible as is given in figure 8.

In both cases an after treatment to remove CO_2 and/or H_2S is necessary if a relative high amount of acid gases is present in the feed stock. This is indicated in the figure symbolic by the DEA after treatment which is a common practise to sweeten sour gas.

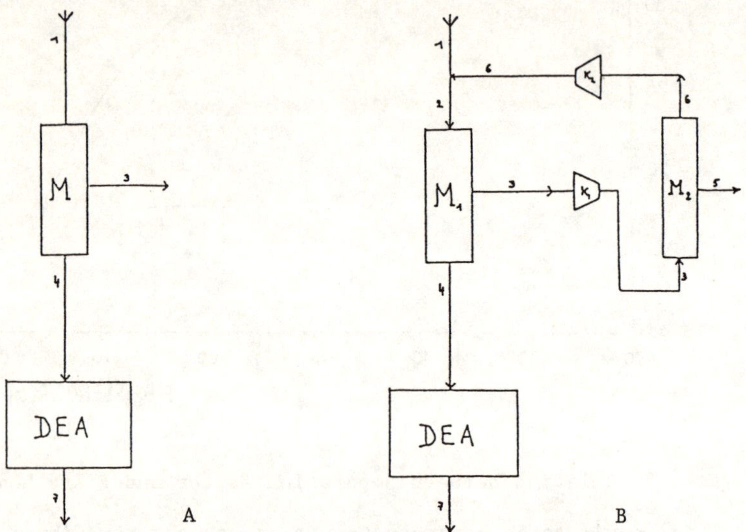

Fig. 8, Diagram of the mixed processes with one membrane
stage (A) or two membrane-stages (B)

In the system of figure 8B the permeated gas is treated another time to separate the sour components from the permeated gas. Natural gas can then be recirculated and the sour components can be treated whenever necessary (e.g. H_2S in the Claus process).

In the absence of H_2S the selection between a dual membrane system and a mono membrane system, both in combination with a DEA after treatment, is based upon economical considerations. Here the value of lost CH_4 is an important parameter.

Conclusions

Extensive studies in the comparison of these two processes have shown that mixed membrane processes with mono membrane systems are always more economical than dual mixed membrane systems (even when full commercial costs of lost CH_4 are taken into account) and that the dual mixed membrane system is less economical than the existing processes (like DEA-absorption).

Considering these conclusions, it could be deducted that in the
case when H_2S is a component in the sour natural gas, the
necessary mixed dual membrane systems will have less
opportunities when the economy is considered compared to the
existing treatment with only DEA etc.

6.3. Techno -Economical Evaluation (iso-cost curves)

Depended on the feed composition there is always an optimal
composition of the retentate to feed the DEA absorption process.
It is necessary to obtain the most economical situation in the
combination of both processes.

Studying this system the general conclusion could be drawn that
improvement of the permeability of the membrane will bring more
economical advantages than improvement of the selectivity
provided that a certain threshold value on selectivity is
obtained as is shown in figure 9. This figure presents lines
with equal separation costs.

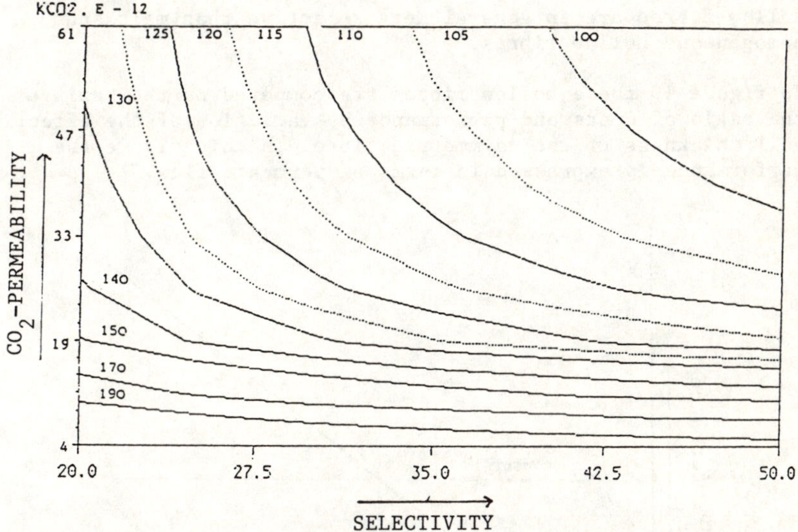

Fig. 9, Equal costs of produced gas as function of the effective
permeability and the selectivity

The same aspect can also be illustrated by using Iso-cost
curves. The Iso-costs curves represent a ratio of the costs of
the produced gas of the mono membrane, mixed process compared
with a DEA-process (see figure 12). The graphs can be used for a
fast and easy positioning of polymer types.

These figures are valid for typical melt spun, homogeneous
membranes and typical wet spun asymmetric membranes as given in
paragraph 2.3. and discussed in chapter 7.

7. <u>PRODUCTION COSTS</u>

The production costs of the final membrane module were analysed.
The costs and performance of the different types of hollow
fibres have been compared before hand.

7.1. <u>Comparison of hollow fibres</u>

The asymmetrical hollow fibres, made by a wet spinning process
and the homogeneous hollow fibres, made by a melt spinning
process were compared on the basis of costs and performance. The
criteria for the performance were the selectivity and the
permeate flux.

To compare the selectivities of the different hollow fibre types
the dimensionless number of Gorissen (see paragraph 6.1.) was
used. Fibres with typical dimensions, as given in paragraph
2.2., were compared, while the effective wall thickness of the
asymmetric hollow fibre was varied.

The conclusion from this evaluation is that wet spun asymmetric
hollow fibres are in general more selective than melt spun
homogeneous hollow fibres.

In figure 10 these hollow fibres are compared on the basis of
the ratio of costs and performance by variation of the effective
wall thickness of the asymmetric fibre. In this figure the
performance is expressed in terms of permeate flux.

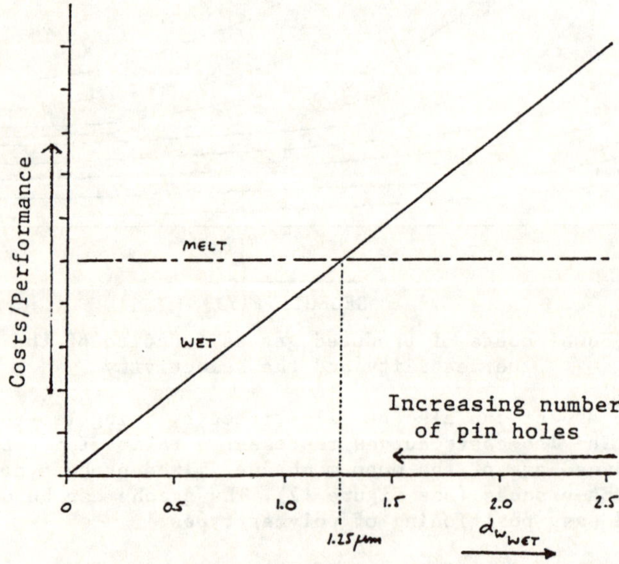

Fig. 10, Costs/Performance comparison of hollow fibres as a
function of the effective wall thickness of the wet
spun membrane.

Figure 10 shows clearly the preference for wet spun asymmetric
hollow fibres already at relatively large effective wall
thicknesses. Only, if pinholes can not be avoided (at very small
wall thicknesses) and the selectivity will decrease the choice
for an asymmetric wet spun membrane is worse.

As the cost of the membrane area is strongly influenced by the
outer diameter of the asymmetric hollow fibre wet spinning
development has to be done to reduce the outer diameters of wet
spun hollow fibres. So far, the hollow fibres produced by wet
spinning still have a too large outer diameter.

7.2. Analyses of final module costs

Analyses of the cost of the final module are presented in the
table of figure 11. The final module consists of a pressure
resistant housing and a set of compact membrane discs. This
module is provided with flanges and ready for installation. It
has a disc diameter of 0,50 m and the separation unit has about
10.000 m^2 of membrane area of homogeneous hollow fibres.

Cost components	Melt spinning	Wet spinning
Polymer material	7%	6%
Spinning costs	7%	5%
Package manufacturing	16%	17%
House	66%	67%
Assembly	4%	5%

Fig.11, build up of the costs of the final module

This figure shows there are hardly differences in the costs
build up between melt and wet spun fibres to produce membrane
modules. Even the total costs of the final module are about
equal.

The costs of the final module are mainly determined by the costs
of the housing. The costs of the compact membrane disc are
therefore relatively unimportant (within certain limits). This
means that e.g. the costs of the polymer material are of minor
importance. This has opened the way to explore other, rather
complicated, polymer materials (see chapter 8).
It also means that wet spun hollow fibres are very competitive
to melt spun hollow fibres. The outer diameter of asymmetric
hollow fibres is an important factor in the costs of the
modules. Reduction of this dimension is therefore of importance.

8.   Reorientation of research activities

The above mentioned conclusions have resulted in a reorientation
of the research activities in that sense that synthesis of new
polymer types is necessary to find the right solution for this
gas separation application.

Polymers with fluorine groups have been synthesised with
promising results and a variation of the fluorinated polymer
segments show variation of selectivity and permeability as is
given in figure 12.

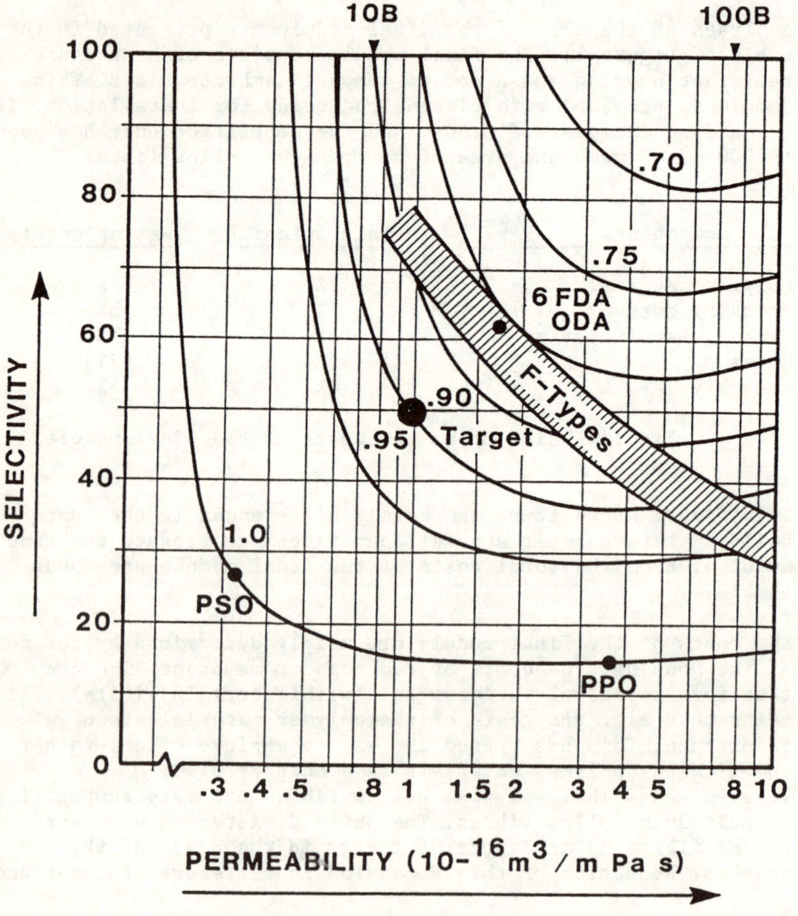

Fig. 12,   Permeability and selectivity of the fluorinated
polymers compared to the original polymers, target
and iso-cost curves

An example of such a polymer is given in the next figure
(figure 13):

6FDA-ODA

Fig. 13, Example of the structure of a new synthesised polymer

It is clear that these polymer groups would fulfil the
requirements as formulated in the target. However the selectivity
decreases at high pressures. The research activities on
membranes are concentrated to find a solution for this decrease
in selectivity.

Although these polymers will be most probably expensive this will
not be an obstacle as the polymer cost are of minor importance
with respect to the feasibility (as shown in paragraph 7.2.).

9. MARKET OPPORTUNITIES

From market studies it is shown that the market for sour gas
sweetening is rather limited.
A bigger market opportunity for these membranes lies in the
production of CO_2 for Enhanced Oil Recovery. This is a future
market and dependent on the oil price.

10. <u>Discussion and Conclusions</u>

The starting point of the project to develop the desired
membranes by modification of existing polymers like PSO and PES,
has been based on the following assumptions:

- the polymer material should be relatively cheap to produce
  complete gas separation modules economically
- the starting material should be shaped at ease into hollow
  fibres by a melt spin technology, also for production
  costs reasons
- chemical or physical modification of the available polymers
  could have brought the desired result with respect to
  permeability and selectivity in a short period
- the well known characteristics of these polymers could have
  brought a quick result in obtaining the desired pressure
  resistant hollow fibres.

In previous chapters it has been clearly demonstrated that the
desired result could not be obtained starting from the above
mentioned assumptions.

The economical evaluation based on intensive modelling has shown
clearly that the polymer price is very moderately influencing
the cost of the final module. This conclusion is any how valid
for this type of gas separation at high pressures.

It has also been demonstrated that the general idea that
improvement of selectivity will bring better gas separation
results is disputable. For this project improvement of
permeability instead of selectivity has to be preferred.

The generality of these two conclusions should be justified for
every other gas separation process using membranes.

It has also been demonstrated that wet spun hollow fibres are
more economical than melt spun hollow fibres when compared on
the basis of costs and performance. This is putting much more
challenge to further development of asymmetric hollow fibres
especially in the direction of reducing the outer diameters.

These conclusions about wet spun hollow fibres and polymer costs
have opened the way to start the development of new polymers,
even expensive ones, adequate for this type of gas separation.

As the project will not be continued for mainly lack of market
opportunities in the coming years, this new polymer development
can not be continued in the scope of this project.

DEVELOPMENT OF GASEOUS PERMEATION MEMBRANES ADAPTED TO THE PURIFICATION OF HYDROCARBONS

A. DESCHAMPS (1), R. DICK (2), C. LECOMTE (3)
(1) Institut Francais du Pétrole
(2) Institut National de Recherche Chimique Appliquée
(3) Commissariat à l'Energie Atomique

Summary

This research is carried-out within the framework of a BRITE programme involving TOTAL/CFP, IFP, CEA, IRCHA, GASUNIE. The objective is to develop gaseous permeation membranes adapted to natural gas dehydration and sweetening, especially for offshore application. Characteristics of membranes adapted to this application are discussed through chemical engineering and economic evaluation studies. A specific polymer belonging to the aromatic polyimides family has been selected for the manufacture of membranes meeting the requirements of the objective. Hollow fibers have been produced and tested in form of minimodules of a few hundred of fibers. The results obtained on pure gases and simulated natural gas are discussed.

1. INTRODUCTION

The objective of this research is to study and develop gaseous permeation membranes for the treatment of a raw natural gas to meet pipe line specifications for transportation.

In order to decrease the exploitation cost of gas fields located in places presenting difficult access, such as offshore, it is necessary to limit the treatment operations realized on the production site to the strict minimum enabling the transportation of the gas by pipe line. Final treatments to meet gas distribution specifications can be performed at a lower cost in more convenient areas and particularly onshore.

This minimum treatment consists in dehydrating (at least partially) the gas, to avoid corrosion and hydrates formation in the pipe or to make possible their control (by injection of additives for examples). Even if not strictly necessary the partial removal of acid gases, H_2S and CO_2, can be advantageous in some cases.

The usual methods for gas treating implement absorption processes using solvents or amines solutions for acid gases removal and triethylene glycol for drying. Although they have been the subject of many studies and improvments, these processes present several disadvantages particularly within offshore operations : high investment costs, heavy and cumbersome equipments inducing high platform costs, furnace requirement for regeneration (safety problem), problems of operation needing a permanent presence of operators on the site.

In comparison, a membrane process looks very attractive for this application :

- it is a passive process using partial pressure difference as driving force,
- weight and dimensions of equipments are reduced,
- possibility of automation and telecontrol,

- modular and consequently evolutive system.

In spite of these potential advantages, membranes processes are not yet used in this field because commercial membranes characteristics are not adapted to natural gas dehydration, both from the point of view of stability and permeabilities/selectivities range.

2. <u>MEMBRANES CHARACTERISTICS FOR NATURAL GAS DEHYDRATION</u>

The problem of natural gas dehydration by permeation in offshore production has already been analyzed by FOURNIE and AGOSTINI [1] by means of chemical engineering and economic evaluation studies. From this previous work and complementary studies performed within this project, the following conclusions have been drawn :

- the main advantage of the permeation system by comparison with the traditional gas treatment processes lies in its simplicity. The membrane process scheme must then avoid heavy and expensive peripheral equipments such as compressors, heat exchangers, condensers,.... wich could penalize it.

In consequence a one stage permeation scheme (FIGURE 1a, b), without recompression of the permeate for upstream recycling, will be generally considered. The permeate can be used for on-site energy production or is sent to the flare and then has to be minimized.

If the pre-treatment of gas is light, the membranes will have to withstand very severe and unusual operating conditions especially concerning the presence of water and heavy hydrocarbons vapours. So, a special attention has to be paid to the chemical and mechanical stability of membranes.

- The membranes permeabilities to methane and water vapour must be adjusted so as to reach a good compromise between the membrane area to be used (which conditions the weight and the cost of the unit) and the quantity of lost gas.

As an example, FIGURE 2 presents the results of calculations performed for a natural gas dehydration unit equiped with membranes of different characteristics (Flowscheme of FIGURE 1b). The water content of the gas has to be reduced from about 600 ppm vol. (saturation at 70 bars, 30 °C) to 145 ppm vol. A sweeping gas is used on the low pressure side of the membrane (BP = 1,5 bar) with a flow rate of 5 % of that of the feed gas. Obviously, a lower sweeping gas flow rate can be used. From these results it appears that membranes presenting a high H_2O/CH_4 separation factor and a low methane permeability ($< 10^{-6}$ cm^3(STP)/cm^2.s.cmHg) lead to a low methane loss but to a very high membrane area. At the opposite, a low H_2O/CH_4 separation factor (< 200) and a high methane permeability ($> 10^{-4}$ cm^3(STP)/cm^2.s.cmHg) result in a low membrane area but a high methane loss. From this first approach the following characteristics were defined for the membrane to be developed in the project :

CO_2/CH_4 separation factor $>$ 50
H_2S/CH_4 separation factor $>$ 50
H_2O/CH_4 separation factor $>$ 200
CH_4 permeability between 10^{-4} and 10^{-6} cm^3(STP)/cm^2.s.cmHg

Good stability in presence of natural gas containing CO_2, H_2S, water vapour and hydrocarbons vapours.

Broad operating temperatures range (20 to 80 °C).

Mechanical properties allowing a pressure difference up to 100 bars.

Later, taking into account the membranes manufacturing constraints encountered, two target membranes were defined more precisely :

- membrane M_1 ($GCH_4 \simeq 5.10^{-5}$ cm^3(STP)/cm^2.s.cmHg, α $H_2O/CH_4 \simeq$ 300 which can be used without sweeping gas on the low pressure side (scheme of

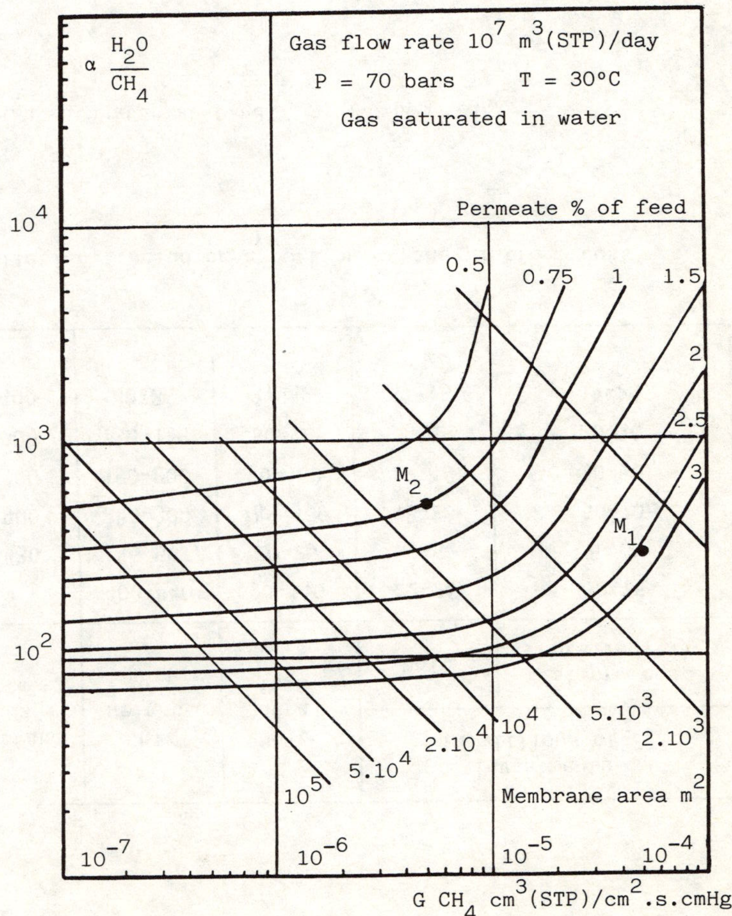

Figure 1 : Natural gas dehydration by permeation

Figure 2 : membranes characteristics for natural gas
dehydration

TABLE I
COMPARISON OF NATURAL GAS DEHYDRATION PROCESSES

Case	Membrane	MP gas % of gas treated	LP gas % of gas treated	Membrane area in m^2	Unit weight in t	Unit area in m^2	Investments in millions of FF	
							Unit	Platform cost entailed (1)
Glycol	-	-	0.5	-	400-440	90-110	22-26	70-75
1	M1	0	4.2	1,430	45-50	20-25	3-5	8-12
2	M2	0	2.8	8,900	280-300	140-150	18-26	50-70
3	M2	1	2.9	5,900	190-200	90-100	12-18	30-50
4	M2	2	3.3	4,000	130-135	60-70	8-12	20-30
5	M2	5	5.9	2,500	80-85	37-42	5-7.5	14-20

(1) Additional investment for an off-shore installation, including a portion of the platform cost

Characteristics of the feed gas

- Flow rate : 10^7 NM^3/day
- Composition (simplified) Mole % dry CH_4 95, CO_2 5

  H_2O saturated at Pand T : 1040 ppm Vol.

- Pressure : 70 bars, T : 40 °C

Water content of treated gas : 170 ppm Vol.

FIGURE 1a),
   - membrane M_2 (($GCH_4 \simeq 5.10^{-6}$ $cm^3(STP)/cm^2.s.cmHg$, α $H_2O/CH_4 \simeq 500$)
for which the use of a sweeping gas on the low pressure side is generally
necessary to obtain reasonable membrane areas (scheme of FIGURE 1b).

   The main results of an economic evaluation study of a natural gas
dehydration process using these two types of membranes are presented in
TABLE I. They show that the investment cost for the membrane unit is lover
than that for a conventional glycol installation in four of the five cases
studied. The advantage of the membrane process is even greater when the
additional costs entailed by the installation on a platform are taken into
account (gains in weight and space). In the case of membrane M_2 it is also
shown how a compromise between the investment cost and gas loss can be
found.

   The main objective of the experimental work conducted in this project
is to develop membranes presenting such optimized properties for natural
gas dehydration. Obviously the best CO_2/CH_4 and H_2S/CH_4 separation factors
compatible with this target are also researched.

3. EXPERIMENTAL TECHNIQUES FOR MEMBRANES MANUFACTURE AND CHARACTERISATION

3.1. Membranes manufacture

   In a first step dense films and flat asymmetric membranes were
prepared by usual methods to evaluate the ability of several polymers for
the manufacture of membranes meeting the objectives (intrinsic
permeabilities, chemical and mechanical stability,...).

   In a second step hollow fibers were produced from preselected
polymers using a dry-wet spinning process including phase inversion of the
polymer solution. The polymer solution is extruded through the annular
orifice of a spinnerette in the form of a filamentary stream which is
introduced under tension into a first coagulating bath. The extrusion is
carried out with simultaneous injection of a core fluid through the
central orifice of the spinnerette in order to obtain the fiber canal. The
hollow filament is directed through guide rolls to a second coagulating
bath, a washing section, and then, is rolled up on a winder. A stretching
of the hollow filament takes place due to the speed difference between the
extrusion of the spinning solution and the rolling up of the filament.
Many parameters as temperature and composition of the spinning solution,
of the coagulating bath and of the core fluid... can be modified to adjust
structure and characteristics of the hollow fibers. Post-treatments
including heating in controlled atmosphere and coating with polymers are
also applied to improve membranes resistance and selectivities.

   For testing hollow fibers, bundles of 50 to 200 fibers are potted at
each end in epoxy tube sheets and mounted in a stainless steel vessel
(FIGURE 3) having pressure resistance up to 100 bars. The effective length
of the fiber bundles is about 30 cm.

3.2. Membranes characterization

   A first characterisation of flat membranes and hollow fibers is
performed by measuring their permeabilities to the main components of
natural gas in presence of pure gas. Usual methods are implemented for
permanent gases (high pressure 2 to 40 bars, low pressure 0 to 1 bar,
temperature 20 to 70 °C). For water vapour a specific apparatus (FIGURE 4)
in which the permeated water is collected in a cold trap (77 K) and
weighted after a period of time, is used.

   Performance and ageing tests are then performed on selected hollow
fibers using synthetic gas mixtures saturated in water and liquid

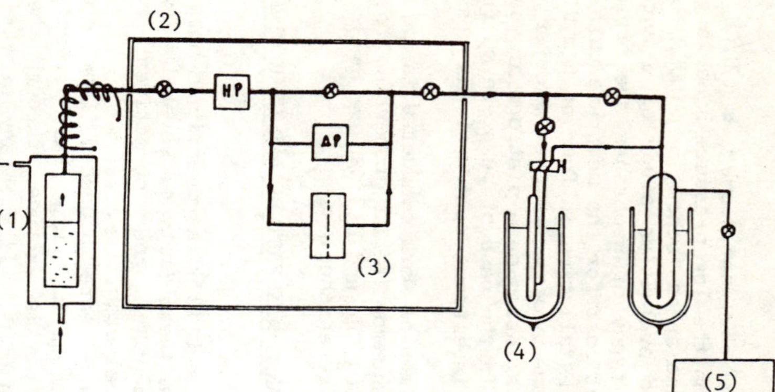

(1) Water vapour generator
(2) Stove
(3) Membrane cell
(4) Cold trap
(5) Vacuum pump

Figure 4 : Water vapour permeameter

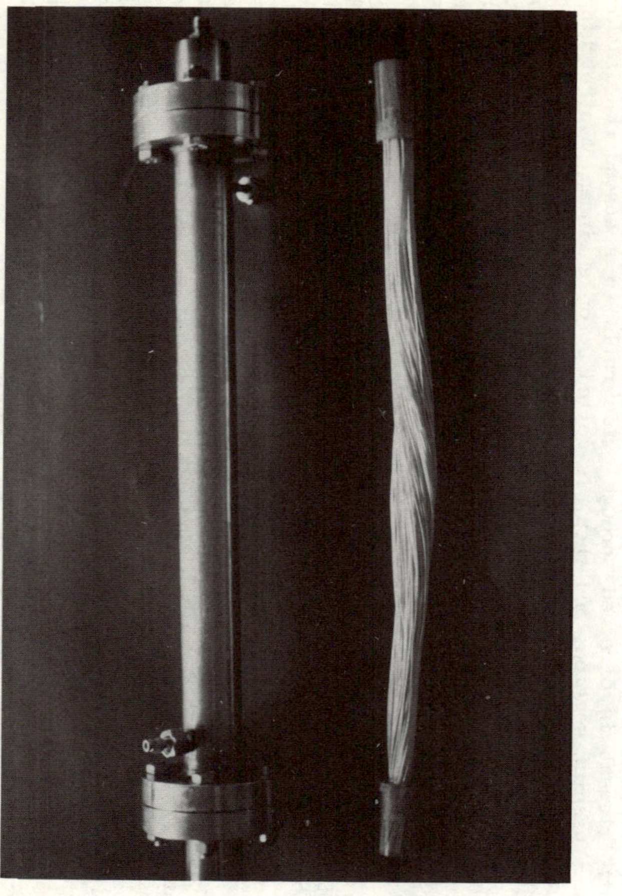

Figure 3 : Hollow fiber minimodule

hydrocarbons under controlled temperatures. The test facility (FIGURE 5) allows to determine membranes permeabilities to the different components in presence of a sweeping gas on the low pressure side of the fibers. Typical composition of the "simulated natural gas" used for ageing tests is as follow :

	CH_4	CO_2	H_2S
% Vol. (dry)	79	20	1

H_2O and heptane : saturated at 40 °C, 30 bars.
Module temperature : 50 °C

Structure and stability of membranes are also characterized using different techniques such as scanning electron microscopy, IR, mechanical tests.

4. RESULTS AND DISCUSSION

4.1. Polymer selection

To elaborate membranes meeting the requirements of the objective, a special attention has been given to polymers belonging to the aromatic polyimides family, due to their high thermal and chemical stability, their good mechanical strength as well as their potential gas separation properties.

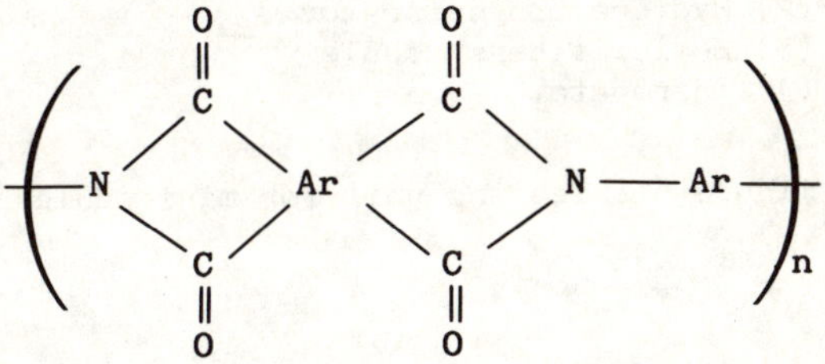

Different commercial polyimides have been evaluated first in the form of dense and flat asymmetric membranes and later in the form of hollow fibers. Although some of them have shown good separation factors as dense films, they have been excluded from further evaluation because fibers spun from their solutions were too brittle.

To obtain hollow fibers presenting required permeabilities and mechanical strenght, a specific polymer having a pronounced hydrophilic character and a sufficiently high molecular weight has been synthesized. This polymer has intrinsic separation factors of about 200 000 for H_2O/CH_4 and 80 for CO_2/CH_4. After preliminary tests it has been selected for membranes development.

4.2. Membranes characteristics and performances

The main results concerning the separation H_2O/CH_4 obtained with flat asymmetric membranes and hollow fibers prepared from this polymer are presented on FIGURE 6 (permeabilities measured on pure gases).

They show that raw membranes obtained directly through the technique of phase inversion presents a H_2O/CH_4 separation factor which increases from about 10 to 10 000 or more when methane permeability decreases from 10^{-3} to 10^{-7} $cm^3(STP)/cm^2.s.cmHg$. This behaviour can be simulated with a model assuming that the total flux through the membrane is the sum of a pure permeation flux and of a diffusion flux, of Knundsen or Poiseuille

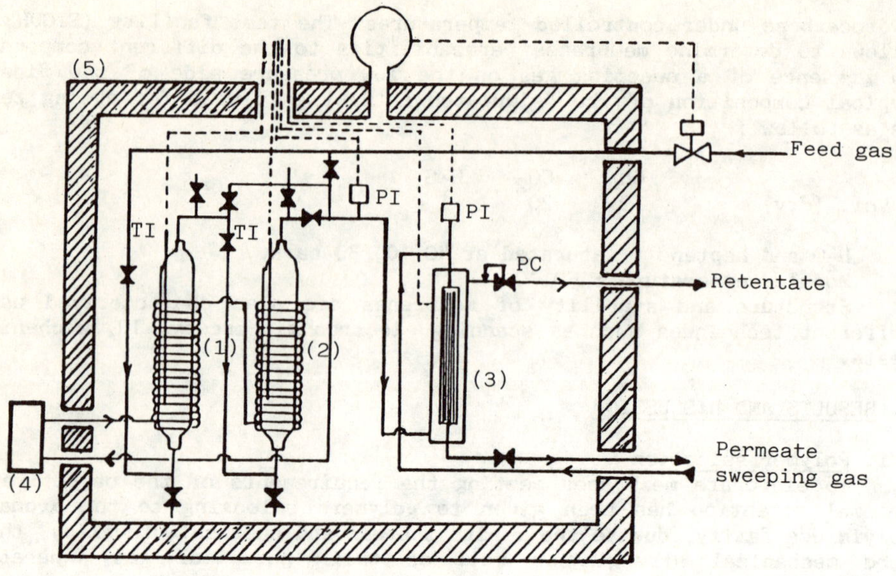

(1) H_2O saturator
(2) Hydrocarbon saturator
(3) Hollow fibers module
(4) Thermostat

Figure 5 : Testing unit for minimodules

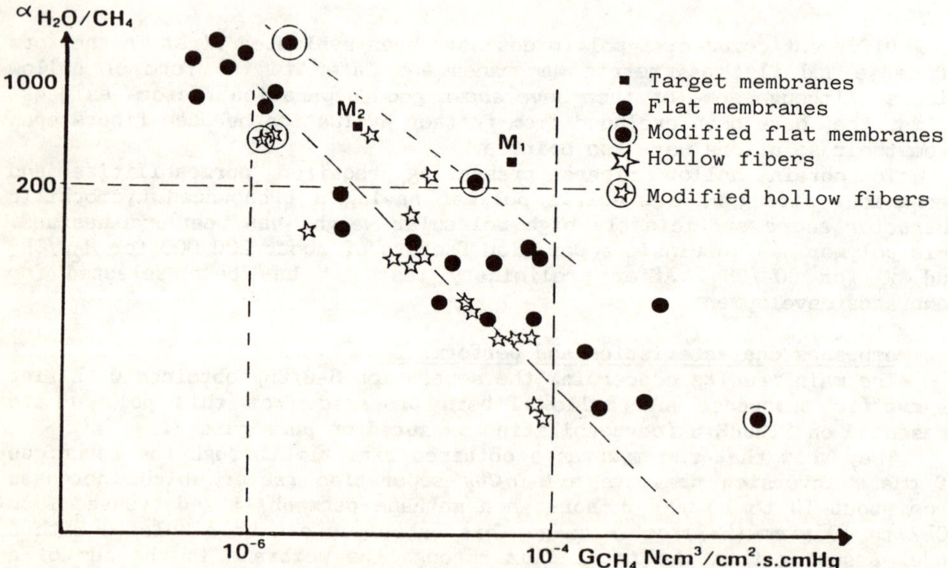

Figure 6 : Permeability and selectivity of membranes

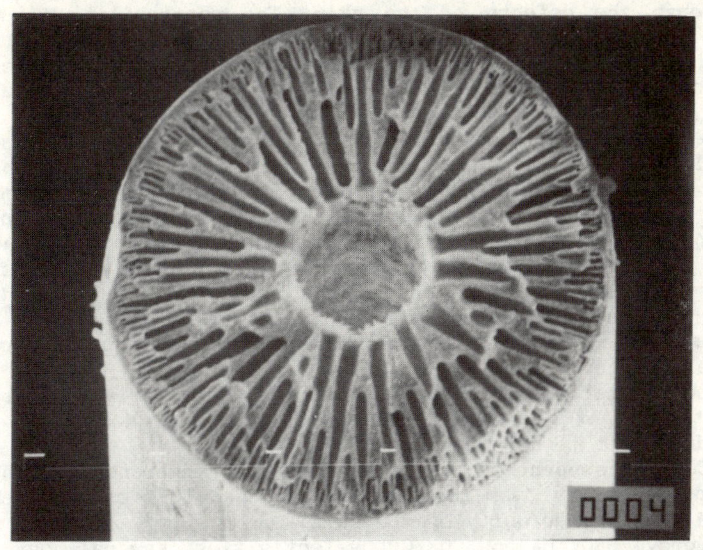

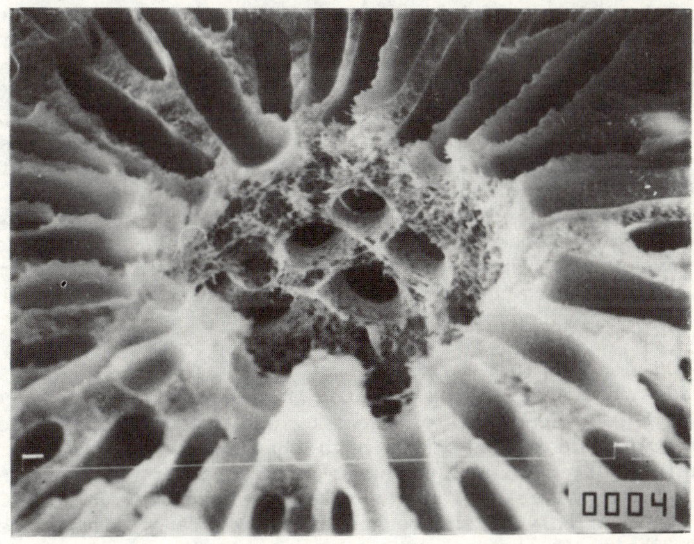

Figure 7 : Sem photographs of a hollow fiber

type, through the defects of the skin. The permeability to methane is primarily governed by this later one. Due to the very high intrinsic permeability of the polymer to water vapour, the membrane permeability to this component is much less influenced by the residual porosity of the skin.

These different characteristics of permeability/selectivity correspond to different pore structures of the membranes. By changing the preparation parameters of hollow fibers it was possible to obtain five different types of fibers structures. FIGURE 7 presents SEM photographs showing one of these structures which is characterized by a thin outside skin and a fiber wall of high porosity, comprising many finger like macrovoids disposed radially around the fiber hole and emerging in it. This structure is particularly adapted to the use of a sweeping gas on the low pressure side of the membrane. Others structures presenting a spongy part are also interesting.

Hollow fibers presenting permeabilities characteristics very close of those of the target membrane M_2 have been produced. Membranes of M_1 type are more difficult to obtain directly.

The CO_2/CH_4 separation factor of these raw membranes is generally low when the permeability to methane is higher than 10^{-6} $cm^3(STP)/cm^2.s.cmHg$, due to the residual porosity of the skin.

To improve H_2O/CH_4 and CO_2/CH_4 selectivities, several post-treatments have been applied on flat membranes and then on hollow fibers. Coating with silicones or other polymers allows to decrease the permeability to methane by a factor of 10 to 100 without lowering significantly the permeability to water and, in some cases, to carbon dioxyde. The membranes selectivities can be also strongly modified by heat treatments. This effect depends on the initial structure of the membranes and on the residual solvent content. The permeability to water is generally slightly reduced but the stability of membranes is highly improved, especially in presence of liquid water and hydrocarbons. There are still some difficulties to obtain simultaneously high CO_2/CH_4 separation factors and required properties for natural gas dehydration.

Concerning the stability of membranes, ageing tests lasting up to two months have been performed on simulated natural gas. Raw and heat-treated membranes have shown stable performances. Some evolution of permeabilities has been noticed for coated membranes ; it seems mainly due to the presence of heavy hydrocarbon vapours.

5. CONCLUSION

Membranes characteristics adapted to natural gas dehydration have been defined through chemical engineering and economic evaluation studies. A specific polyimide presenting a high chemical and thermal stability, a good mechanical strenght and adequate gas separation properties has been selected as basic material for membranes manufacture. Hollow fibers were produced from this polymer and tested at the scale of minimodules of a few hundred of fibers. They have shown characteristics close to those of the target membranes.

However, more work is necessary to optimize their properties and to study their behaviour in representative conditions. Some tests on real natural gas are planned for the beginning of 1989.

REFERENCES
[1] FOURNIE, F. and AGOSTINI, J.P. (1984). Permeation : A new competitive process for offshore gas dehydration. 16th Annual Offshore Technology Conference, Houston, May 7-9 1984 (OTC 4659).

OPTIMIZATION OF AN UF PILOT PLANT FOR THE TREATMENT
OF RADIOACTIVE WASTE

I W CUMMING AND A D TURNER
CHEMICAL ENGINEERING AND MATERIALS DEVELOPMENT DIVISION
HARWELL LABORATORY, UKAEA, ENGLAND

Summary

A crossflow filtration process has been developed to treat the
Harwell site low level active liquid waste. A pilot plant was then
constructed and operated to determine the α and $\beta\gamma$ removal from the
effluent. The unit has been used to assess the long term
performance of ultrafiltration and microfiltration membranes. The
reliability of the pilot plant equipment has also been demonstrated
over periods of between 6,000-10,000 operating hours. Tests have
shown that the process gives optimum α removal at pH5 down to 1.5
mBq/ml and that addition of copper ferrocyanide and zirconium
phosphate to the effluent can significantly improve $\beta\gamma$ removal.
Direct electrical Membrane Cleaning has been shown on a
laboratory scale to enhance crossflow filtration performance at
conductive UF and microfiltration membranes by improving their
performance at reduced crossflow velocities. Pilot scale units
based on Tech Sep M4 and MA1/S sintered stainless steel membranes
have been designed and constructed for longer term evaluation in the
UF pilot plant. Tests on the former have recently begun.

1.   INTRODUCTION

A new process for the decontamination of radioactive effluent based
on crossflow ultrafiltration has been developed and tested at pilot plant
scale using a radioactive liquid waste. The process combines the
simultaneous removal by filtration of insoluble particulates already in
the waste and finely divided inorganic ion-exchange material added to the
effluent to absorb soluble radionuclides.

The main objective was to develop a process which could offer
substantial advantages over a conventional waste treatment such as a floc
based process. The benefits offered by crossflow filtration are that not
only is a better decontamination factor achievable because of a more
effective separation of colloidal solids from the waste liquor, but that
as only small quantities of additional ion-exchange materials are added
to the waste, the volume of sludge arising from the treatment is
minimized. The process also produces a pumpable slurry which can be
handled by subsequent waste treatment steps such as cementation.

The pilot plant was constructed for long term testing of crossflow
filtration to demonstrate the decontamination factors and the throughputs
achievable on a real waste. It was also used as a test bed to assess
equipment suitable for use in an active plant. In particular, the
performance of both ultrafiltration and microfiltration membranes was
determined over appreciable periods of time. The low active waste used
for these tests arises in considerable volumes ($\sim$ 150m³/d) from various
operations on the Harwell site. The effluent contains α, β and γ

activity, all at levels of ~ 0.4 Bq/mℓ. At present, the effluent is treated by precipitation of ferric hydroxide at pH 10.5, and sedimentation. .

The pilot plant is now being used to test a novel system of enhancing plant throughput known as Direct Membrane Cleaning (DMC). This technique relies on the application of an electric current pulse through the waste suspension between a counter electrode and an electrically conducting membrane.

2.   DEMONSTRATION ULTRAFILTRATION PILOT PLANT

The pilot plant was initially designed to be fitted with a Tech Sep M4 membrane ultrafiltration module. This contains 37 parallel tubes each 1.2m long with a 6mm internal diameter. These consist of a zirconia membrane supported on the inside surface of porous carbon tubes. The molecular weight cut-off of the membrane is quoted at 20,000 - equivalent to a nominal pore size of ~ 2nm.

A schematic diagram of the pilot plant is shown in Figure 1. The waste feed is prefiltered by a coarse 0.5mm strainer and pumped to a 150 litre feed tank, where the pH of the waste is adjusted automatically by the addition of either sodium hydroxide or nitric acid. Also, any precipitant used to enhance the removal of soluble radionuclides can be dosed into this tank. The effluent is then pumped at pressure into the crossflow filtration circuit. The flow around this loop is maintained by a centrifugal pump at a crossflow velocity of ~ 4.5m/s while the pressure can be varied between 2-5 bar(g). The operating conditions of the circuit are monitored by a flowmeter and two stainless steel diaphragm pressure gauges, which measure the module inlet and outlet pressures. Concentrate is removed from the circuit by the opening of a discharge valve for a few seconds twice an hour. The control of the plant is carried out by a CBM8032 microcomputer connected via an analogue to digital converter and a switch sensing/relay output interface to the plant instrumentation, pumps and valves. Changes in the plant operating conditions are made via the microcomputer which also controls the plant 24 hours a day with only daytime operator attendance.

The plant has now been operated for over 10,000 hours, of which the first 4,500 hours used a single Tech Sep M4 module which was still operating satisfactorily when removed from the plant. During the rest of the operating time, various microfiltration modules have been tested which are described in a previous paper [1]. The equipment used in the pilot plant has generally given very satisfactory performance. However, some useful experience on key items - as detailed below - has been gained over the substantial operating time.

Flowmeters

Initially, the flow around the filter circuit was monitored by both doppler and time of flight ultrasonic flowmeters. Neither of these instruments gave reliable readings so the plant was then operated for a substantial period of time using only the pressure drop between the inlet and outlet of the module to give an indication of flow. This was kept between fixed upper and lower limits - which if exceeded would initiate a plant shut-down. However, an electromagnetic flowmeter with an internal ceramic liner is now being used, and this has given very reliable operation over the last 6000 hours.

Pressurizing pump

The plant was originally designed to operate at a fixed volumetric throughput. The effluent was pumped into the filtration circuit by a

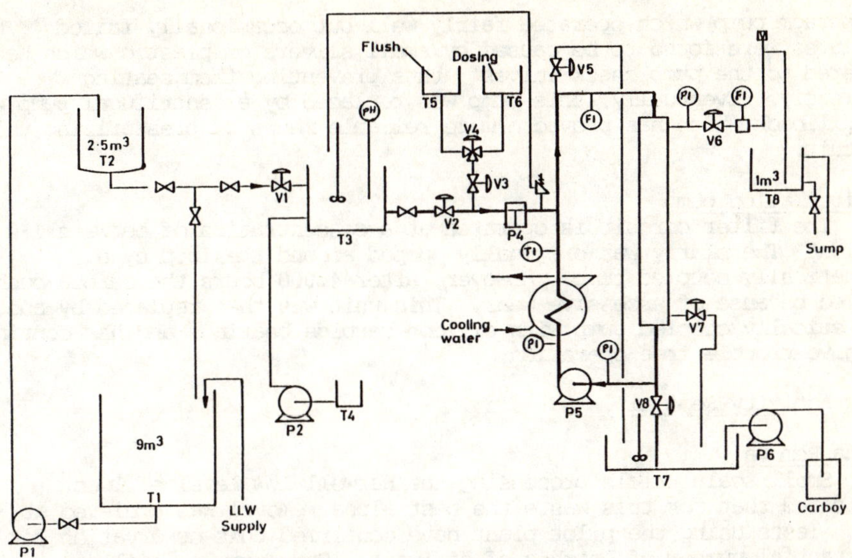

Figure 1 Schematic diagram of Harwell LLW pilot plant

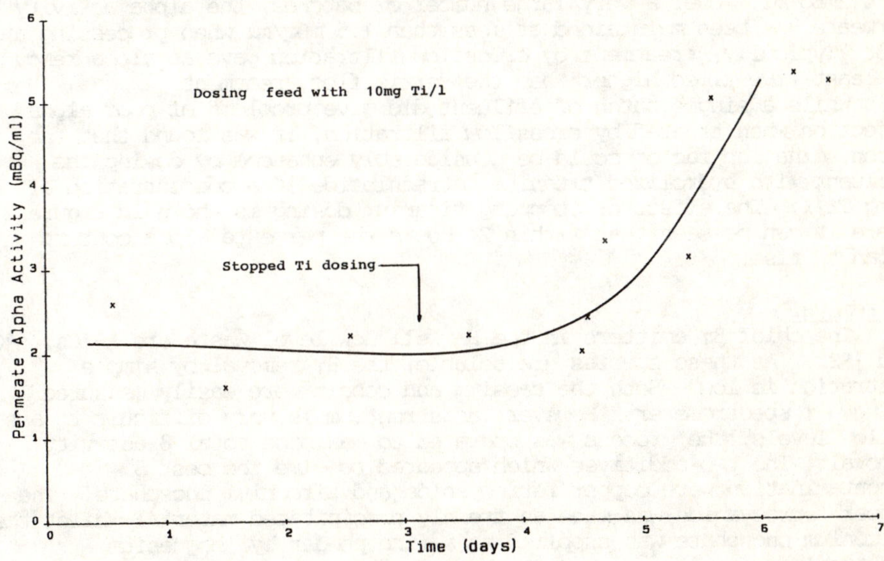

Figure 2 Effect of titanium hydroxide on alpha removal
from Harewll LLW.

diaphragm pump which operated fairly well but occasionally failed. These failures were found to be caused by small slivers of plastic which had adhered to the pump check valves - thus preventing them seating correctly. Eventually, this pump was replaced by a centrifugal pump with a spillback which has proved a more reliable means of pressurizing the circuit.

Recirculation pump

The filter circuit is operated at a concentration of between 1-2 wt% solids. The slurry was originally pumped around the loop by a magnetically coupled pump. However, after 4,000 hours the carbon bushes failed because of excessive wear. This unit was then replaced by another magnetically coupled pump with silicon carbide bearings and has continued to give trouble-free operation.

2.1 Activity Removal

Alpha Removal

Small scale trials processing the Harwell low level effluent indicated that for this waste the best alpha removal was achieved at ~ pH 4.5. Tests using the pilot plant have confirmed this observation for a substantial number of batches of effluent. For example, effluent containing a very low level of alpha activity of 6.8mBq/ml was processed for one day at pH5, a second day at pH7 and a third day at pH 9.3. At pH5, the permeate activity was below the detection limit of 1.5mBq/ml; on the second day at pH7 it rose to 8.7 mBq/ml (activity was being released in the filter circuit) and on the third day at pH9 it declined to 4.2mBq/ml. Over a very large number of batches, the alpha activity in permeate has been maintained at less than 1.5 mBq/ml when processing at pH5. Typically, treatment by crossflow filtration gave an alpha removal at least five times higher than the ferric floc treatment.

While a single batch of effluent did give problems of poor alpha rejection when treated by crossflow filtration, it was found that the decontamination factor could be considerably enhanced by dosing the effluent with hydrolized titanium tetrachloride at a concentration of 10mg Ti/ℓ. · The effect of stopping titanium dosing is shown in Figure 2 where it can be seen that within 24 hours the permeate alpha content began to rise.

βγ removal

The chief βγ emitters in the Harwell low level waste are ^{137}Cs, ^{60}Co and ^{90}Sr. As these species are soluble, the βγ removal by simple filtration is low. Both the caesium and cobalt were easily measured using a γ spectrometer. However, as strontium is very difficult to assay at low levels, the process was operated to maximise total β activity removal. The two additives which appeared to give the best β decontamination were copper ferrocyanide and zirconium phosphate. The copper ferrocyanide was made as freshly precipitated material whilst the zirconium phosphate was supplied as a fine powder by 'Magnesium Elektron'.

The removal of ^{137}Cs from the effluent could be achieved by injecting very small quantities of copper ferrocyanide into the filtration circuit. The efficiency of removal of ^{137}Cs from the effluent was found to improve with dosing time. This would suggest that the completeness of ^{137}Cs absorption increased with the circuit copper ferrocyanide concentration. Figure 3 shows the improvement in caesium decontamination factor with dosing time at different effluent additive

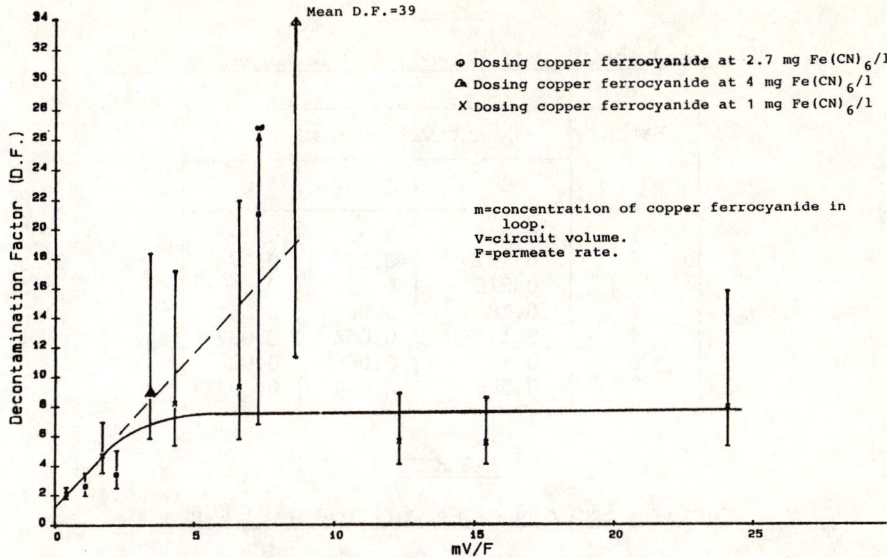

Figure 3 Caesium removal by copper ferrocyanide.

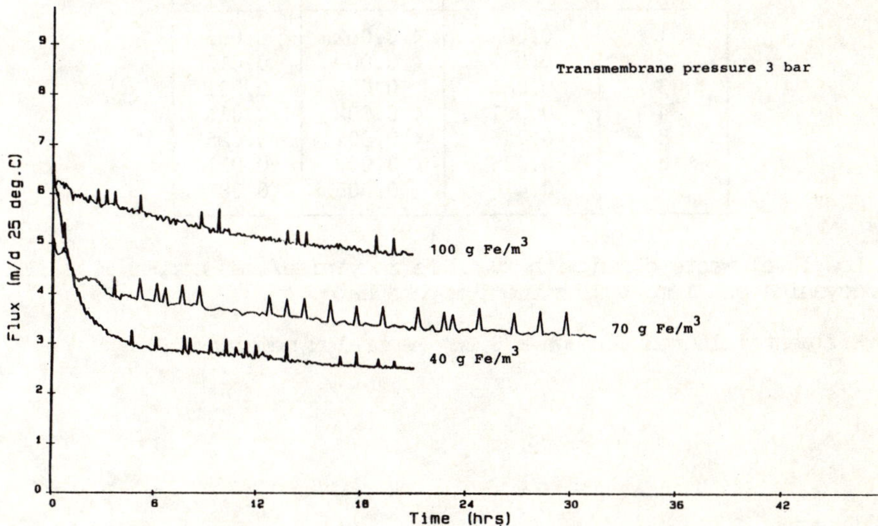

Figure 4 Flux profile for Membralox module processing LLW
dosed with ferric nitrate at pH 5.

<u>TABLE I</u>

<u>Feed activity of Harwell low level waste</u>

Batch	Activity, Bq/ml		
	β	^{137}Cs	^{60}Co
1	0.394	0.23	0.12
2	0.131	0.177	0.07
3	0.215	0.098	0.06
4	0.18	0.06	0.041
5	0.143	0.054	0.037
6	0.15	0.066	0.028
7	0.585	0.034	0.49

<u>TABLE II</u>

<u>Permeate activity of Harwell low level waste</u>

Batch	Activity, Bq/ml		
	β	^{137}Cs	^{60}Co
1	0.06	< 0.002	0.055
2	0.045	< 0.002	0.012
3	0.065	0.002	0.03
4	0.081	< 0.002	0.014
5	0.046	< 0.002	0.025
6	0.035	< 0.002	0.011
7	0.47	0.005	0.38

1.   Low level waste dosed with 20 mg ferrocyanide/ℓ as copper ferrocyanide and 10mg/ℓ of zirconium phosphate.

2.   Effluent filtered through a 0.2μm Membralox membrane.

concentrations of 1, 2.7 and 4mg ferrocyanide/ℓ. As can be seen from Figure 3, at a dosing concentration of 4mg ferrocyanide/ℓ of waste, the ^{137}Cs content in the permeate had declined to near the detection limit.

The completeness of ^{60}Co removal from the effluent, however, was found to be more variable. Tests on three different batches of waste indicated that as the copper ferrocyanide concentration was increased to 10mg ferrocyanide/ℓ of effluent, the ^{60}Co removal improved. Increasing the copper ferrocyanide further did not appear to give any additional enhancement. Unlike the absorption of ^{137}Cs, the build up in concentration of the copper ferrocyanide in the filtration circuit did not give any further improvement in ^{60}Co removal.

The pilot plant was also used to test seven different batches of low level waste with the feed effluent dosed with preformed copper ferrocyanide (20mg ferrocyanide/litre of effluent) and powdered zirconium phosphate (10mg ZrP/litre of effluent). The β, ^{137}Cs and ^{60}Co activities in the feeds are given in Table I and the average activity of the permeates are given in Table II. These tests indicated that the ^{137}Cs removal was consistantly high, while the ^{60}Co removal varied from batch to batch. The β content of the permeates was low, with the exception of batch 7 which had a relatively high ^{60}Co content in the feed and a very poor ^{60}Co removal. It was thought this variation in performance was due to the effect of complexing agents in the effluent.

2.2 The effect of ferric ion concentration on plant throughput

The pilot plant has also been used to test the effect on membrane flux of ferric hydroxide dosed into the low level waste. A Membralox module containing 29 tubes of 7mm inside diameter and 0.75m length with an active filtration area of 0.48m² was used for these tests. The membrane had a nominal pore size of 0.2μm.

The low level effluent was dosed with ferric hydroxide precipitated in the plant feed tank by the simultaneous addition of ferric nitrate and sodium hydroxide to maintain a constant pH. The flux decline at dosing levels of 40, 70 and 100g Fe/m³ is shown in Figure 4. As can be seen here, increasing the ferric concentration from 40 to 100g Fe/m³ approximately doubled the membrane flux. Reducing the ferric concentration below 40g Fe/m³ did not cause any further decline in flux, while increasing it above 100g Fe/m³ did not give any additional enhancement. This experiment demonstrated that appreciable quantities of a dosing reagent were necessary to change the membrane flux performance.

2.3 Membrane cleaning

The membrane was at first cleaned by recirculating an enzymatic detergent solution around the filtration loop for 1 hour at 50°C. This cleaning procedure gave a good recovery of the membrane permeability but produced 50 litres of effluent containing detergent. This could only be treated by a ferric floc precipitation to remove its now soluble alpha activity. As this meant a separate process would be required for the treatment of this secondary waste, the cleaning process was changed. Sodium hydroxide solution (1M) was recirculated around the filtration loop for one hour at 50°C followed by 1M nitric acid also for one hour at 50°C. This procedure gave an improved recovery of membrane permeability and produced only 30 litres of waste which was a neutral solution of sodium nitrate. This effluent was suitable for treatment in the crossflow filtration plant. Effective membrane cleaning has also been shown to be achieved by recirculating 0.5 M sodium hydroxide followed by 0.5M nitric acid for 1 hour each at 50°C.

2.4 <u>Conclusion</u>

These tests have shown that the treatment of a real radioactive waste can be carried out effectively in a crossflow filtration plant. Both inorganic microfilters and ultrafilters have given much better α removal than a conventional floc treatment and an improved β removal. The pilot plant has been shown to be capable of reliable operation over a substantial period of time with very little maintenance required for instrumentation, pumps and valves. The process is capable of very flexible operation, in that feed pH can be easily changed as can be the dosing reagents for absorbing soluble radionuclides. Technology based on these process developments has been chosen by British Nuclear Fuels plc to treat some of the liquid wastes arising at the Sellafield site [2].

3. <u>ELECTRICAL FILTER CLEANING</u>

3.1 <u>Introduction</u>

As an additional development of crossflow filtration technology, Direct electrical Membrane Cleaning (DMC) is currently being evaluated at the pilot scale - having been successfully demonstrated in the laboratory. As has been described in a previous paper [1], this technique can enhance permeation rates at lower crossflow velocities by the periodic removal of surface fouling from a conductive membrane through the in situ electrolytic generation of microscopic bubbles by the application of a short current pulse. This flows through the waste suspension to the membrane itself from an anodic counter electrode. The dislodged material is carried away in the crossflow stream. The effectiveness of cleaning is enhanced by a transient reduction in transmembrane pressure during current application, as this maximises the net velocity of particle movement away from the filter surface, both through the rate of bubble growth under reduced external pressure and also by minimising the permeate flow in the opposite direction as a result of filtration. In order to evaluate the effectiveness of DMC under larger scale waste treatment conditions, modules have been constructed that can be installed in the Harwell UF pilot plant facility. The two membrane systems being tested after laboratory trials are the Tech Sep M4 ultrafilter (currently being used in the UF plant) and a 3 μm sintered stainless steel fibre microfilter. As was described in the earlier paper [1], current densities of 50 mA/cm² applied for only 1 second are sufficient to defoul the Tech Sep M4 membrane, while for the sintered stainless steel fibre membrane, 100-200 mA/cm² is required for 5 seconds. The frequency of cleaning needed to maintain the flux depends on the nature of the particulates and their concentration - up to 12 pulses/hour for high solids - falling to only 1 pulse/hour for low solids. Minimising the pulse duration and current density not only limit the charge passed, thus reducing gas production to an insignificant level, but also enable a larger membrane area to be cleaned for a given size of power supply. Instead of cleaning all of the filter area at once, a smaller power unit can be used continuously - cleaning only a fraction of the total area at a time by switching between filtration modules sequentially.

3.2 <u>Ultrafiltration module</u>

As has been described previously, Tech Sep M4 membranes comprise a microporous graphite tube (5 μm pore size) of 10 mm outside diameter, 6 mm internal diameter. The bore of the tube is coated with a thin layer of ultrafine porous zirconia (1-10 nm pore size) to give it its ultrafiltration properties. Due to the construction of the filter membrane, filtration is from 'in-to-out'. As a result, fouling can build up on the inner surface of the tube despite feed crossflow through its bore. Because of the constraints of this geometry, the counter electrode needed to apply DMC to the inner surface of the tube has to be in the form of an axial wire. As the Tech Sep tube is only stable under cathodic cleaning pulses, the central electrode has to be an anode. The most commonly used commercial dimensionally stable anode systems are based on noble metal coated titanium. The diameter and materials of construction of the central conductor are a compromise between several factors. The smaller the inter-electrode gap, the higher is the resistance to hydraulic flow through the tube, while the electrical resistance between the electrodes falls. In addition, in order to guarantee uniformity of treatment along the entire filter tube, the resistances of the membrane and counter electrode need to be matched.

At the lower crossflow velocities permitted through the use of DMC (~ 1 m/s), the Reynolds number for flow down a Tech Sep tube containing a central 1.5 mm anode is ~ 2,300 - on the borderline between laminar and turbulent flow. As increased solids content and non-Newtonian behaviour will tend to decrease any turbulence, our design calculations have been based on laminar flow. Table III shows the effect of the presence of the central conductor on volume flow rates and crossflow velocity for a given pressure gradient. For an anode of much greater than 1.8 mm diameter, the restriction is becoming significant. In addition, a larger gap is less likely to become blocked by filaments in the feed and reduces the probability of accidental electrode contact. In addition, however, it should be remembered that the effectiveness of DMC in maintaining membrane cleanliness means that crossflow velocities can be significantly reduced, e.g. by a factor of 5. Table IV shows the effect of anode size on voltage drop through the electrolyte and the current density at the anode surface. The latter needs to be kept below 1 A/cm² to ensure a reasonable electrode life. The third factor to be considered is the uniformity of current density distribution over the length of the filtration membrane. This is optimised by matching the electrical resistance of the central anode to that of the membrane and making their connections to the power supply at opposite ends. However, due to the resistance of Tech Sep membranes being as high as 1.3 Ω/m, and industrially available anodes being based on commercially pure titanium of significantly lower resistance, the membrane has been shunted with a stainless steel connector over its length to balance the resistances. As a result, the variation in current density over the length of the membrane is < 4% - even for a low conductivity feed. An additional benefit of this design is that the overall cell resistance is lowered, thus enabling a 36 V power supply to be more than adequate.

A three tube module was therefore constructed on these principles (Figure 5) such that the tubes could either be cleaned singly or in parallel. The total membrane area of the three 1.2 m membranes was 680 cm². A volumetric recirculation rate of ~ 0.3 m³/h is required to give a crossflow velocity of 1 m/s. At a current density of 50 mA/cm², each tube requires a cleaning current of 11.3 A at a voltage of 20.5 V for low level waste. Cleaning the three tubes simultaneously is therefore well within the 100 A/36 V capabilities of the Hewlett Packard HP6456B power supply. The cleaning pulses are activated by the pilot plant microcomputer management system by a relay to the remote control panel of the power supply.

As shown in Figure 6, the UF plant recycles feed through the filter unit maintained at pressure (e.g. 2 bar) by a feed pressurising pump, while the permeate emerges at atmospheric pressure. As from previous

TABLE III

Effect of the presence of an axial electrode on the laminar fluid flow down a 6 mm diameter filtration tube for a constant pressure gradient.

Central conductor diameter (mm)	Annular cross-section area (%) *	Volume flow rate (%) *	Crossflow velocity (%) *
2.4	84	20.4	24.3
2.1	88	25.2	28.7
1.8	91	30.4	33.4
1.5	94	36.2	38.6
1.2	96	42.6	44.4
0.9	98	49.6	50.7

* all normalised to the unrestricted tube.

TABLE IV

Effect of axial anode diameter on electrical power required during DMC at a Tech Sep M4 filter tube.

Central conductor diameter (mm)	Voltage drop in electrolyte* (V)	Power dissipation (W)	Current density at anode surface (mA/cm²)
2.4	11.2	126	125
2.1	12.8	144	143
1.8	4.6	165	167
1.5	16.9	191	200
1.2	19.6	221	250
0.9	23.1	261	334

* Typical low level waste feed ρ = 810 Ω cm.
Current density at cathode 50 mA/cm².
Filter tube diameter 6 mm, length 120 cm.

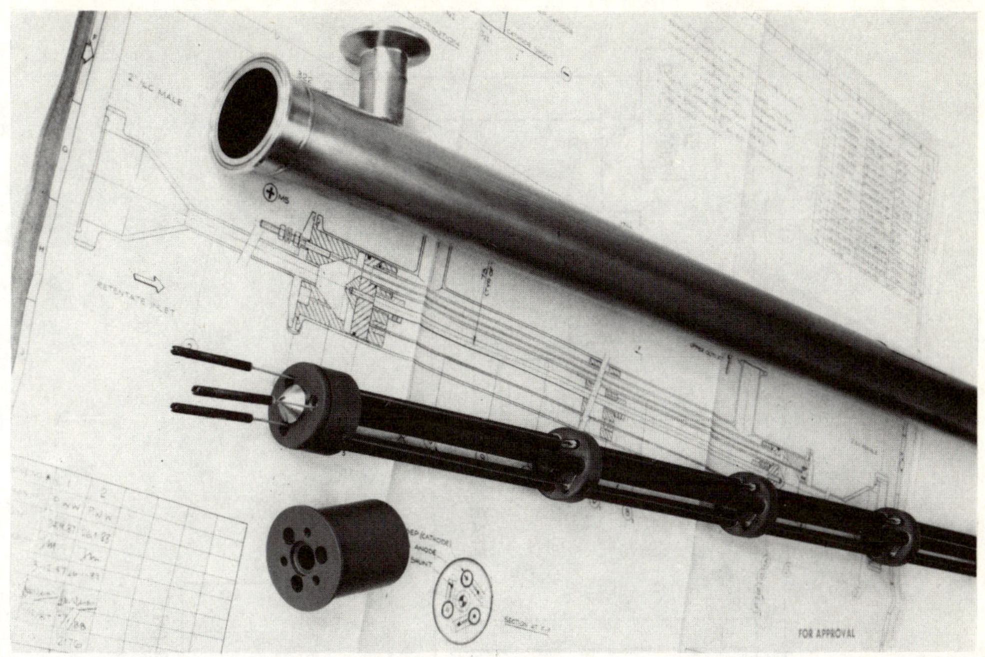

View from the anode connection end

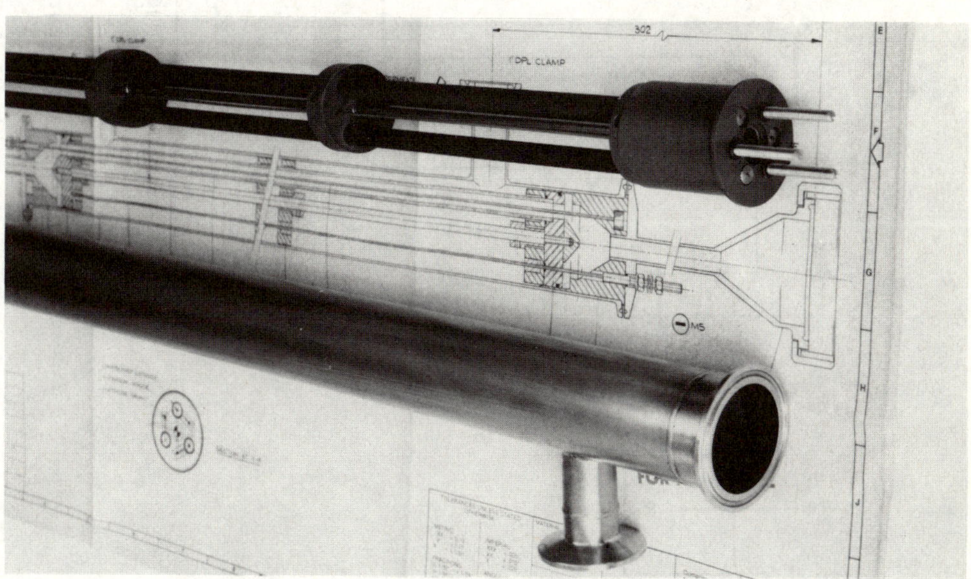

View from membrane connection end

Figure 5   Three tube DMC-UF pilot plant module

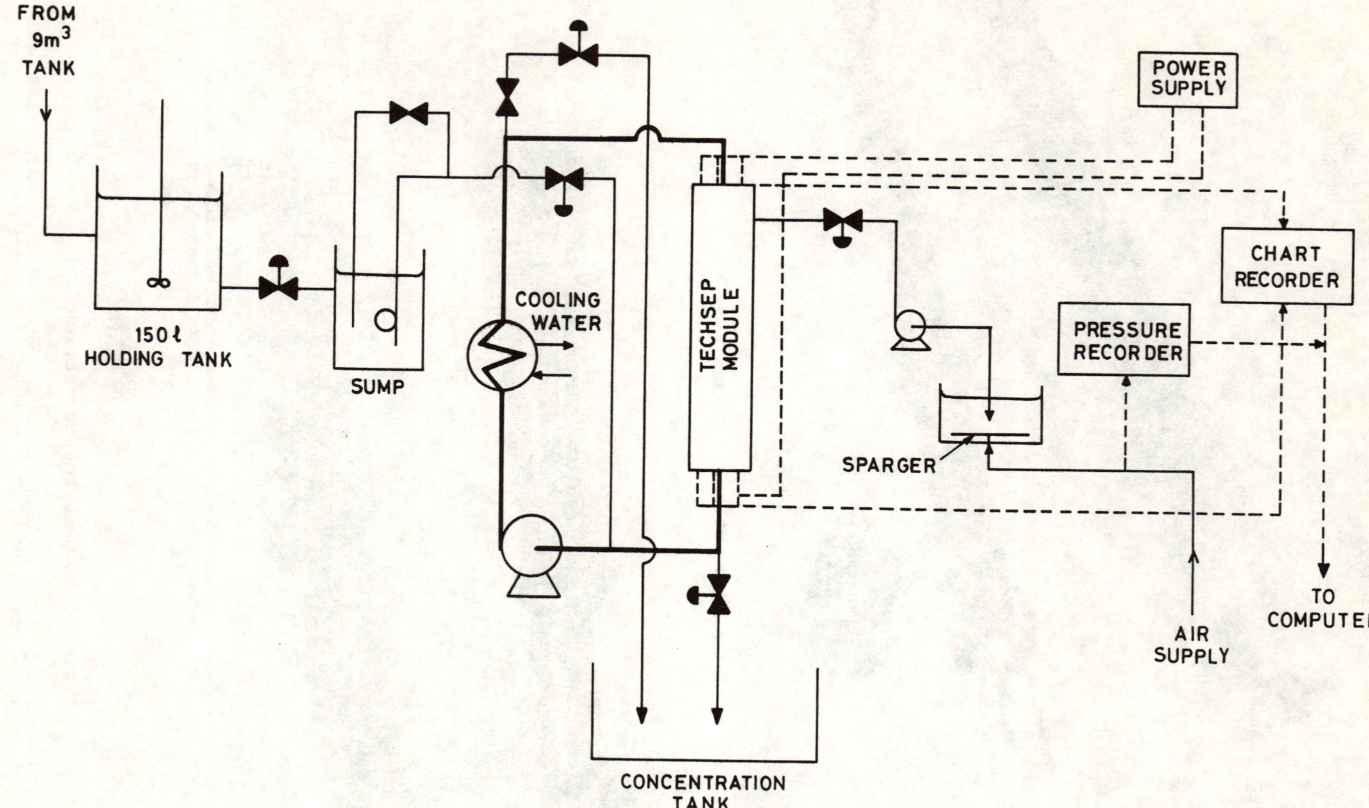

Figure 6 UF pilot plant layout for the treatment of LLW

work [1] it has been shown that DMC pulses are more effective at a transiently low transmembrane pressure, the loop is depressurized during pulse application by the synchronized opening of the concentrate removal valve. However, so as to give an acceptable volume reduction factor, the feed pressurization pump is inhibited during 11 out of 12 such operations. Unless this occurs, the feed added during the 4 seconds that the concentrate removal valve is open will displace an equal volume of concentrate. With four DMC pulses/h, concentrate will be removed every 3 hours.

Calculations of the composition and volume of off-gas produced as a result of DMC have been carried out to confirm that no gas should accumulate in the high pressure recirculated feed loop as bubbles but rather should pass out in the dissolved state with the permeate to be released in the permeate tank. The off-gas comprises 27% N_2 (from the increase in temperature produced by the pump desorbing dissolved air), 36% O_2 (from electrolyis and desorbing dissolved air) and 37% H_2 (from electrolysis). This is diluted to below 2% by a 1 ℓ/h air purge through this vessel.

Tests are currently in hand to demonstrate the 2-4 fold enhancements in permeation rate produced by the removal of surface fouling deposits by DMC, and to confirm the long term maintenance of process DF achieved with conventional UF.

3.3 Microfiltration module

The second module to be evaluated is based on a Fairey Microfiltrex MA1/S sintered stainless steel fibre membrane, which has a 3 μm absolute cut-off. A similar philosophy to that used for the UF DMC system has been adopted in the design of this unit. Tubular membranes have the advantage of minimising the length of seal for a given membrane area, thus reducing the risk of leakage. 'In-to-out' filtration not only preserves the simplicity and uniformity of flow patterns, but reduces wear by avoiding any changes in direction of the feed suspension. In addition, the use of an axial counter electrode minimises the area of electrode needed. However, due to the greater difficulty experienced at the time of the design study in producing a 6 mm i.d. tube from sintered stainless steel fibre membrane, an 18 mm bore design has currently been adopted. This is located inside a perforated thin walled stainless steel tube - both for mechanical support and also to enhance the conductivity of the filter. A 60 cm length was chosen to give an area of 340 cm², so that the HP6456B power supply could deliver up to 200 mA/cm² if the tubes were cleaned singly, or up to 100 mA/cm² if all the tubes in a three tube module of total area 0.1m² were cleaned simultaneously.

In order to optimise hydraulic resistance to flow and electrical resistance through the feed suspension, the central anode size had to be increased from the wire used for the UF unit to a thin walled titanium tube of 10 mm o.d. with its exterior surface coated with a thin layer of platinum to ensure a long life as an anode. Due to the larger filter diameter, this design has a longitudinal pressure drop three times less for the same crossflow velocity when compared to the UF unit. As Harwell Low Level Waste typically has a resistivity of about 500 Ω cm due to the chemicals added to generate ~ 25 ppm (Fe) iron hydroxide floc, 40 V is required to give a cleaning current density of 150 mA/cm². If the external diameter of future replacement filter tubes are reduced to 16 mm, this gives an inter-electrode gap of 3 mm, thus enabling a cleaning current density of 191 mA/cm² to be delivered from a 36 V power supply. Although the narrower cell gap does increase the longitudinal pressure drop by a factor of 1.8, this is still 3.5 times less than the Tech Sep M4 design.

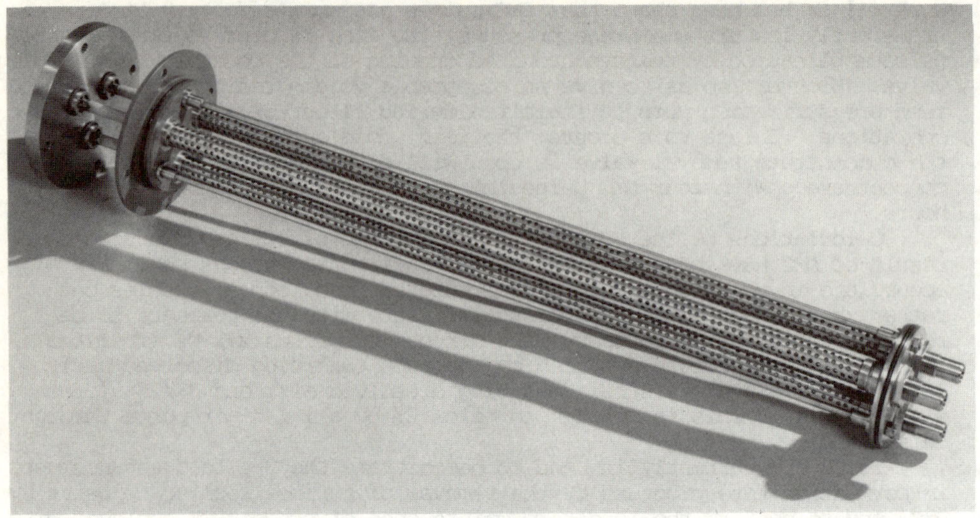

Internal filter tube arrangement

Assembled filter module

Figure 7   Three tube sintered stainless steel fibre microfiltration pilot plant unit.

A three tube module has been constructed on these principles (Figure 7) with replaceable filter elements. As with the UF unit, the tubes can either be cleaned singly or in parallel. In order to give a crossflow velocity of 1 m/s, a volumetric recirculation rate of 1.9 m³/h is required. Calculations on the off-gas produced show that due to the raising of the current density to 150 mA/cm² and the pulse duration to 5 seconds, the volume liberated in the permeate tank increases to 1.05 ℓ/h – of which 37% is H_2, 28% N_2 and 35% O_2. However, as the filter throughput will typically be at least 120 kg/h (0.6 m/h/bar at 2 bar through 0.1 m²), as before all the gas will pass in the dissolved state into the permeate vessel. The hydrogen content can be diluted to < 2% here by a 20 ℓ/h air purge.

3.4 Plant scale-up

Further scale-up of either of these two systems can be achieved by increasing the number of tubes/module and by increasing the number of modules operating in parallel. Each filter unit could be cleaned independently of the others by isolating it with remotely operated valves during DMC pulse application. In this way, plant operation is not interrupted and a small power supply can be used more or less continuously as it cleans modules sequentially. The optimum module size is affected not only by the reduction in engineering costs with larger units, but also by the cost of an increasing size of power supply. In particular, for the microfiltration membranes further developments in membrane structure and materials may lead to reductions in current density and pulse duration required for cleaning, which will allow an increase in membrane area/module, either by increasing tube length and/or the number of tubes.

3.5 Conclusions

In considering the economics of DMC, a range of factors need to be evaluated. Due to the extra axial electrode and the power supply required to drive the process, the capital cost of a module of given membrane area is increased. However, this is off-set by the enhancement in performance in increased permeation rates (by factors of at least 2-4), in enhanced filtration module availability (resulting from obviating the need for backflushing or chemical washing to restore the filtration flux) and in reduced crossflow velocities. The former two factors lead not only to a reduction in the number of filtration modules necessary, but also could make savings in civil and engineering costs arising from a more compact plant. The reduction in crossflow velocity not only reduces plant wear resulting from the abrasive nature of the suspended particulates thus reducing the cost of a plant with a guaranteed operating life, but the pumping energy and pump sizes are also considerably reduced. For a UF plant operating at 2 bar, reduction of crossflow velocity from 5 m/s to 1 m/s results in a reduction of recycled volumetric flow by a factor of 5.3 (the extra results from the presence of the central wire). This not only influences the size of pump but also its running costs. The reduction in crossflow velocity also enables more viscous suspensions to be treated, as although the presence of the central electrode increases the pressure gradient along the tube by a factor, this is more than off-set by the flow reduction – thus leading to an overall saving of 1.9 in pressure drop. Further improvements in both enhanced permeation rate and reduced crossflow velocities may result from work currently in progress on improved membrane materials.

However, in common with conventional crossflow UF and microfiltration, due to the continuous shear applied to the concentrated product, it remains an easily pumped slurry compatible with subsequent immobilization techniques such as cementing. In contrast, rotary drum and plate and frame filters give rise to a cake product, which is significantly more difficult to handle. From plant operating experience on UF and laboratory evaluation of DMC, the overall process lends itself to automation and should have a low maintenance requirement.

<u>Reference</u>

[1] Cumming, I.W. and Turner, A.D., Recent Developments in Testing New Membrane Systems at AERE Harwell for Nuclear Applications, presented at CEC conference on Future Applications for Membrane Processes, Brussels, 6-7 December 1988.

[2] Howden, M and Moulding T.L.J., Progress in the Reduction of Liquid Radioactive Discharges from the Sellafield site. International conference on Nuclear Fuel Reprocessing and Waste Management. "RECOD 87".

<u>Acknowledgement</u>

This work has been part funded by the Commission of the European Communities in the framework of its research programme on radioactive waste management. In addition, the work has also been carried out with financial support from the U.K. Department of Environment, U.K. Department of Energy, British Nuclear Fuels plc, and Ministry of Defence. In the D.o.E content, the results will be used in the formulation of Government Policy, but at this stage they do not necessarily represent Government Policy.

USE OF MEMBRANE TECHNOLOGY FOR PROCESSING WASTE WATER FROM OLIVE-OIL PLANTS WITH RECOVERY OF USEFUL COMPONENTS

L. MASSIGNAN, P. DE LEO, D. TRAVERSI
Centro Ricerche Bonomo - Castel del Monte - Andria (Bari)
A. AVENI
Istituto Ricerche Breda - Bari

Summary

Pre-treatment tests with enzymatic and chemical methods were carried out in order to overcome the membrane fouling during the reverse osmosis process of vegetation waters. Treatment with pectolytic enzymes do not improve the water filterability while the use of calcium chloride solution at room temperature induces an effective separation of the gelling substances. Natural ageing for 4 months also allows to obtain a liquid that can be processed in reverse osmosis plant. Treatment efficiency was tested by a semi-pilot reverse osmosis unit with flat membranes. Permeation rate of 380-390 $l/m^2 d$, COD rejection of 98% and 80% recovery factor were observed both for calcium pre-treated and naturally aged waters. The concentrated fraction can be used as component for mixed animal-feeding after the removal of polyphenolic substances.

1. INTRODUCTION

The olive-oil production generates a very important environmental problem in various Mediterranean Countries. Husks and waste-waters (named vegetation waters (VW)) are the principal wastes of the olive-oil industries (Fig. 1). Husks are normally used for producing husk-oil by solvent extraction while VW are accumulated and discharged after sedimentation and recovery of superficial oil.

Depending on the different processes (continuous or discontinuous), for each ton of oil produced, 2 to 5 cubic meters of VW are discharged. Due to the high content of organic substances (sugars, nitrogenous compounds, fats, organic acids, polyalchools, polyphenols, pectins, etc.) and inorganic components (mainly potassium and phosphorus salts), the pollution effect is very strong: the COD and BOD_5 contents vary respectively in the range 30-200 and 20-60 kg/cubic meter.

The pullution problems are very strong in Apulia, where, as pointed out by studies carried out by Centro Ricerche Bonomo and Istituto Ricerche Breda (1), about 2000 oil-mills are operating with an annual VW production of 800000 cubic meters, that result in polluting load equivalent (during the operating season) to the waste water of a town of about 2.5 millions of people.

At the present state of the VW treatment technology, industry has found no interest in supporting on a wide scale any traditional process (thermic, chemical or biological): this because of the high investment and operational costs, the shortness of the production period (3-5 months) and the small size of the olive-oil mills (2,3,4).

The treatment methods based on membrane technologies (reverse osmosis, ultrafiltration, microfiltration) seem to be interesting and to admit further developments (5,6,7): they are suitable to concentrate the organic substances and allow the recovery of some valuable components. These membrane processes can be used in modular plants and therefore it becomes easy to plan depuration projects either for individual or associated olive-oil mills. Besides the application of these technologies depends on the possibility of the economical recovery of the concentrate, whose composition must be controlled and, if necessary, modified according to the final destination. For example, if the concentrate is used as animal feeding, the presence of polyphenols can cause some reduction of the protein bioavalaibility; on the other hand, the recovery of polyphenols can give economical benefits as they can be used as natural antioxidants and pigments (8,9).

Nevertheless the application of membrane processes involves some technological troubles related to the presence in the VW of gelling substances, like pectins, that give rise to fouling phenomena strongly reducing the membrane efficiency (10,11). Therefore the removal of these substances before the treatment of VW with membrane processes seems to be absolutely necessary.

This report shows the results of some experiences of enzymatic, chemical and chemical-physical pre-treatments of VW (12) carried out in order to overcome the membrane fouling during the reverse osmosis process. Tests have been performed using a commercial pectolytic preparation, normally used in the oenological industry, and calcium chloride solutions. Calcium chloride was selected as flocculation agent because of its capacity to interact with pectic substances and to modify the protein solubility ("salting out" phenomenon). These pre-treatments have been compared with the natural ageing of the VW that was described as a method for improving the membrane process performance (5).

2. MATERIALS AND METHODS

2.1 BATCH PRE-TREATMENTS

Pre-treatment tests with enzymes and calcium chloride have been performed using vegetation waters drawed not from the settling tank but directly from the centrifuge of a discontinuous oil-mill. In this way we collected waters with a solid concentration higher than the one normally measured in wastes of the olive-oil plants but we avoided any external pollution. The main characteristics of the VW are reported in Table I.

The VW were stored at room temperature (16-18°C) in an open tank; samples for the pre-treatment tests were taken out every 24 hours for 10 days in order to verify the ageing effect on the response of VW to the enzymes and chemicals. All treatments were carried out at room temperature, using 500 ml aliquots of vegetation water.

The enzymatic treatment was performed using a commercial pectolytic preparation (Ultrazym 100 - Esseodue Novara): the enzyme (100 ppm) was added by gently stirring to the VW adjusted to pH 3.7 with HCl 37%.

For the calcium treatment an exahydrate calcium chloride solution (143.5 g/l of Ca++) was added by stirring to the VW (1:100 v/v).

Combined enzymatic and chemical treatments were also tested at the same conditions above described.

Tests were carried out daily with the following experimental design:

TREATMENT

1	Control (non treated VW)
2	Enzyme (100 ppm)
3	$CaCl_2$ (1435 ppm Ca++)
4	Enzyme (100 ppm) + $CaCl_2$ (1435 ppm Ca++)

All the samples were left settling in graduate cylindres and the volume of settled sludge was detected after 24 hours. The sludge was analyzed for dry matter content. The kinematic viscosity at 38°C (Cannon-Fenske capillary viscosimeter) and the capillary suction time (with 18 mm cylinder) of the supernatant were also measured.

2.2 REVERSE OSMOSIS (R.O.) TESTS

A semi-pilot continous unit was utilized for the R.O. tests. The unit was equipped with a DDS (De Danske Sukkerfabrrikker) module supporting flat membranes DDS-HR95 with nominal NaCl rejection of 95%. The total membrane area was 0.252 m^2.

All tests have been carried out with a feed rate of 600 l/h at constant temperature (30°C). The temperature was controlled in the feed tank by recirculating cold water in submerged coil. A high pressure cartridge filter was placed between the volumetric pump and the R.O. module.

The VW for R.O. tests have been drawed from the settling tank of a continuos oil-mill. Three samples of VW, each of 150 l, have been stored in open tanks. The first (control) was fed to R.O. unit whitout any treatment; the second sample was treated with $CaCl_2$ as described before, then filtered through a cotton cloth (120 l of filtered liquid has been collected) and fed to R.O. unit; the third one had been left settling for 4 months in the open tank at room temperature, then filtered and fed to R.O. unit. The VW characteristics detected before and after the pre-treatments are reported in Table II.

Experiences with enzyme pre-treated VW have not been planned since the tested enzyme showed low efficiency in improving the VW filterability.

3. RESULTS

Table III shows the results of the measurements performed during the batch pre-treatment tests.

The total dry matter of the untreated VW decreases during 10 days by about 14% due to the natural fermentation of the organic substances; this has a remarkable effect on the response of the VW to the treatments. For the untreated VW the viscosity and the CST of the supernatant do not change during the time of the test; the addition of enzyme causes a remarkable decrease of the viscosity without a decrease of the CST values.

Both untreated and treated VW show an increaese of the sediment volume (Fig. 2) after 3-4 days of ageing; a real jump was observed after·8 days in the samples treated with $CaCl_2$. This phenomenon was joined to an halving of the supernatant viscosity and a very high reduction of CST.

The treatment of 8 days aged VW with $CaCl_2$ gives rise to an evident separation of sludge (35% v/v) with a dry matter content of about 16%; 46% of the original dry matter is then removed from the VW before it is fed to R.O. process.

Fig. 3 shows the flux rate of permeate observed during the R.O. tests at different operative pressures. The detected flux rate for the untreated VW was very low (about 100 $1/m^2d$) and did not increase by increasing inlet pressure; this confirms the presence of a "gel-layer" on the membranes. On the contrary the samples pre-treated with $CaCl_2$ or aged 4 months showed a linear response to the increasing pressure; the permeation rate was more than 400 $1/m^2d$ at 5000 KPa with slightly higher values for the naturally aged VW.

The flux rates measured in continuous testes during 6 days, for the 3 samples, at a pressure of 5000 KPa are reported in Fig. 4. For the untreated VW a drop of flux was observed during the first few hours of run. For both the pre-treated samples, only a little decrease was registered in the first two days; this may be ascribed to a slight fouling: periodical washing with tap water allowed to restore the starting flux. It was extimated that the washing time and the water volume caused a loss of plant productivity of about 5%. During the tests with pre-treated VW the flux rates did not change for different values of recovery factor (RF) until 60% RF was reached, then a noticeable decrease was observed, and at 80% RF a flux of about 150 $1/m^2d$ was measured.

The COD rejection was quite constant for all RF values ranging aroud 98%.

4. <u>CONCLUSIONS</u>

The results of the VW pre-treatment tests and of the reverse osmosis experiences provide some suggestions that can be summarized as follows:
- the enzymatic treatment whit a commercial pectolytic preparation allows a VW viscosity reduction up to 50% without, however, any filterability improvement. This pretreatment seems then to be not suitable to avoid the R.O. membrane fouling.
- better results can be obtained by chemical treatment with $CaCl_2$ solutions. The tratment of 8-10 days aged Vw leads to a physical separation of most of gelling substances. The supernatant obtained after sedimentation or centrifugation shows a low viscosity and a good filterability and can be

fed to a R.O. process allowing permeate flux rates 4 time higher than in untreated VW.
- almost the same results can be obtained by natural ageing of VW for at least 4 months. The ageing time may vary in relation to the ambient temperature and the presence of wild yeasts in the settling tank.
- the pre-treated VW can be processed in a continuous R.O. plant operating at 5000 KPa with a permeate flux rate of 380-390 l/m^2d and a 80% recovery factor. The permeate obtained shows a residual COD of about 1000 ppm.
- the R.O. concentrate, mixed with the sludges obtained can be used as component for mixed animal-feeding after having removed polyphenols. Furthermore the recovery of these substances can find useful applications in the natural antioxidants field.

5. FINAL CONSIDERATIONS

The results of these tests showed that it is advantageously possible to use the membrane technology for the VW treatment with recovery of useful products. However, the same tests have pointed out the need to perform further studies in order to improve knowledges on the following items:
- optimization of treatment methods mainly related to the use of concentrate
- selection of high efficiency membranes and modules
- definition of process parameters and membranes washing cycles
- examination of the different enzymatic methods for reducing the fouling problems on the membranes
- examination of the feeding aspects in order to prepare mixtures, integrated with VW concentrates, suitable for individual animal species and their physiological conditions.

On the basis of these considerations a BRITE research programme has been prepared by a research team, whose partners are Italian and Spanish Industries and Research Centers involved in waste treatment and by-products recovery. The research foresees the development of membranes processes according to the olive-oil waste water treatment needs and the experimental examination of the possibility to extend the membranes use for the improvement of low-quality vegetable oils.

REFERENCES

(1) CENTRO RICERCHE BONOMO (1982). Indagine sulla consistenza quantitativa e sulla distribuzione delle acque di vegetazione dei frantoi in Puglia. Technical Report n° 1.
(2) GARRIDO FERNANDEZ A. (1975). Tratamiento de las aguas residuales de la industria del alderezo. Metodo para su eliminacion o reacondicionamiento para su posterior empleo. Grasas y aceitas, 26 (4), 237.
(3) BOARI G., BRUNETTI A., PASSINO R., ROZZI A. (1984). Anaerobic digestion of olive oil mill waste waters. Agricultural Wastes, 10, 161.
(4) ARPINO A., CAROLA C. (1978). Lo smaltimento delle acque di vegetazione

provenienti dagli impianti di estrazione dell'olio di oliva. Nota II.
Riv. It. Sost. Grasse, 55, 24.

(5) POMPEI C., CODOVILLI F. (1974). Risultati preliminari sul trattamento di depurazione delle acque di vegetazione delle olive per osmosi inversa. Sci. Tecn. Alimenti, anno IV, 363.

(6) RAMPICHINI M. (1987). New applications of ultrafiltration: effluents from synthetic fibers and olive oil production. Chimica Oggi, 4, 21.

(7) JEMMETT M.T., CARRIERI C., DE LEO P., KAPSIOTIS G.D. (1983). Esperienze di osmosi inversa ed ultrafiltrazione di acque di vegetazione delle olive con moduli a membrane piane. Scienza dell'Alimentazione, anno 12, 1, 37.

(8) CAMURATI F., FEDELI E. (1982). Attività antiossidante di estratti fenolici di acque di vegetazione delle olive. Riv. It. Sost. Grasse, 59, 623.

(9) CAMURATI F., LANZANI A., ARPINO A., RUFFO C., FEDELI E. (1984). Le acque di vegetazione della lavorazione delle olive: tecnologie ed economie di recupero di sottoprodotti. Riv. It. Sost. Grasse, 61, 283.

(10) VIGO F., GIORDANI M., CAPANNELLI G. (1981). Ultrafiltrazione di acque di vegetazione da frantoi di olive. Riv. It. Sost. Grasse, 58, 70.

(11) VIGO F., DE PAZ M., AVALLE L. (1983). Ultrafiltrazione di acque di vegetazione da frantoi di olive. Esperienza gestionale in impianto semipilota. Riv. It. Sost. Grasse, 60, 267.

(12) MASSIGNAN L., DE LEO P., CARRIERI C. (1985). Depurazione mediante osmosi inversa di acque reflue da oleifici. Scienza dell'Alimentazione, anno 14, 6, 421.

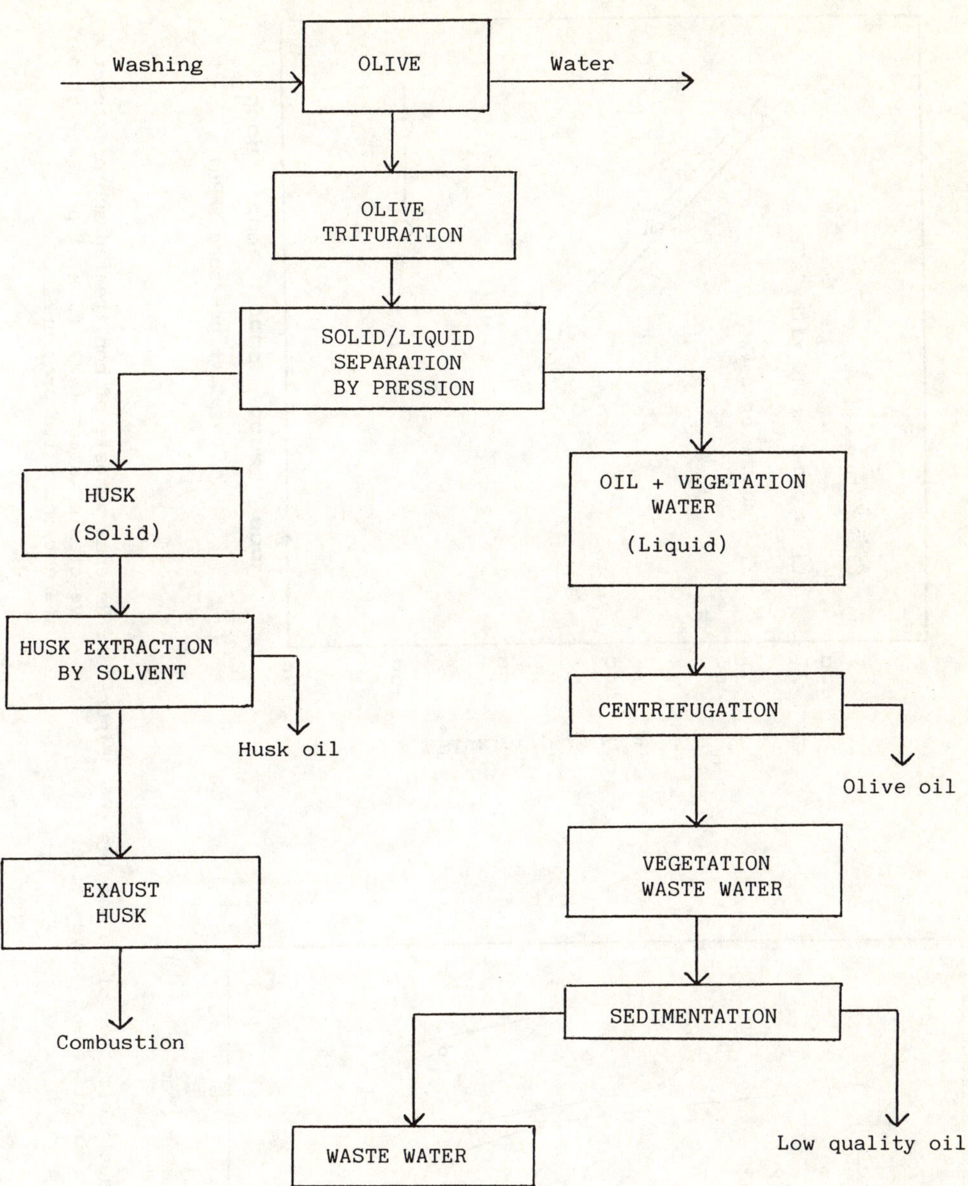

Figure 1 : Scheme of the olive-oil traditional process

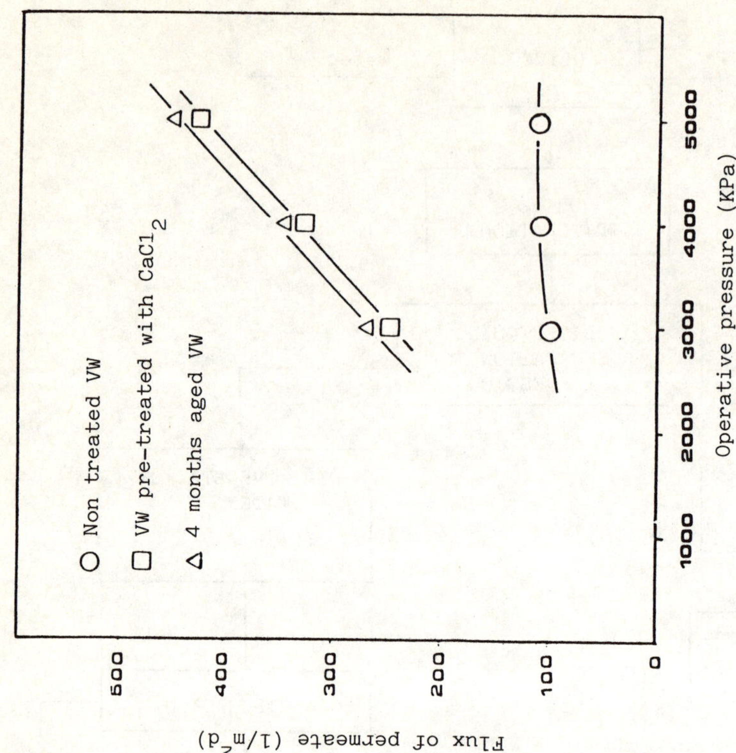

Figure 3 : Reverse osmosis of non treated and pre-treated vegetable waters (VW). Flux of permeate at different operative pressures

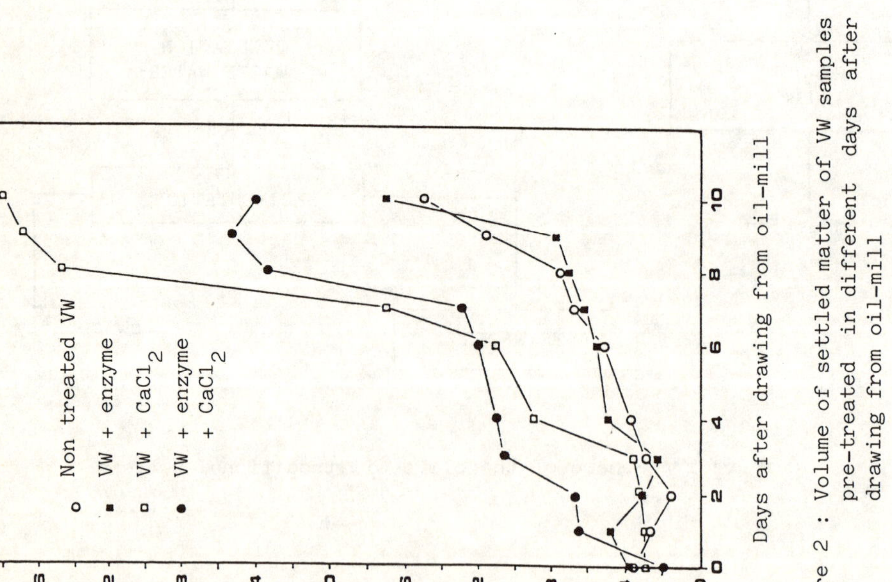

Figure 2 : Volume of settled matter of VW samples pre-treated in different days after drawing from oil-mill

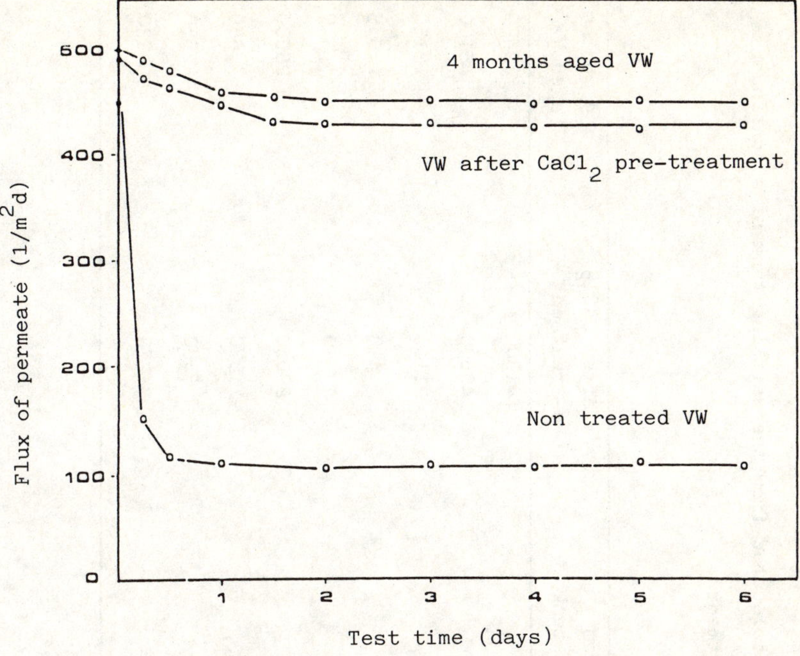

Figure 4 : Reverse osmosis of non treated and pre-treated VW.
Flux of permeate during the test time

Table I : Characteristics of vegetable waters used for batch pre-treatment
tests

Total dry matter (ppm)	131700
COD (ppm O$_2$)	172600
BOD$_5$ (ppm O$_2$)	71500
pH	5.40
Specific gravity	1.046
Suspended dry matter (ppm)	13600
Volatile dry matter (ppm)	115200
Ca^{++} (ppm)	1250
Na$^+$ (ppm)	80
K$^+$ (ppm)	8750
N$_{tot}$ (ppm)	780

TABLE II – Characteristic of vegetable waters collected from the settling tank of a continuous oil-mill, before and after pre-treatments.

	VW from settling tank	VW after $CaCl_2$ pre-treatment	4 months aged VW
Total dry matter (ppm)	55100	41300	39400
COD (ppm O_2)	65300	48500	46100
BOD_5 (ppm O_2)	27800	21100	13150
pH	5.1	4.8	5.5
Suspended dry matter (ppm)	7030	0	0
Volatile dry matter (ppm)	44300	29050	28500
Ca^{++} (ppm)	720	1830	680
K^+ (ppm)	2750	2500	2550
N_{tot} (ppm)	450	275	220

TABLE III — Effects of different pre-treatments on sedimentation and filterability of vegetable waters (VW)

		Days after drawing from oil-mill										
		0	1	2	3	4	5	6	7	8	9	10
Total dry matter %		13.17	13.15	12.96	12.61	12.12	n.d.	12.10	12.10	11.76	11.56	11.38
Non treated VW	S	3.7	2.7	1.6	3.0	3.7	"	5.1	6.7	7.5	11.5	15.0
	R	5.56	4.86	2.96	5.30	6.71	"	8.21	9.74	9.24	13.69	16.58
	V	2.54	2.40	2.39	2.41	2.41	"	2.52	2.97	3.14	3.28	2.92
	CST	400	400	400	400	400	"	400	400	400	400	400
VW pre-treated with enzyme	S	3.5	4.9	3.3	2.3	5.0	"	5.5	6.1	7.0	7.8	17.2
	R	7.72	10.91	7.94	6.71	14.23	"	14.12	12.75	15.47	15.45	25.68
	V	1.05	1.10	1.10	1.12	1.07	"	1.06	1.14	1.02	1.01	0.98
	CST	400	400	400	400	400	"	400	400	160	110	50
VW pre-treated with CaCl$_2$	S	3.2	2.8	3.1	3.6	8.8	"	11.0	17.1	34.8	37.0	38.0
	R	5.40	5.26	5.69	5.80	12.99	"	15.36	24.00	46.16	49.80	50.89
	V	2.62	2.57	2.61	2.75	2.86	"	3.78	4.28	1.18	1.14	1.09
	CST	400	400	400	400	400	"	400	400	18	10	12
VW pre-treated with enzyme + CaCl$_2$	S	1.9	6.4	6.7	10.6	11.0	"	12.0	12.8	23.4	25.4	24.0
	R	3.77	10.12	10.63	15.96	17.70	"	17.95	18.61	36.21	39.37	37.11
	V	1.46	1.61	1.57	1.50	1.26	"	1.98	2.15	1.02	0.99	1.03
	CST	400	400	400	400	400	"	400	50	15	10	10

S = Sediment after 24 hours (% V/V); R = Removed dry matter after sedimentation (% of total VW dry matter);
V = Kinematic viscosity of supernatant (centistokes); CST = Capillary suction time of supernatant (sec.).

A STUDY ON MEMBRANE TECHNOLOGY FOR THE TEXTILE INDUSTRY: MEMBRANE PREPARATION AND CHARACTERIZATION

S.N. GAETA, M. SOGLIANO, and E. PETROCCHI
Separem S.p.A., Biella, Italy
E. DRIOLI
Università di Calabria, Arcavacata di Rende (CS), Italy

Summary

Have been developed special membranes and modules adequate to recover water and auxiliary chemicals from textile effluents. The membranes can be classified as loose-reverse-osmosis membranes. They were developed starting from polyamides of the polypiperazine family produced at Separem.

The membranes were prepared by using a technique of phase inversion from solution. The parameters to prepare the membranes have been optimized. The membranes are characterized by intermediate values of rejection (85-90%) and an average flux of 55 lt/h m^2. They are extremely resistant to pH (1-13) and oxidizing species (at least 25 ppm of Cl_2, continously).

The spiral wound modules developed with these membranes, properly designed to meet the fluid-dynamics requirements to minimize fouling, have been tested with textile effluents. The optimal conditions to operate the reverse osmosis plant have also been developed. The results indicate that the modules are adequate to economically treat textile effluents. Preliminary evaluations indicate that the cost of permeate water is less than 500 Lit/m^3 and that the pay-off time for the reverse osmosis plant is about 2 years.

1. INTRODUCTION

Textile industries traditionally use a huge quantity of water which is normally discharged after expensive treatments to decrease the pollution load within legal limits. Due to the increased costs of pollution control, to the lack of water, and to the valuable chemicals contained in the effluents, it is advantageous to apply the membrane technology to recover water, auxiliary chemicals, and energy. Therefore, the traditional "money-wasting" process of pollution control can be converted in a profitable operation. Moreover, by recycling the waste-effluents are avoided all the environmental problems connected with the discharge of polluted effluents.

Due to the harsh environment of the textile effluents in which the membrane have to operate, it is necessary to have membrane modules characterized by excellent chemical and mechanical resistances. In fact, this reduces the cost of pretreatment and cleaning operations by allowing the use of more aggressive detergents which are normally less expensive than detergents specifically developed for less resistant membranes. Moreover, the life of membranes is presumably longer.

By using the innovative membranes described in this paper, reverse osmosis processes, rationally integrated in textile plants, can easily be used to reclaim water and auxiliary chemicals by allowing their recycle and approching a zero-discharge system. This is schematically illustrated in Figure 1.

A conventional cycle of a textile process, showing the relative amounts of energy, water and chemicals used, is reported in Figure 2. The process is relative to a cotton dyeing plant, but can be generalized to other processes. In Table I are reported the typical characteristics of waste-waters from a textile dyeing process and their range of variation. The potential species which can be recovered are reported in Table II. Traditionally the effluent is disposed in equalizing tanks and thereafter treated in situ or in municipal waste-water treatment plants.

2. RESULTS AND DISCUSSION
2.1 MEMBRANE

The novel membranes developed for textile applications were prepared at Separem by using proprietary polymers of the polypiperazineamide family. The general formula and main characteristics are reported in Figure 3. They are produced by Separem by using an interfacial polycondensation between a carbonilic acid and a piperazineamide. X and R_i are different radicals depending on the desired polymer properties. These polymers are soluble and suitable for preparing membranes.

The membrane prepared are homogeneous, asymmetric and supported. They are prepared by a technique of phase inversion from solution. The general procedure is reported in Figure 4. The polymer is completely dried and then dissolved is a strong organic solvent. The solution is filtered, deaerated, and casted on a non-wooven support. The phase inversion is started in air and completed in a non-solvent bath. The membrane after a post-treatment is housed in spiral wound modules. The membrane casting parameters have been reported elsewhere(1).

In Figure 5a is illustrated a SEM picture of the cross-section of the membrane. The skin layer, responsible for the transport properties, is about 1 micron thick, the supporting sponge-like structure is about 70 microns thick. In Figure 5b and 5c are shown enlargements of the bottom and top part of the membrane, the asymmetry is evident.

In Table III are reported some characteristics of the membrane. The membranes are characterized by intermediate NaCl rejection and by high fluxes. These good transport properties are due to the rigid and regular polymeric molecules which allow good intramolecular interactions and a higly porous structure. The membrane can withstand a pressure as high as 6 MPa without severe compaction. These mechanical properties are due to the high molecular weight and low molecular weight distribution of the polymer.

The pH resistance is illustrated in Figure 6, where the variation of flux and rejection is reported versus pH. These results confirm that the membrane can be used safely at any pH between 1 and 13. The Chlorine resistance resulted excellent. In Table IV are summarized the results of specific tests reported elsewhere (2). The membranes can be exposed to chlorine up to 200.000 ppmCl$_2$-hours without variation of transport properties. The life of the membrane is remarkably longer if the operating temperature is lower than 25°C.

2.2 TEXTILE APPLICATION

To prevent fouling and module damage due to clogging is necessary a preliminary removal of fine suspended solids and colloids from effluents. Therefore the effluent needs to be pretreated before being processed in the reverse osmosis plant. However, due to the excellent resistance of the membrane used it is not necessary any pretreatment to prevent chemical deterioration of the membrane.

In Figure 7 is reported the fouling index for the modules when treating a textile effluent from a cotton dyeing plant as a function of pressure, axial flow rate and temperature. The bigger the fouling index the lower the permeate flux per module, i. e., the more expensive the process. The data reported in Figure 7 led us to identify as optimum operating conditions for a module the following: Pressure=4 MPa, Axial Flow Rate=2m^3/h; Temperature=30°C.

These results were utilized to operate a system to recover water and Na$_2$SO$_4$ from a 14 m^3/day effluent of a textile industry. The reverse osmosis was integrated with traditional systems as reported in Figure 8. The biological treatment reduces the chemical oxygen demand and the surfactant content of the effluents. The physico-chemical unit, followed by sand and cartridge filters, lessens the fouling load to the reverse osmosis module. These prefiltration steps are fundamental to enhance the life of membranes. The results of the separation obtained are reported in Table V. The permeate flux is reported in Figure 9. The permeate has always been reused for washing and dyeing operations. The quality of the fabrics obtained has always been judged commercially acceptable (either dark or white shades). The concentrate, after decoloration, can be reclaimed to the dyeing bath. The cost of the operation is reported in Table VI.

3. CONCLUSIONS

The membranes developed by Separem specifically for textile applications are adequate to treat textile and, generally, industrial effluents. They are extremely resistant to chemical and oxiding species.

The process developed by integrating a reverse osmosis unit with a conventional waste treatment system is valid to recover water and auxiliary chemicals from textile effluents.

The work also demonstrates that for reverse osmosis treatment of industrial effluents, besides having stable membrane modules, a thourough study of the engineering aspects of the membrane process is necessary. In fact, by optimizing the condition of the prefiltration and of the reverse osmosis process the separation obtained and the life of the membrane modules have been greatly enhanced.

The cost of the water recovered is about $500 \, Lit/m^3$. The potential recovery is about $1300 Lit/m^3$. The impact that this technology can have on the textile industry is even greater for these areas where there is a lack of water or for rapidly expanding industries where the existing pollution control system are unadequate to meet the increased demand (quantitative and qualitative).

4. ACKNOWLEDGEMENTS

The authors wish to tank Dr. P.K. Bhattacharya, Dr. U. Fedele, and Dr. E. Vigliani for their valuable contribution during the experiments with textile effluents.

The work reported was partially funded by the Commission of the European Communities under BRITE Contract No. RI-1B0073-I(S).

5. REFERENCES

(1) DRIOLI,E., GAETA,S., and SOGLIANO,M. (1987). Italian Patent No. 21852 A.

(2) GAETA,S., PETROCCHI,E., and DRIOLI,E., (1988). Preparation and characterization of Novel Chlorine Resistant Reverse Osmosis Polyamide Membranes. Submitted for publication in J. Membrane Sci.

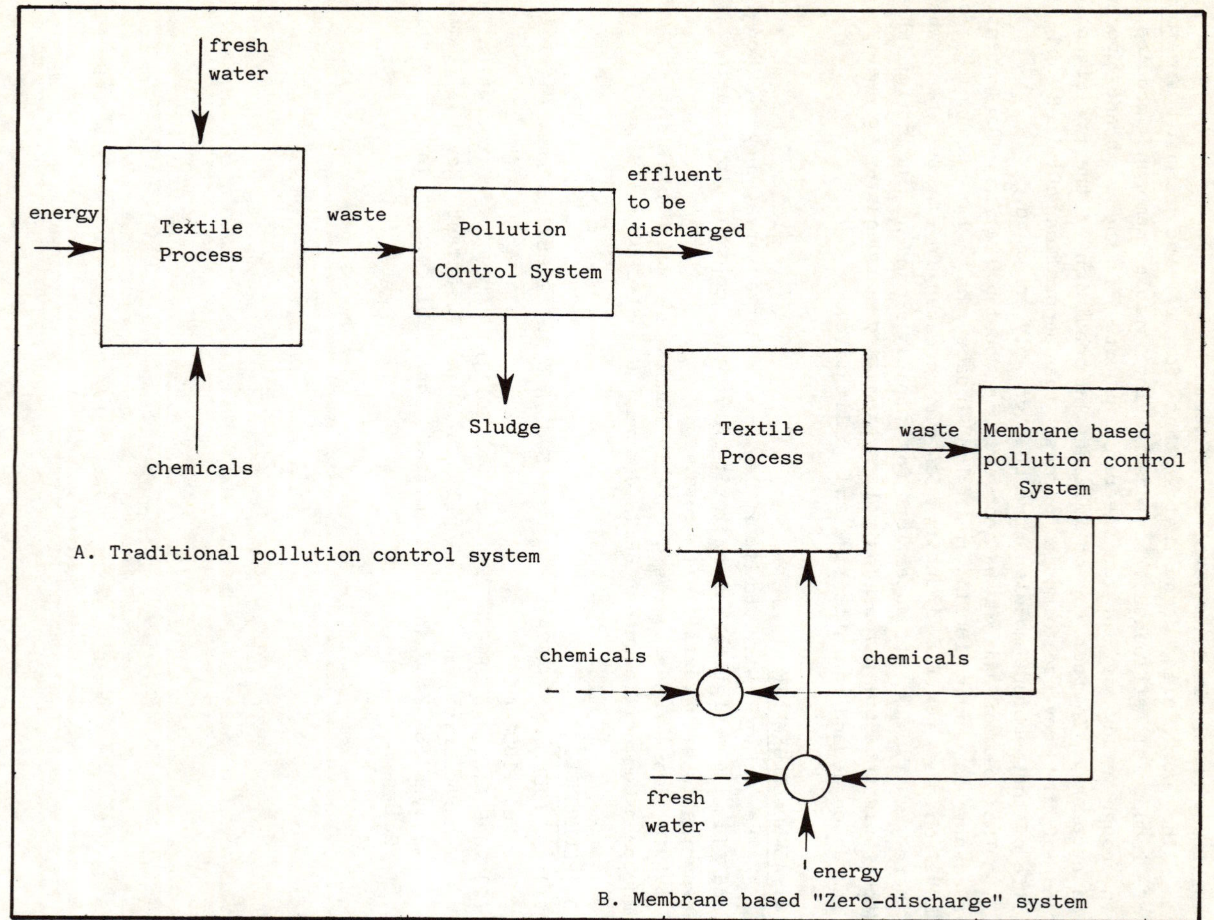

Figure 1. Membrane based versus traditional pollution control systems.

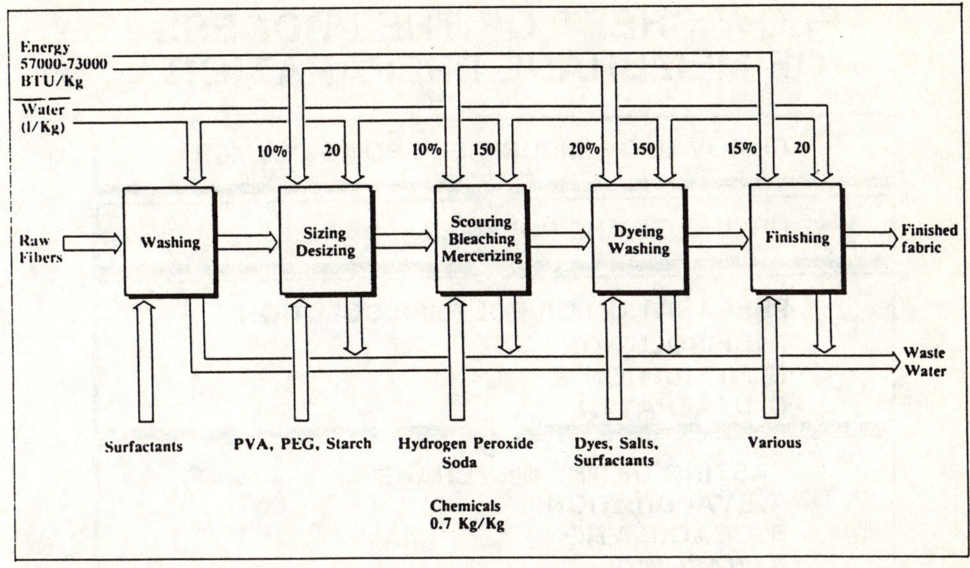

Figure 2 : Schematic diagram of a process for dyeing and finishing cotton
fabrics

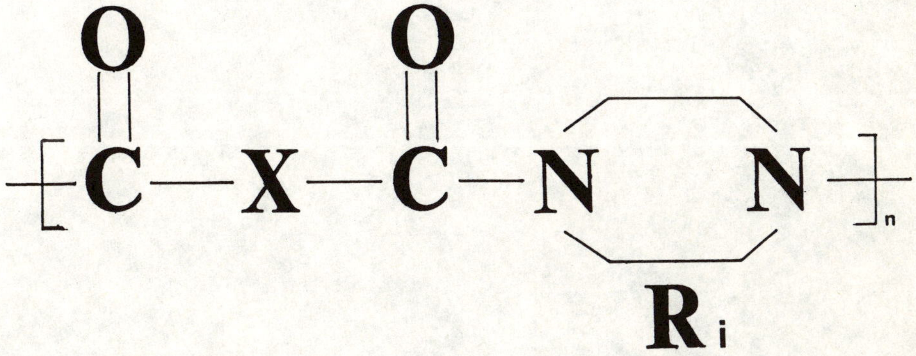

Poly (Piperazinamides)

Tg > 230°C   Modulus of Elasticity > 30,000 Kg/cm^2

Amorphous   Heat Deflection Temperature > 300°C

Figure 3 : General formula and properties of the polymers used to produce
Separem S.p.A. membranes and modules

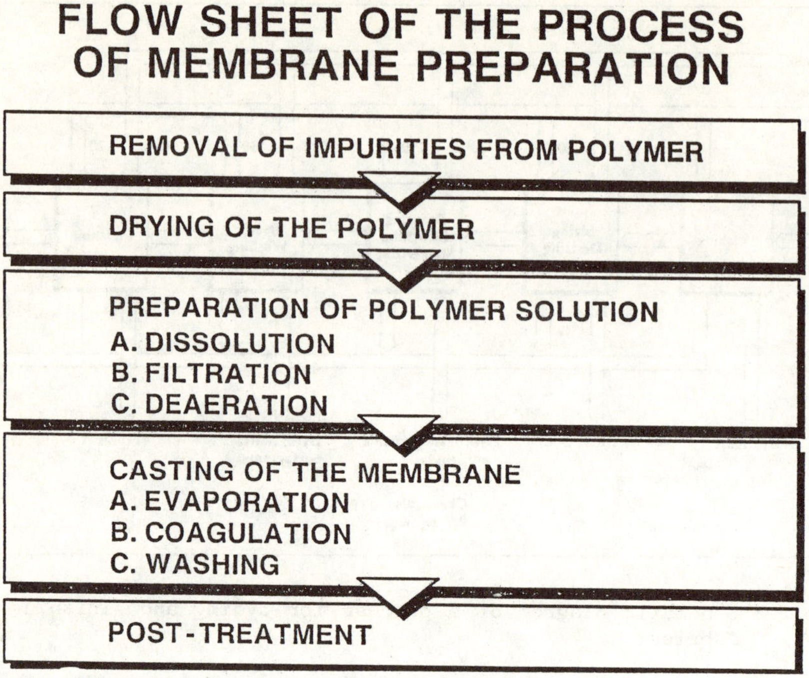

FLOW SHEET OF THE PROCESS
OF MEMBRANE PREPARATION
REMOVAL OF IMPURITIES FROM POLYMER
DRYING OF THE POLYMER
PREPARATION OF POLYMER SOLUTION
A. DISSOLUTION
B. FILTRATION
C. DEAERATION
CASTING OF THE MEMBRANE
A. EVAPORATION
B. COAGULATION
C. WASHING
POST-TREATMENT

Figure 4 : Schematic diagram of the method to prepare membranes by phase inversion

10µm 30.1kV 1.25E3 0800/88 SNIA

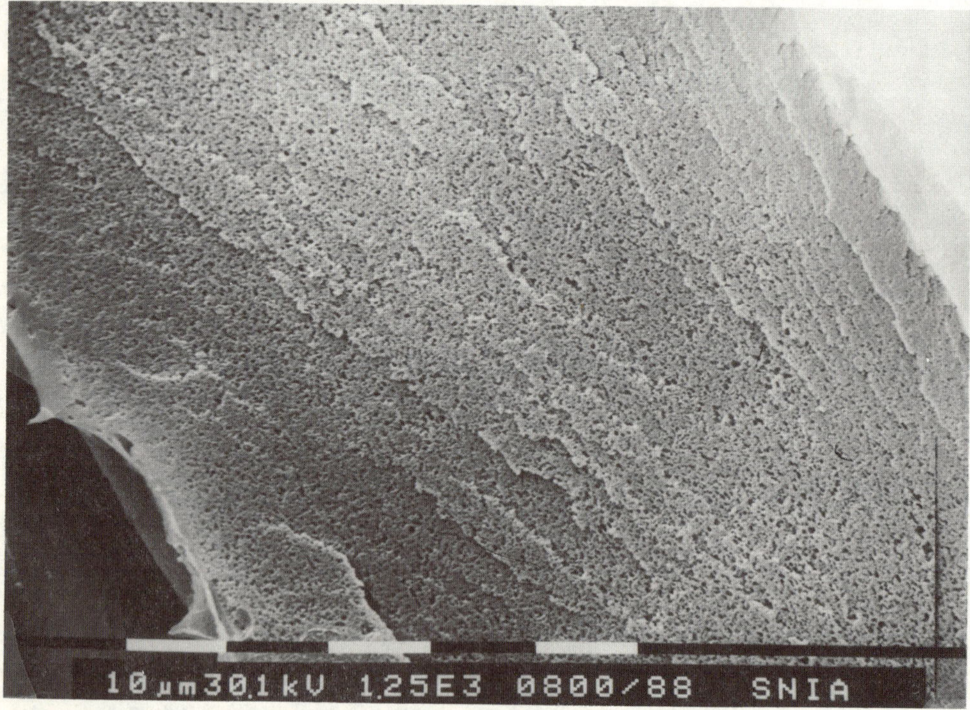

Figure 5a : Cross section of the membrane ; SEM picture

Figure 5b and 5c. Enlargements of the top and botton part of the membrane

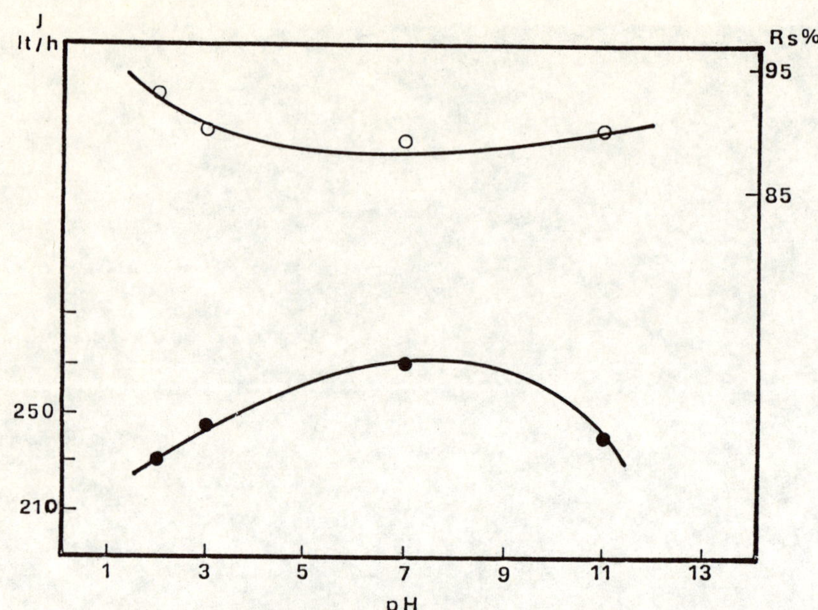

**Variation of flux and rejection of Separem's Membranes with pH.
T= 25 °C, P= 3MPa, C= 2000 pp of NaCl**

Figure 6 : Flux and rejection versus pH

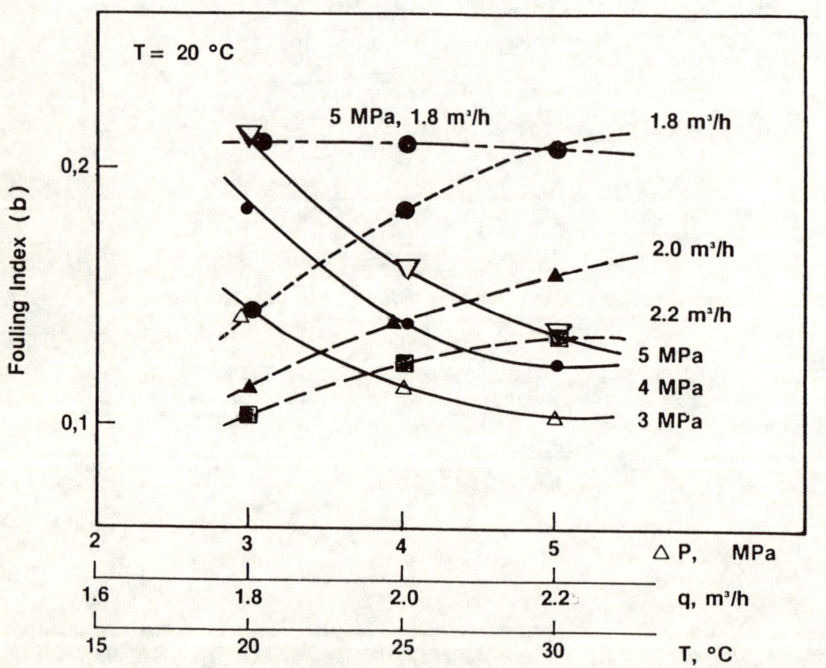

Figure 7 : Fouling index versus pressure (---), axial flow rate (——), and
temperature (——–——–—)

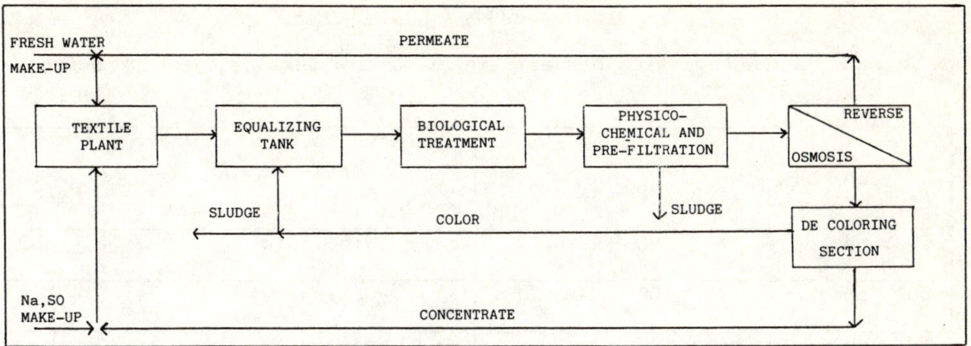

Figure 8 : Schematic of the process to treat an effluent from a textile dyeing plant to recover water and auxiliary chemicals

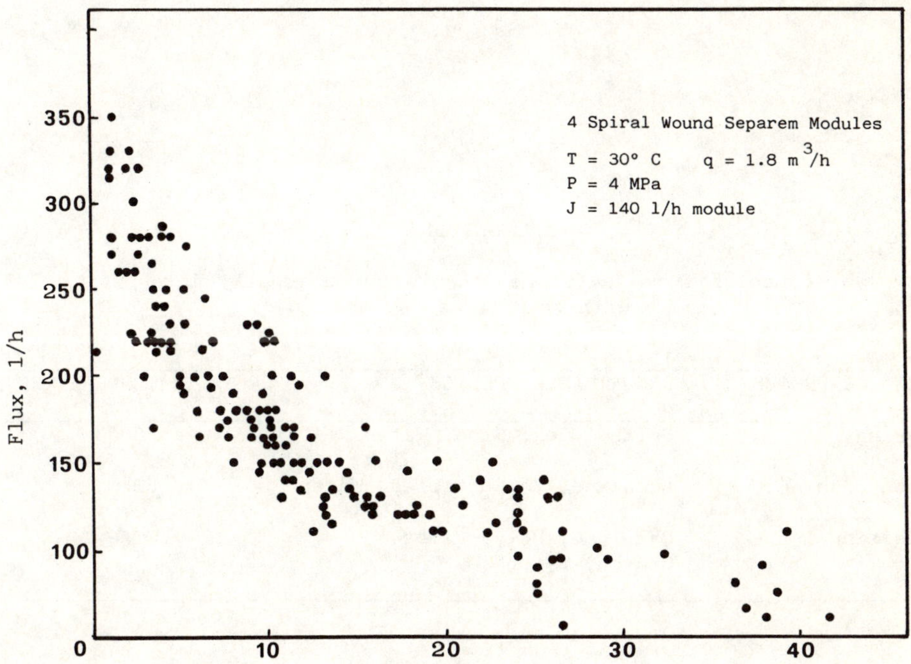

Figure 9 : Flux versus Na_2SO_4 concentration. Na_2SO_4 concentration, g/1

TABLE I

Typical characteristics of waste-water from a textile dyeing process

		Surfactants, ppm	
T, °C	30 – 80	Anionic	0 – 15
COD, ppm	80 – 2000	Non-ionic	2 – 350
BOD, ppm	50 – 1000	Cationic	0 – 15
TSS, ppm	40 – 100	Cr^{+3}, ppm	0 – 3
Color	visible, 1:40	Cr^{+6}, ppm	0 – 2
		NH_3 , ppm	0 – 30
		Phenols, ppm	0 – 12
		pH	2 – 10

TABLE II

Recoverable species from textile effluents by rationally integrating R.O. with traditional systems

TEXTILE PROCESS	RECLAIMABLE SPECIES	WASTE TEMPERATURE
Dyeing	Salts, Dyes, Soda, Water	50°C
Wool washing	Lanoline, Combustibles, Water	50°C
Desizing	PVA, PEG, Polyacrilates Starch, Water	80°C

TABLE III

Average characteristics of the membranes

TYPE   HOMOGENEOUS, ASYMMETRIC AND SUPPORTED

Thickness, microns
     Skin   0.8 ÷ 1.5
     Sponge layer   70 ÷ 80

Transport properties
     Flux*   50 - 60 lt/h m^2
     Rejection*   92 - 95 %
     Compaction**   0.01

Resistance :
     Pressure   up to 6 MPa
     Temperature   up to 40°C
     pH range   1 - 12
     Biological   excellent
     Chlorine   25 ppm, continously
     (at any pH)   higher than 100, shock treatment

(*) T = 25°C, P = 3MPa, C = 2000 ppm NaCl
(**) Defined as m = In J/Jo/In (t-to) at T = 25°C, P = 3 MPa,

TABLE IV

Chlorine resistance of the membranes developed

Concentration		Exposure time *	
pH	4	7	10
Chlorine concentration (ppm)			
2	10 years	1.4 years	4.6 years
100	1800 hr	220 hr	400 hr

* Time during which the membrane maintains 99 % of transport properties

TABLE V

Results of the concentration operations

	Fresh feed	Concentrate	Permeate
Suspended solids, ppm	<1	43	none
COD, ppm O_2	50	2010	none
Surfactants			
. anionies (MBAS)	0.35	10	none
. non ionies (BIAS)	6.8	260	<1.0
pH	7.8	8	7
Color	visible	none*	none
Sulphates, ppm	1000	40.000	600
Hardness, °F	4	177	<1
Flow rate**, m^1/h	3.5	1.5	2

(*) After decoloration   (**) Average value

TABLE VI

Economical evaluation

1. OPERATING COSTS, LIT/M		
Pretreatment		180
Energy		45
Chemicals	Total	225
Reverse osmosis		
Energy		500
Cleaning		50
	Total	550
2. RECOVERY *		
Water		700 (based on municipal cost for treating waste effluents)
Chemicals		600 (based on 95 % recovery of Na_2SO_4)
	Total	1300

(*) Besides, it is also possible to recover surfactants and energy (hot water and power from the high pressure R.O. streams)

APPLICATION OF ELECTRO-OSMOSIS TO RADIOACTIVE WASTE PROCESSING

A.D. TURNER
Materials Development Division,
Harwell Laboratory, UKAEA, England.

Summary

Electro-osmotic dewatering (EOD) is an attractive remotely
controllable process for the final stage of slurry concentration
after gravity settling or filtration prior to immobilization in
cement. It operates by using an electric field instead of pressure
as the driving force across a microporous non-conducting membrane –
partitioning the solid and liquid components by charge separation.
The high solids retention factor (> 99.99%) yields Decontamination
factors > 1000. Dewatering of a wide range of suspensions can be
accomplished at a variety of membranes – including fabrics (eg
cotton) for LLW, and ceramics or glass for MLW. Volume reduction
factors of 10 can be achieved for low conductivity streams
(0.01–1 S/m) at high processing rates (0.2–1.5 m/h) to give products
of 30–40% solids. These are kept fluid by crossflow recirculation at
$\sim$ 1 m/s, so that they can be subsequently immobilized simply by
cement powder addition. Higher conductivity feeds can be treated
after simple washing at a microfilter.

Planar and tubular membrane designs have been considered for compact
full-scale plants (< 20 dm^3 cell size for a 2.6 m^3/d sludge
throughput). Cost-benefit analyses indicate that, in common with
other techniques, large savings on immobilization and disposal costs
(> 87%) can be made by this further dewatering stage after gravity
settling prior to cementing. Estimates for EOD of capital and
running costs amount to 2% and 1.3% respectively of the savings made
over 10 years, with electrical power contributing only 0.15%. These
arise from the compactness of the plant and the energy efficiency of
the process to produce a fluid product suitable for direct
cementation.

1. INTRODUCTION

Traditional methods of solid/liquid separation used in the nuclear
industry include such techniques as gravity settling, centrifugation and
filtration. Although a Rotating Drum Vacuum filter is capable of
producing a product of $\sim$ 30% solids, this is in the form of a solid cake
and is therefore difficult to incorporate in cement during immobilization
without the addition of more water – thus degrading the overall volume
reduction factor and adding to process complexity. Gravity settling, on
the other hand, typically produces a slurry of 1–5% solids content –
somewhat less than crossflow filtration for gelatinous suspensions
($\sim$ 10%). This limit for all filtration processes is because in time a
viscous, partially dewatered fouling layer becomes deposited on the
membrane. As this has a permeability significantly lower than the

membrane itself, the filtration rate falls. Although in crossflow
filtration the thickness of this layer is controlled by the establishment
of a hydrodynamic boundary layer parallel to the membrane surface, the
permeability decreases progressively with increasing feed concentration.
As a result, the filtration rate may become unacceptably low – even though
the feed may still contain > 95% water. The use of Direct Electrical
Membrane Cleaning may increase the extent of dewatering possible by
crossflow dewatering at an acceptable rate, but it is unlikely to be able
to reach 30-40% solids, as at the high pressures normally used and as a
result of the rheological properties of concentrated dispersions, fouling
and blockage become increasingly likely. Concentration to this high
solids content is a desirable target as it leads to significant additional
volume reduction of the product requiring disposal, while leaving
sufficient residual water for cement hydration on immobilization of the
waste by cement powder addition.

Electro-osmotic dewatering (EOD), however, is a process ideally
suited for the thickening of slurries resulting from gravity settling or
crossflow filtration, as it operates by using an electric field instead of
pressure as the driving force across a microporous membrane – partitioning
the solid and liquid components by charge separation.

2. ELECTRO-OSMOSIS

The process operates by the pores of a non-conducting microporous
membrane support becoming coated with the particulate being separated –
thus imparting a small surface charge (characterized by the zeta
potential), which is matched by ions of equal and opposite charge
dispersed into the adjacent solution. This is illustrated in Figures 1-3
for an iron hydroxide colloid produced at pH 1.35 by boiling a 10 mM
$FeCl_3$ solution. Figure 1 shows electron probe X-ray microanalysis [1]
across the thickness of a range of microporous membranes fabricated from
mixed cellulose acetate/nitrate after soaking for 20 minutes. The
colloidal particulates can be seen to have absorbed uniformly throughout
the membrane for pore sizes greater than 0.1 μm. Below this critical
level, the particles were too big to penetrate the structure and occlude
the internal surface. For the successfully coated membranes, as the
absorbed particulates are immobilized, when an electric field is applied
across the membrane between external electrodes, the solution containing
the local excess of counter ions is transported along the pore. This is
shown schematically in Figure 4. By normal solution viscosity effects,
the complete fluid content of the pore is dragged through as well, through
the movement of only a very small electric charge. Figure 2 shows the
flow rate to be a linear function of current density for successfully
coated membranes, while the uncoated very fine pore size medium passed
virtually no permeate. In addition to this effect, the dispersed
particles are transported away from the membrane surface by
electrophoresis as shown in Figure 4. This not only inhibits membrane
fouling, but also significantly reduces the fraction of highly charged
fine particles passing through, thus giving solids retentions > 99.99%,
and therefore a high process DF. This is illustrated in Figure 3, where
for a series of three membranes of different pore size, increasing the
cell voltage reduces the fraction of fine particles passing through.

The electro-osmotic flow as the result current flow is described by
the Smolouchowski equation [1]

$$V = \frac{I \, \varepsilon \, \zeta}{4 \pi \, \eta \, K_o},$$

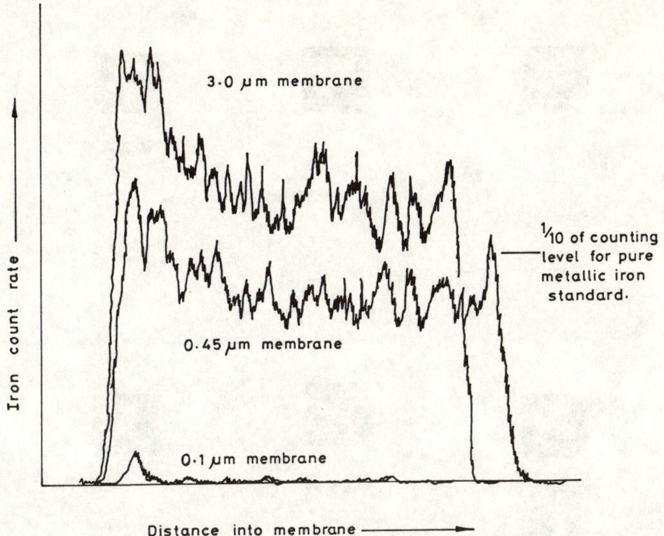

Figure 1   Electron probe X-ray microanalysis of microporous membranes of different pore size after soaking in an iron hydroxide colloid.

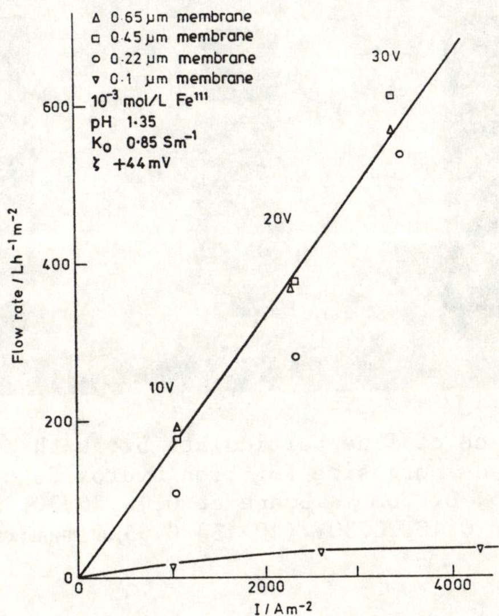

Figure 2   Electro-extraction from iron hydroxide colloid as a function of current density for a range of microporous membranes.

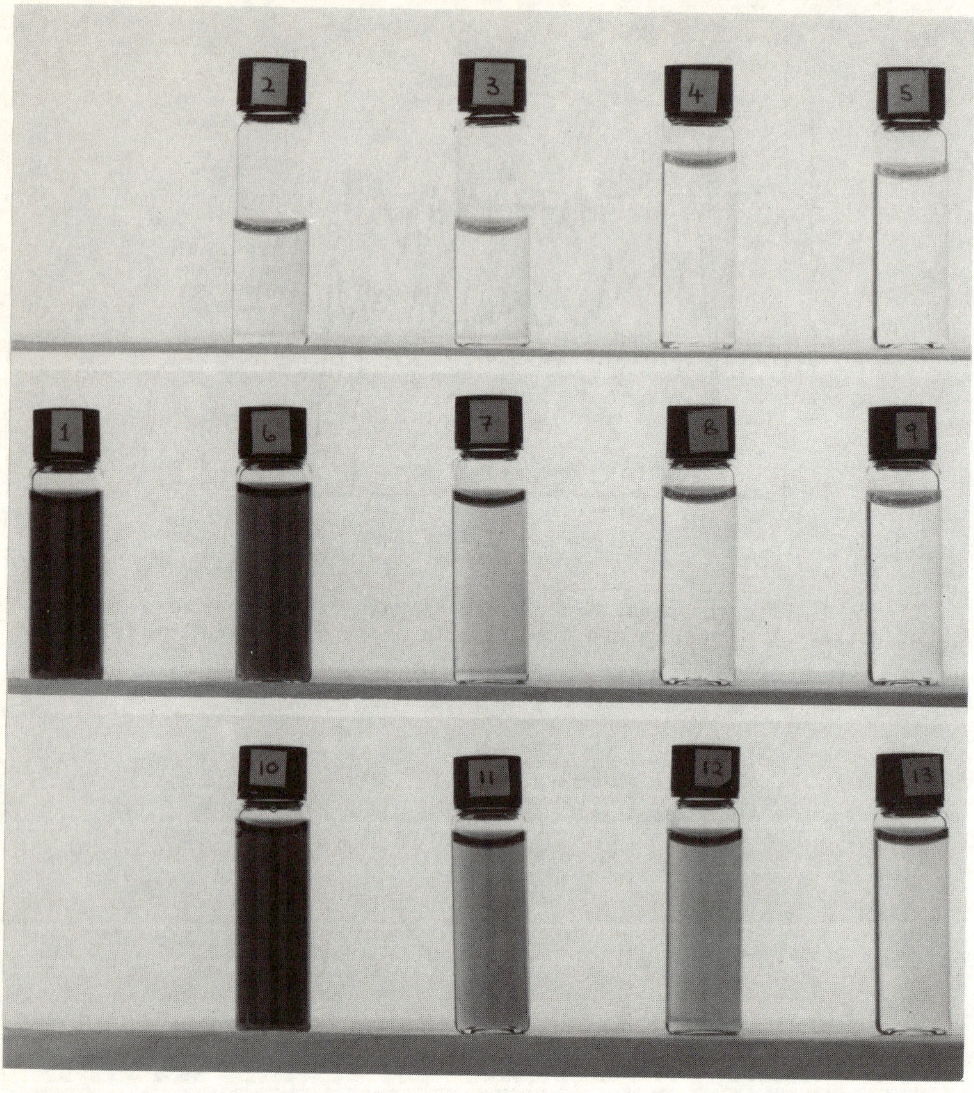

Figure 3 Dependence of fine particulate break-through on applied voltage and membrane pore size for iron hydroxide colloids. (1) sample being treated.(2-5) 0.22μm membrane at 0,10,20,30V respectively;(6-9) 0.45μm membrane at 0,10,20,30V;(10-13) 0.65μm membrane at 0,10,20,30V.

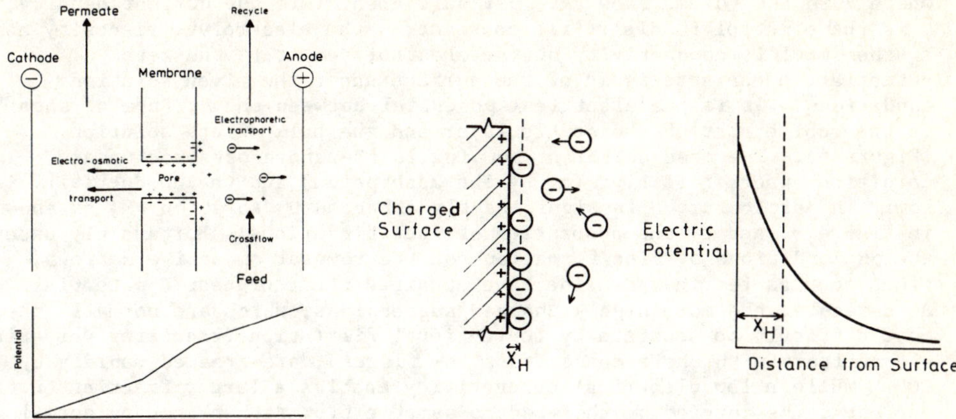

Figure 4 Electro-osmotic dewatering at a microporous membrane

Figure 5 Electric double layer and potential distribution at a charged surface

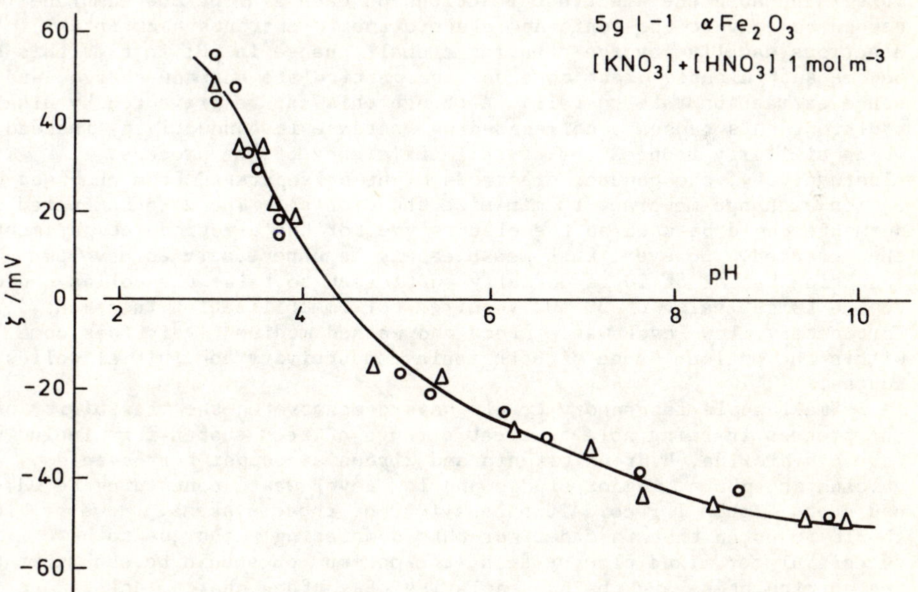

Figure 6 Variation of zeta potential of dispersed α-Fe_2O_3 as a function of pH.

where V is the volume flow rate per unit area, I is the current density, ε is the electrolyte dielectric constant, η the electrolyte viscosity and K_0 the specific conductivity of the electrolyte. ζ is the zeta potential, a characteristic of the surface under the given solution conditions. It is the electrical potential between the surface of shear in the mobile part of the double layer and the bulk of the solution (Figure 5). The most efficient pumping is therefore obtained under conditions where ζ is high (eg by ion adsorption) and the conductivity is low. In particular, ζ is significantly affected by solution pH, as shown in Figure 6, and by the adsorption of specific anions. Fortunately under the pH conditions of floc formation for the removal of active cations, flocs such as ferric hydroxide have attained their highest ζ potential. As a result, the more highly charged suspensions, which are normally the most difficult to separate by conventional filtration (requiring very fine ultrafilters with their concomitant low fluxes), are treated rapidly by EOD. While a low electrical conductivity enables a larger fraction of the current to be carried by the electro-osmotic flow rather than by normal ionic conduction, there is an ultimate limit imposed by the voltage of power supply required. Optimal operation is therefore found in the range 0.01-1 S/m.

Also, as can be seen from the Smolouchowski equation, the dewatering rate is proportional to the impressed current, thus enabling a fine degree of process control to be exercised. In a practical system the product is kept fluid by crossflow recirculation at $\sim$1 m/s. The overall volume reduction factor offered by EOD, though, is normally limited to $\sim$10. This is because the electrode reactions on each side of the membrane needed to convert the ionic and electrokinetic currents back into electrons usually involve creating a small change in pH; in time this may become sufficiently great to cause the particulate surface charge, and hence extraction rate to fall. Although this can be prevented by alkali addition, this causes a corresponding increase in conductivity instead - which similarly reduces the overall efficiency of the process. Alternatively, the counter electrode might be separated from the feed by an ion-exchange membrane to minimize the problem. The decontaminated permeate could be used as the electrolyte for the electrode compartment thus created. However, such measures may be unnecessary as a volume reduction factor of 10 is normally sufficient to raise the solids content to the target value of 30-40% required for immobilization in cement. Fortunately, low level waste flocs and washed medium level flocs come within the optimum range of both ionic conductivity and initial solids content.

Small scale laboratory trials have demonstrated the flexibility of the process in being able to treat a range of feed suspensions including ferric hydroxide, hydrous titania and zirconia, copper ferrocyanide, calcium phosphate, magnox sludge and low level waste containing $Fe(OH)_3$ and soot. Table I records the behaviour of these systems. However, it should be noted that in order for this dewatering technique to be successful for mixed floc systems, the optimum pH should be such that all the particulates have the same polarity of surface charge, unless one component is in a sufficient excess to occlude the other materials. For active species absorbed on the particulates (eg Co, actinides on $Fe(OH)_3$), process DFs were very high - equivalent to virtually quantitative retention.

Table I

Zeta potentials of waste suspensions and flocs

Suspended Solid	pH	K_9 (S/m)	ζ (mV)
$Fe(OH)_3$	1.35	0.85	44
	10.0	0.69	−59
$Fe(OH)_3$ low level waste	9.25	0.2	−55
$Al-Fe(OH)_3$ (washed)	8.0	0.6	16
$Zr(OH)_4$ (washed)	9.5	0.5	−26
$Ti(OH)_4$	11.5	0.08	−47
$Mg(OH)_2$	10.5	0.05	15
$Cu_2Fe(CN)_6$	10	0.09	−8

3. MEMBRANE SELECTION

A range of membrane materials has also been examined [1-3] - both from the point of view of their effectiveness and also their radiation resistance. These included cambric cotton, carbon cloth, cellulose acetate/nitrate, woven polypropylene, glass fibre, zirconia cloth, and microporous stainless steel. Of these, the polypropylene, carbon and stainless steel materials proved to be unsatisfactory electro-osmotic membranes. The relatively coarse porosity of the woven synthetic polymer was less effective than cotton in forming the pre-coat of absorbed particulates at which electro-osmosis subsequently takes place, due to the absence of fine fibres to assist in trapping the floc particles. Electronically conducting materials also proved unsuitable [2]. If they were in electrical contact with the working electrode, then some gassing was able to disrupt the pre-coat responsible for the electro-osmosis, thus reducing its effectiveness. If these membranes were electrically isolated, then degradation was observed to occur by a bipolar cell action - with oxidative corrosion ultimately leading to a breach of the membrane. All the remaining insulating materials performed effectively as EOD membranes. The fact that the nature of the membrane had little effect on dewatering performance confirms that they merely act as a mechanical support for a thin pre-coat of the material being separated - as it is this layer which controls the EOD process. Two caveats to this are that (i) membrane materials whose internal pores are sufficiently fine (< 0.1 µm) to prevent coating in this way will have their electro-osmotic behaviour determined by the filter material itself, and (ii) filters with large pores (> 100 µm) may suffer flow reversal along the pore axis due to the pressure gradient induced by electro-osmosis along the walls - and thus have a reduced effectiveness. The mechanical strength of cellulose acetate membranes was adversely affected after irradiation to 10^4-10^5 Gy, although the cellulose acetate/nitrate mixed ester material survived to 10^5-10^6 Gy. Cambric cotton, on the other hand, performed satisfactorily in excess of 10^6 Gy - especially, in solutions less alkaline than pH 13. As a result, this material should be more than adequate for the treatment of low activity slurries. 3 µm glass fibre and 10 µm zirconia cloths, however, are able to withstand significantly higher doses - thus permitting them to treat MLW suspensions.

4. PROCESS DEMONSTRATION

In order to evaluate this process at a larger scale, a 100 cm^2 planar module has been used - as shown in Figure 7 [2,3]. This has been fabricated from perspex in two halves so that when assembled, it consisted of rectangular feed and permeate compartments separated by the electro-osmotic membrane. The working electrode, which was constructed from flattened stainless steel Expamet mesh, was mounted on the permeate side of the membrane and acted as its support. The counter electrode was a platinized titanium mesh fitted to the rear wall of the feed compartment. The inlet and outlet manifolds were of an optimal design that provided a gradation of the fluid stream dimensions between the pipework and the cell while retaining the same cross-sectional area. The loop in Figure 8 was used to recirculate the feed slurry at crossflow velocities of ~ 1 m/s to provide sufficient shear to maintain the concentrate at a low viscosity while minimizing the hydrostatic transmembrane pressure.

Figure 7   100 cm^2 planar crossflow electro-osmotic dewatering cell.

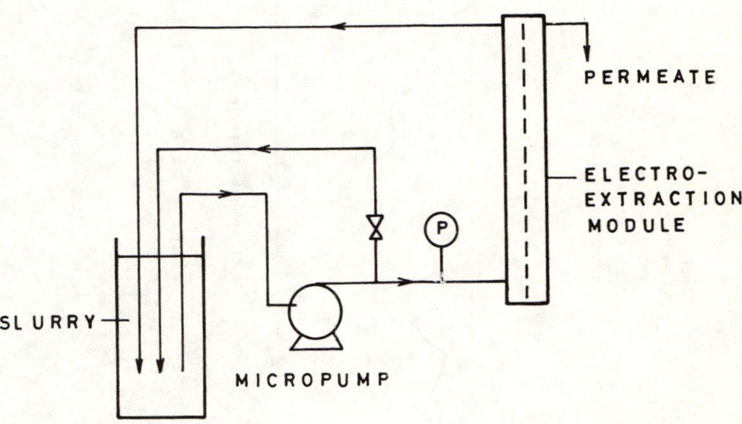

Figure 8   Electro-osmotic dewatering test loop

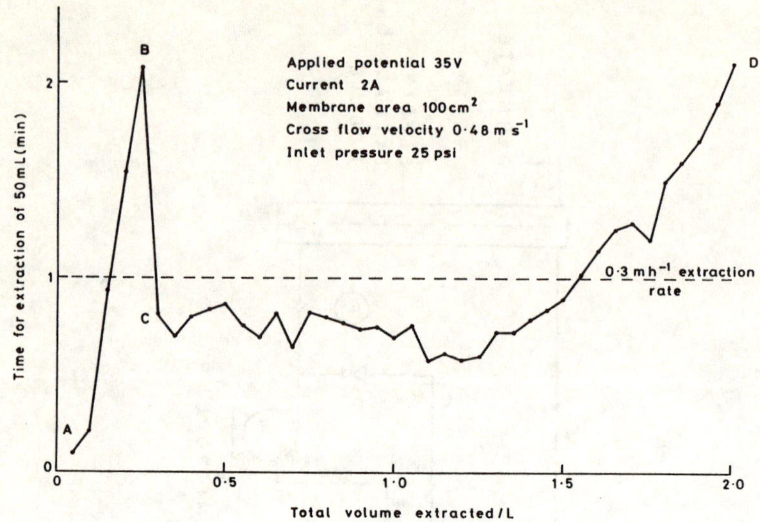

Figure 9   Electro-osmotic dewatering of a 1.6% Fe(OH)$_3$ suspension, initially at pH11 and 0.14 S/m conductivity.

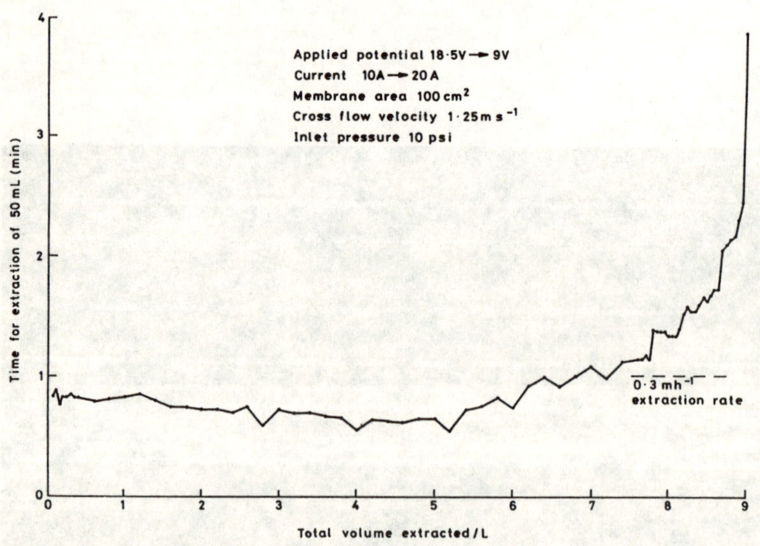

Figure 10   Electro-osmotic dewatering of a 1.2% Fe(OH)$_3$ suspension of 0.39 S/m initial conductivity, maintained at pH10-11 by alkali addition.

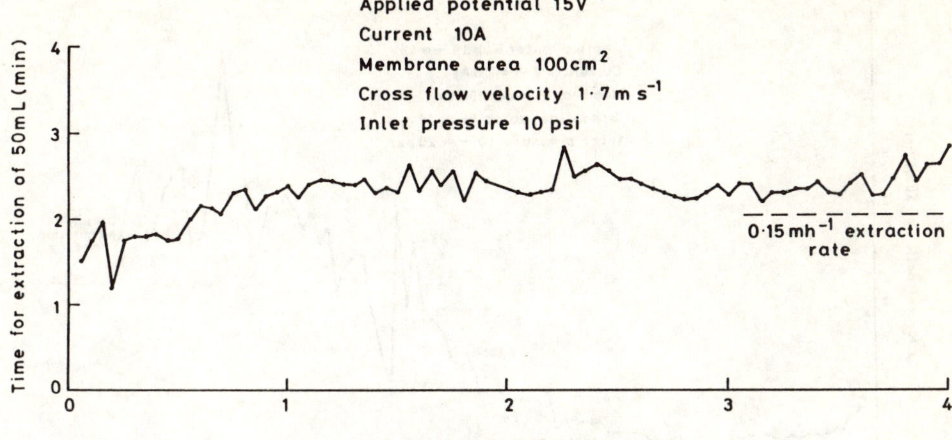

Figure 11 Electro-osmotic dewatering of a 5% solids Harwell low level waste slurry of 0.25 S/m conductivity, adjusted to an initial pH of 10.9. The pH was maintained at 10-11 by alkali addition.

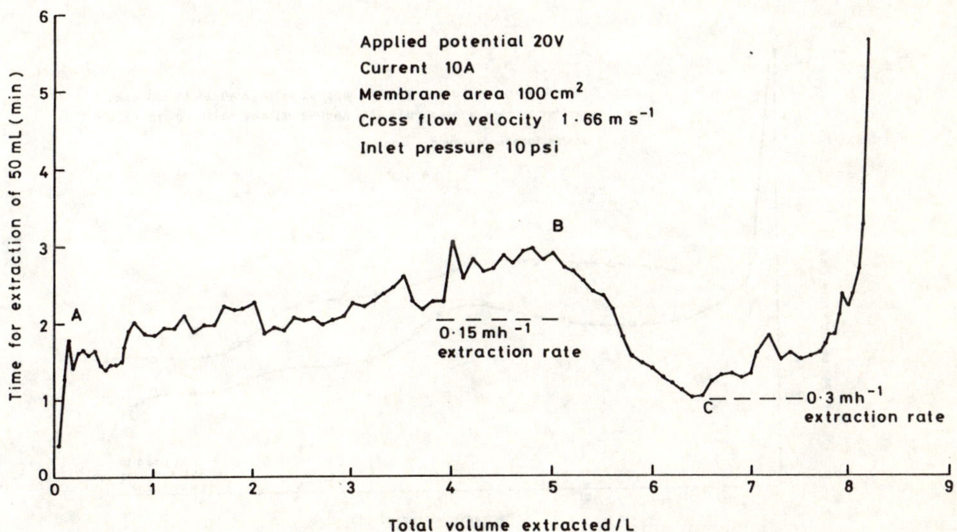

Figure 12 Electro-osmotic dewatering of a 5% solids Harwell low level waste slurry of 0.12 S/m conductivity and adjusted to an initial pH of 10.6. pH maintained at 10-11 from A to B, falling to 7 from B to C, then maintained at pH 7.

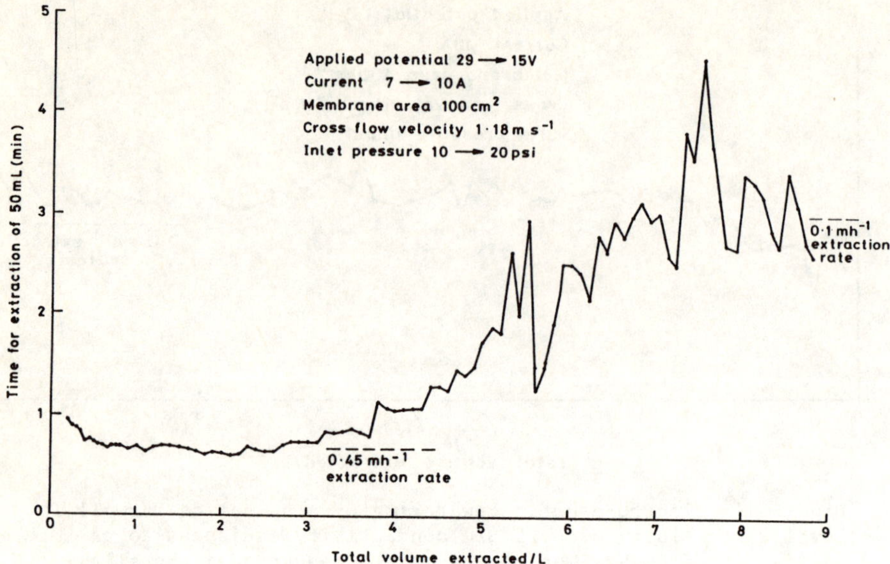

Figure 13 Electro-osmotic dewatering of a 5% solids Harwell low level waste slurry of 0.16 S/m conductivity and initial pH 8.5. pH maintained at 7–8 throughout by alkali addition.

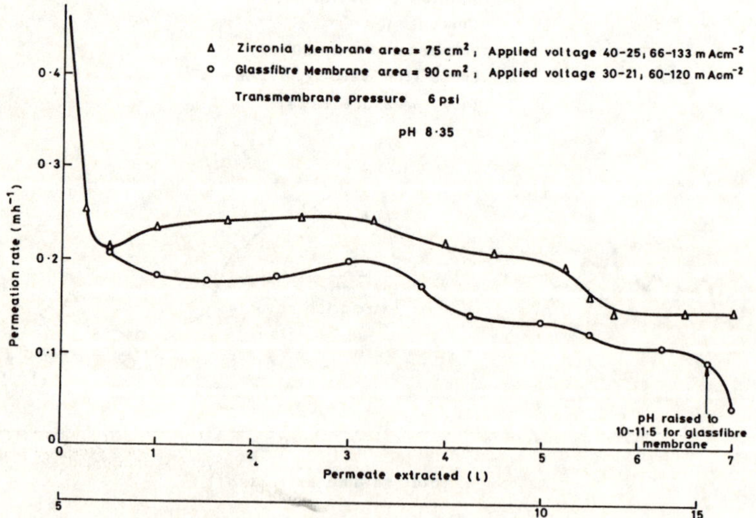

Figure 14 Electro-osmotic dewatering of a 2.6% solids Harwell low level waste slurry of initial 0.12 S/m conductivity and pH8 at zirconia and glass fibre membranes. pH maintained at pH 8.4 by alkali addition.

Figure 15 Electro-osmotic dewatering of Harwell low activity waste slurry.
(1) 5% solids , (2) after concentration to approximately 10% , (3) permeate.

Figures 9 and 10 show the performance of the system with an inactive Fe(OH)$_3$ simulant at a cotton cambric membrane. In Figure 9, initially no current was applied – illustrating the progressive decline in permeation rate due to membrane fouling. However, with the current flowing, not only was the permeation rate significantly increased, but this remained constant up to the end of the run. In Figure 10, a Fe(OH)$_3$ suspension was volume reduced by a factor of 10 to a slurry of 12% solids. From these experiments, batch concentration is recommended as an attractive mode of operation, in that after discharge of the concentrate for immobilization (itself normally a batch process), the next load of more dilute feed to be processed essentially serves the purpose of rinsing out the plant – re-dispersing any deposits and thus preventing progressive blockage.

Figures 11-13 show the subsequent treatment of a genuine ferric floc based active sludge. This also contained substantial quantities of soot from incinerator washings. A slurry arising from gravity settling comprised typically 2.0% Fe(OH)$_3$ together with 3.0% soot and clay. $\beta\gamma$ content varied between 3-10 Bq/mℓ and α 2-6 Bq/mℓ. Figure 11 shows EOD with the feed maintained at pH 10-11 – giving a product of 25%, limited only by the initial feed volume. Figure 12 shows a similar run, but with the pH allowed to fall to 7 part way through the run, at which it was subsequently maintained to give a product of 28%. No α or $\beta\gamma$ activity was detected in the permeate (< 0.01 Bq/mℓ α, < 0.1 Bq/mℓ $\beta\gamma$). Repetition of this run at pH 7-8 throughout (Figure 13) gave a final product of 41.5% solids, with the same quantitative retention of activity observed above.

Figure 14 shows the electro-osmotic dewatering of similar waste streams at a Zircar ZYW30 10 μm porosity yttria stabilized zirconia cloth and a Whatman GFD 2.7 μm glass fibre filter protected between two coarser glass fibre cloths to provide support and protection from abrasion by the feed. Process DFs were > 1600 when the feed pH was maintained at $\sim$ 8. Figure 15 shows the permeate clarity when compared to the initial feed and after concentration to 10% solids.

5. PROCESS SCALE-UP

In order to estimate the size of an EOD unit, reference is made to a large low level liquid effluent treatment plant [2,3] of 100 m^3/h throughput (Figure 16). From the membrane area needed to treat the product from two stage gravity settling, the electrochemical cell would be only < 20 dm^3 in volume. The permeate would be of sufficiently low activity to be discharged, and as the concentrated product is still pumpable, this could be directly immobilized by mixing with cement powder.

An alternative plant design is shown in Figure 17. This uses a combination of a shorter gravity settling stage (eg with an inclined plate separator) with DMC microfiltration, as described in earlier papers. One filter can be used to clarify the supernatant, acting as a guard filter, while the other is used to concentrate the settler underflow (0.5-1% solids) up to 5-10% before EOD to 40%. This strategy minimizes the area of membrane needed in all of these units, thus providing an extremely compact and effective plant. In addition, the power requirement for electro-osmotic dewatering on this scale is seen to be very modest. While for Fe(OH)$_3$ based systems, permeation rates of 0.3 m/h are typical for current densities of 1 kA/m^2, hydrous titania additions (such as might be used for Sr, Co or actinide removal) can give permeation rates as high as 1 m/h for the same current density, thus further reducing the plant size.

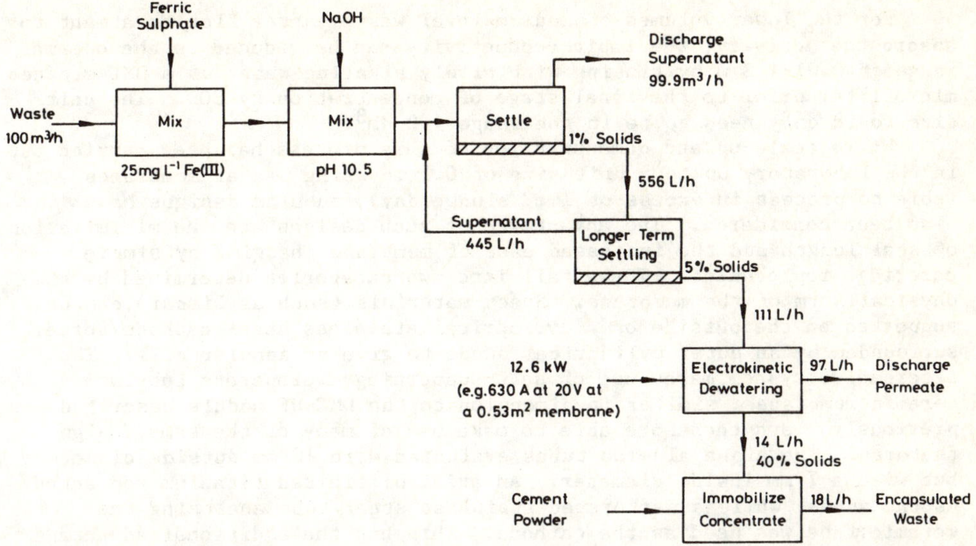

Figure 16 Electro-osmotic dewatering of gravity settled waste treatment flocs.

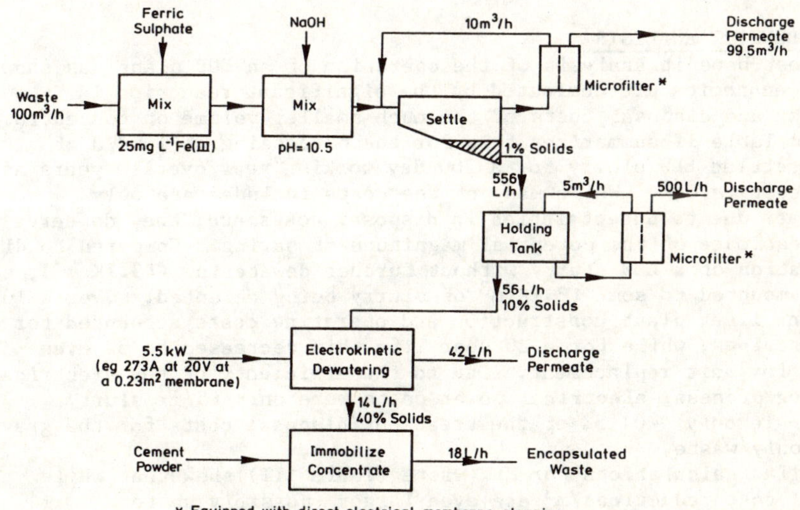

Figure 17 Liquid waste treatment by a combination of micro-filtration and electro-osmotic dewatering.

For the lower volumes of medium level waste, after floc treatment to absorb the activity, the ionic conductivity can be reduced to the optimal range of 0.01-1 S/m by rinsing with mildly alkaline water at a DMC cleaned microfilter prior to the final stage of concentration by EOD. The unit size would only need to be in the range 1-2 dm^3.

While scale-up and demonstration of this process has been carried out in the laboratory up to a unit size of 0.1 m^2 using planar membranes (able to process in excess of 1 m^3 sludge/day), tubular designs have also been considered. The advantages of such designs are the minimization of seal length and the increased ease of membrane changing by simple cartridge replacement. These fall into two categories determined by the physical form of the membrane. Sheet materials (such as Zircar) can be supported on the outside of a cylindrical stainless steel cathode/former, surrounded by an outer cylindrical anode to give an annular cell. The alternative system makes use of self-supporting microporous tubular ceramic membranes, similar in dimension to the DMC-UF module described previously - and therefore able to make use of many of the same design features. The alpha alumina tubes evaluated were 10 mm outside diameter, but with a 7 mm inside diameter. An axial platinized titanium rod acted as the anode, while a perforated stainless steel tube enclosing the ceramic tube was used as the cathode. This has the additional advantage of providing a degree of mechanical support. At current densities of 100 mA/cm^2 and a cell voltage of 20V, a 1% $Fe(OH)_3$ slurry at pH 10 gave permeation rates ranging between 0.23-0.27 m/h for pore sizes 0.2-5 μm, with a minimal transmembrane pressure of 3 psi required for feed recirculation. Full scale units would therefore comprise 20 1.2 m tubes for the treatment of a 5% slurry after gravity settling from a 100 m^3/h plant, or only 9 tubes for the treatment of a 10% slurry after DMC microfiltration.

6. COST-BENEFIT ANALYSIS

A cost-benefit analysis of the operation of an EOD plant has shown that the economics are dominated by the significant reduction in processing and disposal costs of the much smaller volume of concentrated product. Table II summarizes this for the processing of 1 m^3/d of gravity settled LLW slurry for a 200 day working year over 10 years at current cost values. While many of the costs included are only approximate due to uncertainties in disposal costs etc, they do serve to give an estimate of the potential magnitude of saving. Compared to direct encapsulation of a LLW slurry without further dewatering ($£3.7K/m^3$), the savings amounted to some $£3.2K/m^3$ of slurry being cemented. Over a 10 year plant life, plant construction and operating costs accounted for only 3.6% of savings, while for a 20 year life this decreased to 3% even allowing for unit replacement. Due to the efficienty of the electrical dewatering process, electrical power costs were only $£5/m^3$ slurry treated - ie only $\sim$ 0.15% of the treatment/disposal costs for the gravity settled only waste.

Similar calculations for MLW waste (Table III) show that while potential cost reductions/m^3 are even larger (possibly up to $£16.6K/m^3$), overall savings might be somewhat less than for LLW due to the smaller waste volumes being treated. If the volume of MLW is 1/10 that of LLW, over 10 years the operation of the 0.1 m^3/d and 1 m^3/d plants might save £3.3M and £6.5M respectively.

<u>Table II</u>

<u>Cost-benefit analysis for the operation of a 1 m^3/d electro-osmotic
dewatering plant over 10 years at current cost values for
low level waste</u>

	Gravity Settled Waste		EOD Treated Waste	
	Volume (m^3)	Cost $(£k)$	Volume (m^3)	Cost $(£k)$
Gravity settled waste volume	2000		2000	
Waste volume for immobilization	2000		200[+]	
Encapsulation cost ($£2k/m^3$)		4000		400
Disposed volume‡	2600		260	
Disposal cost ($£1k/m^3$)		2600		260
Transport cost ($£325/m^3$)		845		84.5
EOD Plant cost				
Installed equipment				20
Civil engineering				72
Decommissioning*				28
Interest**				30
EOD Plant running cost				
Electricity°				10
Labour				75
TOTAL COST		7445		979.5
COST/m^3 INITIAL WASTE		3.72		0.49

Net Saving: Total £k   6465.5

        £k/m^3 initial waste   3.23

[+] Assuming a volume reduction factor of 10 for EOD operation.
‡ Assuming a volume increase of 30% on encapsulation.
* 30% of plant capital cost.
** 5% of book value over 10 years.
° 0.1 MWh/m^3;   £0.05/kWh.

Table III

Cost-benefit analysis for the operation of a 0.1 m^3/d electro-osmotic
dewatering plant over 10 years at current cost values for
medium level waste

	Gravity Settled Waste		EOD Treated Waste	
	Volume (m^3)	Cost $(£k)$	Volume (m^3)	Cost $(£k)$
Gravity settled waste volume	200		200	
Waste volume for immobilization	200		20[+]	
Encapsulation cost ($£10k/m^3$)		2000		200
Disposed volume‡	260		26	
Disposal cost ($£4.8k/m^3$)		1248		124.8
Transport cost ($£2.8k/m^3$)		748		72.8
EOD Plant cost				
Installed equipment				13
Civil engineering				85
Decommissioning				29
Interest				32
EOD Plant running cost				
Electricity°				1
Labour				75
TOTAL COST		3976		632.6
COST/m^3 INITIAL WASTE		19.9		3.2

Net Saving: Total £k   3343

        £k/m^3 initial waste   16.7

[+] Assuming a volume reduction factor of 10 for EOD operation.
‡ Assuming a volume increase of 30% on encapsulation.
* 30% of plant capital cost.
** 5% of book value over 10 years.
° 0.1 MWh/m^3; £0.05/kWh.

7. CONCLUSIONS

While other solid/liquid separation processes may also be able to provide similar volume reduction factors leading to such savings, EOD does have several distinctive features which give it an advantage over alternative options. In the first place, the continuous crossflow shear maintains the concentrated product (20-40%) in a sufficiently fluid state for subsequent direct immobilization by cement powder addition. Secondly, the plant is very compact, due to the high permeation rates achievable by electrical rather than pressure driven separation (0.2-1.5 m/h). In addition, the presence of the electric field ensures a high process DF even at a microporous membrane. This electrical process is also particularly well suited for automatic remote control, and the energy required is only 1.5-7% of that for evaporation. This, when considered together with the small volumes being treated, is insignificant. It can be seen therefore that electro-osmotic dewatering could have a significant role to play in reducing the cost of active waste disposal by providing a compact and cost-effective volume reduction process for sludges prior to immobilization.

REFERENCES

[1] TURNER, A.D., BOWEN, W.R., BRIDGER, N.J. and HARRISON, K.T. Electrical processes for the treatment of medium-active liquid wastes: a laboratory-scale evaluation. EUR 9522EN (1984).

[2] TURNER, A.D., BOWEN, W.R., BRIDGER, N.J., JUNKISON, A.R. and COX, D.R. Electrical processes for the treatment of medium-active liquid wastes. EUR 10565EN (1986).

[3] TURNER, A.D., BRIDGER, N.J., JUNKISON, A.R. and POTTINGER, J.S. Electrical process for liquid waste treatment. AERE-G3903, DOE/RW/87.126.

ACKNOWLEDGEMENT

This work has been part funded by the Commission of the European Communities in the framework of its second research programme on radioactive waste management. In addition, this work has also been commissioned by the UK Department of the Environment as part of its radioactive waste management research programme. The results will be used in the formulation of Government policy, but at this stage they do not necessarily represent such policy.

APPLICATION OF REVERSE OSMOSIS TO THE TREATMENT OF LIQUID EFFLUENTS PRODUCED BY NUCLEAR POWER PLANTS

Y. HUET[**] - C. MENJEAUD[*] - B. POULAT[**]
FRAMATOME
** Fluid Systems Department Nuclear Engineering Division
* Process Engineering Laboratory of the University of Montpellier

SUMMARY

Radioactive liquid effluents generated during the operation of PWR nuclear power units are currently treated by two independent systems, as a function of their origin (reactor coolant system or other). The effluents from the reactor coolant system are recycled, unlike the others, which, after treatment, are released into the river or ocean that provides cooling water for the unit.
The objective of the treatment of nonrecycled effluents is thus to separate from them as much of the radioactive particles that they contain as possible, so as to be authorized to release into the environment a maximum volume of nonradioactive waste, and to be left with only a minimum volume of concentrated waste, containing most of the initial radioactivity, which must be loaded into casks for storage.
Membrane-based filtration techniques, because they have excellent separation performances, can logically be used for this decontamination of the liquid effluents.
Having developed its own reverse osmosis membrane, Framatome takes the occasion of this symposium to present a possible application in a nuclear power plant, i.e., integration of a reverse osmosis unit into a radioactive liquid effluent treatment system.

1. DESCRIPTION OF A CURRENT NUCLEAR POWER PLANT LIQUID EFFLUENT TREATMENT SYSTEM

The system enables the storage, monitoring, and treatment of liquid effluents that are not recycled for use in the reactor coolant system. At its outlet, consequently, one must only find :
* liquid wastes whose chemistry and radioactivity are compatible with existing regulations, and
* concentrates, containing, in concentrated form, most of the initial radioactivity. After packaging, these concentrates are sent to the appropriate storage sites.

The liquid effluents are mainly collected by the nuclear island vent and drain system. They are sorted into four categories, depending on their chemical quality and radioactivity.
* Equipment drains : borated and aerated effluents of the reactor coolant system wastes type,
* Floor drains : effluents with relatively little radioactivity and low chemically loaded,

* Service drains : effluents with very little radioactivity, and
* Chemical drains : radioactive effluents, chemically loaded.
    The treatment system includes front-end storage, a treatment sub-system, and connections to the systems that handle the final release.
    The front-end storage consists of four groups of identical storage tanks, each being assigned to store a defined category of effluents :
* Equipment drains :   Two 35 m^3 tanks
* Floor drains   :   Two 20 m^3 tanks
* Service drains   :   Two 20 m^3 tanks
* Chemical drains :   Two 20 m^3 tanks.
    Once full, each tank's contents is quickly analyzed. Depending on the results of this analysis, the effluents are either directly sent for release (after chemical neutralization, if necessary) or sent to an evaporator for treatment, if the analysis has revealed a radioactivity greater than 5×10^{-4} Ci/m^3. The role of the evaporator is thus to concentrate the radioactivity, so that the distillate, which in volume is by far the greatest, can be released into the environment. The evaporator has been dimensioned to obtain a decontamination factor of 100. In the existing liquid effluent treatment systems, it is the key part of the system for treating radioactive effluents.

2.   <u>**REVERSE OSMOSIS AND THE TREATMENT OF RADIOACTIVE LIQUID WASTE**</u>
    As described above, the different categories of waste (equipment drains, floor drains, chemical and service drains) collected by the liquid waste treatment system have until now been treated by evaporation. However, for the different reasons listed below, most plant operators now desire to use a different method.
* Experience feedback has revealed a certain number of problems related to the use of an evaporator (generalized corrosion, low availability rate, operating constraints).
* The energy necessary for evaporation makes this process costly.
* The volume-reduction factor is limited, due to the risk of boron crystallization, independently of the radioactivity of the concentrate to be loaded into shipping casks.
    To improve each of these points, Framatome advises adding a reverse osmosis unit. Integrated into the existing liquid waste treatment system, this unit also makes it possible to considerably reduce the annual costs of liquid waste treatment, thanks to a much smaller volume of concentrates to be stored. This advantage is due to the reverse osmosis membrane used, whose selective retention properties enable allowing a large part of the boron to pass through, with very good retention of the radioactive particles in the effluents. Consequently, the evaporator, located downstream, can concentrate the waste to a much higher degree, without risk of boron crystallization.
    Taking advantage of this selective retention characteristic specific to reverse osmosis, Framatome also proposes a new treatment system, which, for the same volume of liquid effluents to be treated, generates three times less concentrated waste requiring storage. The treatment system recommended by Framatome is of the type depicted in Figure 1.
    The overall process design enables the separate treatment technique, which makes the system operation highly reliable and flexible.

3. DESIGN CONSIDERATIONS

The treatment capacity of the reverse osmosis unit, located upstream of the process, must be determined in accordance with the statistics provided by experience feedback, and in agreement with the operating utility's objectives. When the drainage and storage of the different effluents are correct, experience shows that of the volumes effectively treated :
* 90 % of the equipment drains,
* 10 to 30 % of the floor drains, and
* only 2 to 5 % of the chemical or service drains.

On the other hand, it is important to note that the most part of the radioactivity released into the environment is due to nontreated floor drains and that the treated radioactivity represents only 15 % of the released radioactivity.

It is precisely because the floor drains often contain chlorine, however, that the utility avoids to the greatest extent sending these wastes to the evaporator, so they are rarely treated. From the standpoint of reducing the release of radioactive products into the environment, systematic treatment of the floor drains is necessary, and this is now made possible with Framatome's tubular reverse osmosis modules.

4. CHARACTERISTICS OF A REACTOR LIQUID WASTES AND COMPATIBILITY WITH DYNAMIC MEMBRANES

The liquid waste treatment system is by definition the system into which converge all the nonreleasable radioactive liquid wastes. The effluents collected by this system thus have and will always have a highly fluctuating chemistry. For this reason, the methods proposed for their treatment must absolutely tolerate large differences in the chemical composition of the fluids to be treated.

Thus remark shows how difficult it is to define, in the sense of a specification, a representative chemical composition of the future effluents to be treated. Nevertheless, the statistics for different samples taken from the equipment drains and floor drains make it possible to determine the following :
* These effluents have the particularity of being highly loaded with salt, and as concerns the floor drains, they have a high clogging ability.
* Their pH is statistically between 7 and 8.

γ spectrometry of the effluents has made it possible to identify the isotopes responsible for their radioactivity. On the other hand, the average radioactivity of these effluents is, like their volume and their chemical composition, highly fluctuating and greatly variable, depending on the site.

Nevertheless, statistically speaking, the radioactivity of the residual drains can be broken down as follows :

	MINIMUM	MAXIMUM
Co58	67 %	71 %
Co60	1 %	2 %
Sb124	16 %	24 %
Ag110	< 1 %	6 %

along with the sporadic presence of such other isotopes as Mn54, Cs134, Cs137, and Cr51 in the form of trace elements.

Statistically, for the floor drains the radioactivity is distributed as follows :

	MINIMUM	MAXIMUM
Co58	11 %	83 %
Co60	1 %	21 %
Sb124	4 %	13 %
Ag110	1.8 %	6 %
Cs137	–	32 %
Cs134	–	22 %

along with the sporadic presence of such other isotopes as K40, I131, I133, Na24, with relative influence on the overall radioactivity never exceeding 10 %.

5. MEMBRANE CHARACTERISTICS

For such an application, Framatome has developed a reverse osmosis membrane that can be regenerated in situ (see the Annex). With respect to the other membranes available on the market, it presents the following advantages :
1) Being tubular, it is relative immune to clogging.
2) Because it is of the organo-mineral type, it tolerates large acciden-tal variations in the pH (from 2 to 12) without any consequence, and can be used for the treatment of fluids under high pressure (up to 120 bar) and at high temperature (up to 100°C).
3) Dynamic, it can be regenerated in position, which gives this type of membrane a very long useful life.

In the radioactive decontamination application, its isotope-reten-tion properties are as follows :

ISOTOPE	Mn54	Co57	Co58	Co60	Tc99	I131	Cs134
RETENTION FACTOR (%)	63	91	96	96	95	71	41

Boron retention factor at pH7 = 35 %.

6. ECONOMIC ADVANTAGES OF SUCH A PROCESS

The addition of a reverse osmosis unit to an existing liquid effluents treatment system enables making major gains on the annual cost of treating and conditioning radioactive, nonrecycled waste.

For example, the accompanying table (Table 1) presents the savings obtained for the treatment and conditioning of the waste produced annually during the operation of a 900 MWe class PWR nuclear power unit. The costs, expressed in US dollars, refer to the following average French costs :
* Cost of embedding 1 m^3 of concentrate in concrete : $ 5 833
* Cost of embedding 1 m^3 of ion-exchange resins in polymer : $ 35 000
* Cost of transport to the long-term storage site, per m^3 : $ 416
* Cost of long-term storage of waste, per m^3 : $ 500.

7. CONCLUSION

The deposit process, developed in the process engineering laboratory of the University of Montpellier, now makes it possible to industrially produce reverse osmosis membranes. These membranes have the advantages of being able to withstand a high operating temperature and large pH varia-tions, as well as high applied pressure (thanks to the porous stainless-steel support). In addition, their in situ deposition presents numerous advantages from the operational standpoint : the investment is reduced, the operating life is considerably extended, and accidental damage to the mem-brane is without consequence.

PROCESS COST	EVAPORATION	NEW SYSTEM	
Volume to be treated annually	Residual drains Floor drains Chemical effluents	2300 m³ 600 m³ 50 m³	
Volume to be embedded annually	21 m³	7 m³ 0.6 m³	Concentrate Ion-exchanger resins
Volume to be transported annually	126 m³	42 m³ 6 m³	Concentrate Ion-exchanger resins
Annual cost of embedding	$122.00	$40.800 $21.000	Concentrate Ion-exhanger resins
Annual cost of transport	$53.000	$20.000	
Annual cost of storage	$63.00	$24.000	
Total annual cost	$238.000	$105.800	

A N N U A L   S A V I N G S = $ 132.000

T A B L E   1

The reverse osmosis membrane used here can have a multitude of industrial applications. It is already being sold in the USA in the agribusiness field (for fruit juice concentration) and in the nuclear power field. It is being industrially produced in South Africa for use in the textile industry, for retaining the coloring agents in the rinse baths. In Nuclear Power Plants field, Framatome, in association with Metafram, has proven the economic advantages of such a method of decontaminating radioactive liquid effluents, and is now working on the development of a unit capable of treating 3.5 m² per hour.

<u>A N N E X   A</u>
<u>PRINCIPLES OF FORMATION AND CHARACTERISTICS OF A REVERSE OSMOSIS MEMBRANE</u>
<u>THAT CAN BE REGENERATED IN POSITION</u>

<u>WHAT IS THE ADVANTAGE OF IN SITU DEPOSITION OF A REVERSE OSMOSIS MEMBRANE ?</u>

Reverse osmosis is a liquid-phase separation technique by permeation through a selective-permeability membrane, under the action of a pressure gradient. It is generally admitted that separation of the solutes is controlled by the mechanisms of diffusion-solubilization. In other words, the relative affinity of the three constituents, the solvent, the solute, and the membrane, is the most important parameter. The solute can diffuse through the membrane only if it is soluble in the latter.

We can thus immediately state that the idea of a universal reverse osmosis membrane must be abandoned. Hence the idea of a specific deposit in situ on a standard support as the way to adapt the method to the multiple cases to be handled.

Additionally, by using a porous solid support it is possible to deposit several membranes successively on the same tube. We thus hope to simplify the necessary pretreatments during the operation of a unit. In fact, in case of problems, it is always possible to regenerate the membrane.

<u>ADVANTAGES OF A POROUS STAINLESS-STEEL SUPPORT</u>
The porous support made by the Metafram company is composed of sintered type 316L stainless steel. This material resists chemical, thermal, and mechanical attack. In addition, it is chemically inert and suitable for medical or food-related applications.

<u>MEMBRANE FORMATION</u>

<u>PROTOTYPE USED</u> (figure 2)
The prototype used was designed jointly by Framatome and the process engineering laboratory of the USTL, and was manufactured in the Nissan workshops.

It consists of two stainless-steel feed tanks, each of 50 liters capacity, a PMH triplex piston pump, and a pressure damper to attenuate the pump pulsations. The operating pressure is imposed by a needle valve. The fluid circulation system consists of stainless-steel pipes for the high-pressure part and reinforced PVC for the rest.

The physico-chemical parameters, such as the concentrate and permeate flow rates, the pHs, and the concentrate and permeate conductivities, as well as the operating pressure and temperature, are continuously measured and recorded.

All of the fluxes circulate in a closed loop during the formation and characterization of the membranes.

<u>MEMBRANE-FORMATION PROTOCOL</u>
The membrane is deposited inside the porous support, by permeation of the formation solutions, circulating at 10 m/s under a pressure of 70-80 bar and at a temperature of 30-40 °C (figure 3).

The membrane-formation protocol was developed on a module with a surface area of 100 cm², but we are currently working on surfaces with a total membrane area of 2 m². It was obtained on the basis of an operating mode developed in the bibliography by different R & D teams (Johnson, Spencer, Thomas, etc.) :

* Preparation of a solution of zirconium hydroxide at 5×10^{-4} M/l.
* Circulation of this mixture at high speed and an increase of the pressure applied.
* Next, 50 ppm of polyacrylic acid is added to the formation solution.
* The solution is circulated for one hour, then the pH is increased.
* After circulation for one hour at pH = 7, the membrane is considered to be formed.
* The formation solution is then rinsed out with raw water and the membrane is characterized using a saline solution.

Figure 4 shows the evolution of the two characteristics (retention and flow rate) measured during membrane formation. The flow rate mainly drops during the deposit of the first layer, whereas the retention increases especially during the neutralization of the polyacrylic acid.

MEMBRANE CHARACTERISTICS

The characteristics of the membranes made using the small module are included between the following two extremes :
NaCl retention rate = 95 %; Permeate flow rate = 15 l/h per m^2 of membrane
NaCl retention rate = 75 %; Permeate flow rate = 35 l/h per m^2 of membrane.

The membranes have been tested with a 3 g/l solution of NaCl, a pressure of 70 bar, a temperature of 30°C, a tangential circulation speed of 3 m/s, and a pH of between 7 and 8.

REGENERATION OF MEMBRANES AND POROUS SUPPORTS

The main advantage of the method being the possibility of depositing several successive membranes, it is indispensable to be able to remove them easily. For this, we have developed chemical cleaning methods that enable partially or totally regenerating the membrane and the porous support.

PARTIAL REGENERATION

Circulation of a molar soda solution under a pressure of 20 bar causes the removal of the second layer only. A new deposit of PAA then enables regenerating the membrane.

TOTAL REGENERATION

Circulation of a solution of fluoronitric acid enables regaining the original permeability of the porous support. It is then possible to redeposit a complete Zr-PAA membrane.

INFLUENCE OF OPERATIONAL PARAMETERS ON THE MEMBRANE CHARACTERISTICS
(figure 4)
TIME

The flow rate of the permeate decreases and the retention increases significantly during the first two or three hours of operation after the formation. Numerous harsh treatments (cavitation, long waits in the prototype installation or in water samples, etc.) have been inflicted on these membranes and it has been observed after each interruption in their use that an increase in the permeate flow rate and a decrease in the retention are systematically obtained. Nevertheless, it generally suffices to wait about one hour for the operation to return to its pre-interruption characteristics (this is as valid for a one-day interruption as for a two-week one).

This is probably due to the fact that the membrane, less reticulated than conventional membranes, is more sensitive to the phenomena of compacting and swelling.

PRESSURE

Measurements show that there is a proportionality between the applied pressure and the permeate flow rate (slope = 0.5 l/h.m^2/bar). The retention also increases with the applied pressure, but the response is weaker and less reproducible.

TEMPERATURE

The permeate flow rate increases significantly (on the order of 3 % per degree) with the operating temperature, while the retention seems to slightly decrease. This observation is general for all types of membranes. It should be noted, though, that this membrane has been tested statically at 100°C and dynamically at 50°C without any deterioration of any kind. (It was not possible to test it at higher temperatures, because of technical problems having nothing at all to do with the membrane.)

CIRCULATION SPEED

This parameter is adjusted by varying the feed flow rate. The influence of this factor is very small for the small reverse osmosis module ; it is thus not to be taken into account.

PH

Numerous tests have been carried out on different membranes. The response is both highly significant and very reproducible. The permeate flow rate and the retention vary significantly with the pH. The retention is optimum for the pH zone between 7 and 9.5 ; the corresponding flow rate is then a minimum. On either side of this zone, the flow rate increases and the retention drops.

The originality of this membrane lies in the fact that during the pH tests a buffer effect was observed on the water of the acidified system (basified). The membrane reacts strongly to pH changes (Figure 5), but then a neutralization of the solution to be treated as well as a membrane return to its original characteristics are observed. To maintain the solution at low (high) pHs, it is necessary to make several additions of acid (base).

After acidifications (basifications) of very long duration, it may be necessary to neutralize the solution by adding soda (acid). The membrane will then recover its original characteristics.

This membrane thus has an excellent tolerance of large pH variations. (For example, after having being maintained for 450 minutes at a pH near 5.5, approximately one hour after return to neutrality, the membrane recovered its original characteristics.)

TREATMENT OF SEVERAL SPECIFIC COMPOUNDS

CHLORINE (often used in food processiong as a disfinfectant)

A highly concentrated chlorine solution (400 ppm) was treated, with no sign of membrane destruction, unlike the case with the majority of conventional membranes. This compound, furthermore, was particularly well retained.

BORIC ACID (Whose treatment interests the nuclear power sector)

The permeate flow rate and the retention vary with the solution's pH, which makes it possible to adjust this parameter for the desired retention.

COLORING AGENTS (Directly used in the textile industry)

We have systematically obtained 10 % more retention of soluble coloring agents, such as blue acid 161 and red reagent 8, with respect to the retention of NaCl. The permeate flow rate was the same in the two cases.

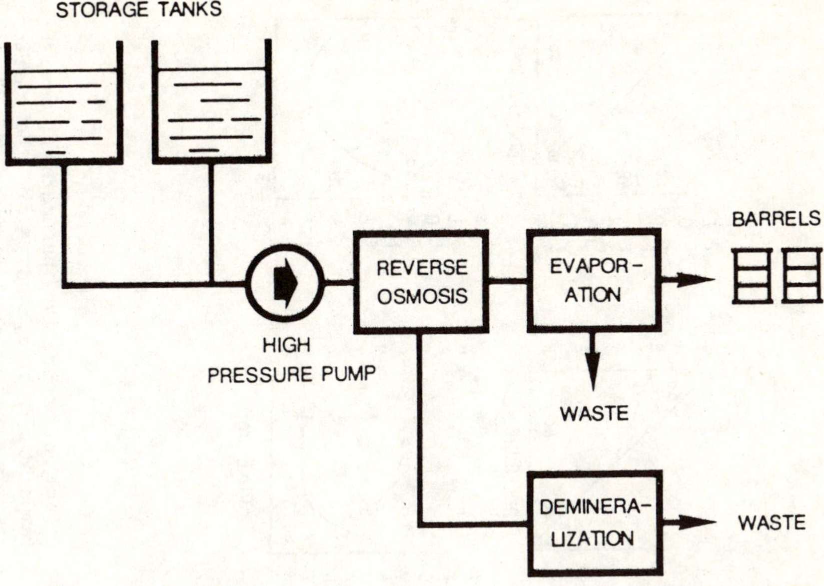

Figure 1 : Reverse osmosis installation

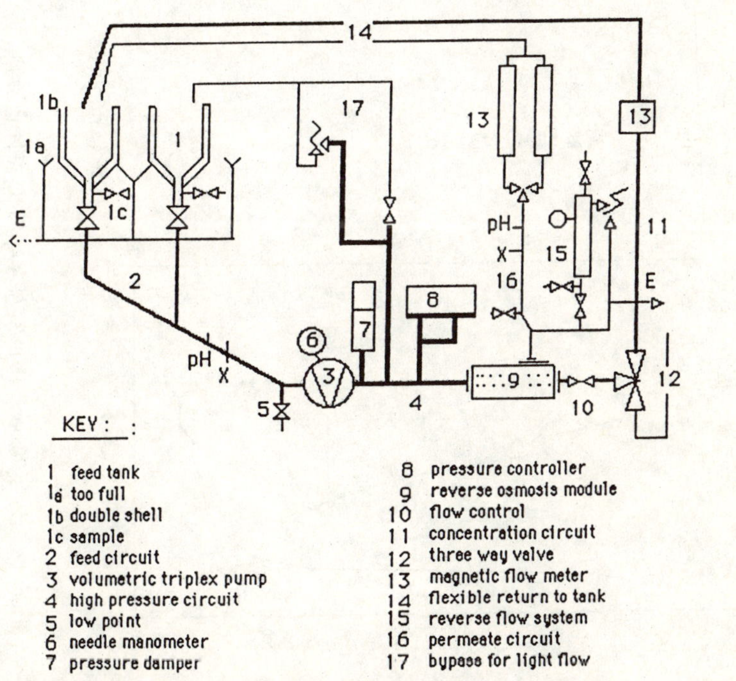

1	feed tank	8	pressure controller
1a	too full	9	reverse osmosis module
1b	double shell	10	flow control
1c	sample	11	concentration circuit
2	feed circuit	12	three way valve
3	volumetric triplex pump	13	magnetic flow meter
4	high pressure circuit	14	flexible return to tank
5	low point	15	reverse flow system
6	needle manometer	16	permeate circuit
7	pressure damper	17	bypass for light flow

Figure 2 : Diagram showing the reverse osmosis prototype allowing the formation and the characterisation of membranes

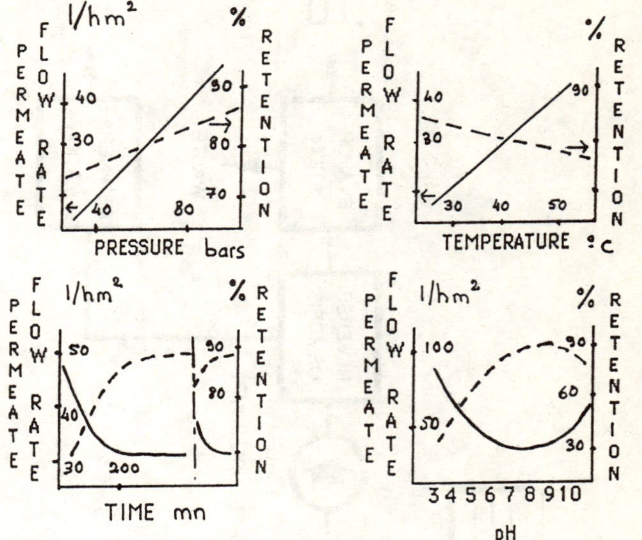

Formation conditions: 70 bar , 10 m/s, APA A, 35 °C, AFM n° 2000

Figure 3 : Curve showing the formation

Figure 4 : Effect of operative parameters (pressure, temperature, time, and pH) on the characteristics (retention, permeate flow) of a reverse osmosis membrane for a solution of 3.5 g/l of NaCl

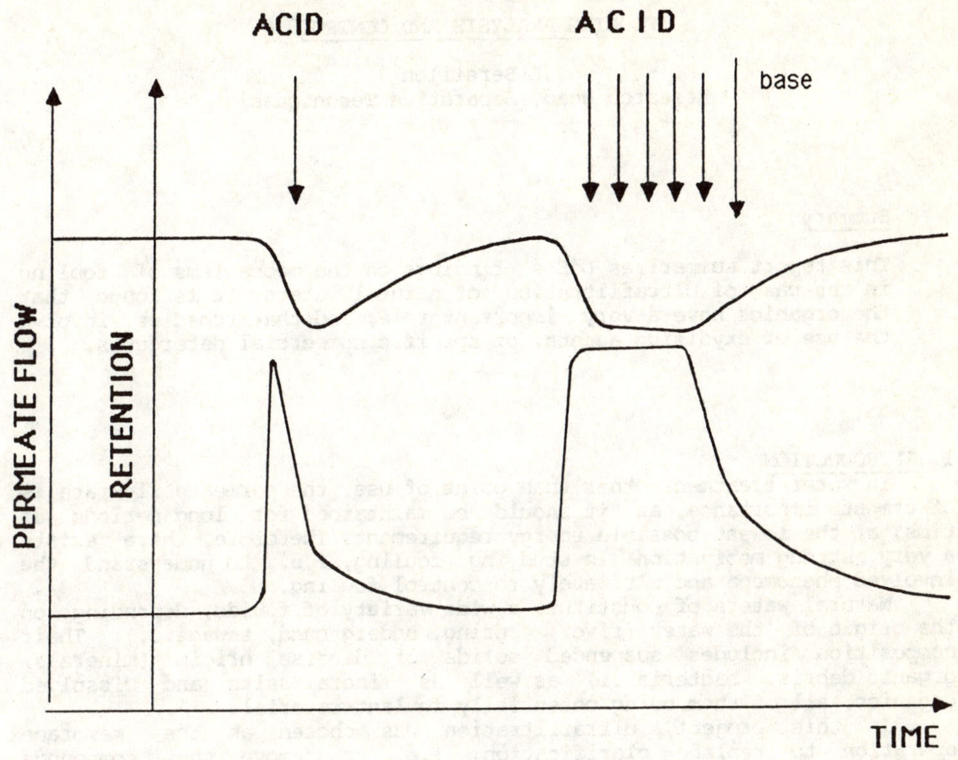

Figure 5 : Effect of pH on retention rate and permeate flow

FOULING ANALYSIS AND CONTROL

J.L Bersillon
Research Head, Separative Techniques

Summary

This report summarizes LdE's findings on the mechanisms of fouling in the case of Ultrafiltration of natural waters. It is found that the organics have a very important role, and that remedies involve the use of oxydizing agents, or specific commercial detergents.

1. INTRODUCTION

In water treatment other than point of use, the permeate flowrate is of utmost importance, as it should be maintained for long periods of time, at the lowest possible energy requirement. Therefore, there exists a very strong motivation in studying fouling, i.e. to understand the involved phenomena and ultimately to control fouling.

Natural waters of constitute a wide variety of fluids, depending on the origin of the water (river, spring, underground, sewage...). Their composition includes suspended solids of diverse origin (minerals, organic debris, bacteria...) as well as mineral salts and dissolved organics, all of them being potentially foulant material.

In this project, ultrafiltration was chosen as the membrane operation to replace clarification, i.e. to remove the compounds responsible for turbidity. These compounds are constituted of suspended material with a particle diameter greater than 0.05 microns ($5\ 10^{-8}$m). Eventhough the suspended solids or the dissolved organics are present in the waters at a relatively low concentration, their influence is crucial, due to the volumes concerned in water treatment. Fouling becomes quickly a major problem. This report summarizes our findings, in terms of identification of the fouling material and the appropriate solutions brought to fouling.

2. FOULING ANALYSIS
2.1 Process analysis

When running an Ultrafiltration module on a water, fouling is responsible for a flux decay as well as in some instances a pressure drop between the inlet part and the concentrate part. The overall permeability decay may be caused by a number of well known phenomena (pore clogging, membrane compaction, curing, formation of a cake), whereas, the pressure drop along the membrane is always due to cake formation.

Therefore, applying hydrodynamic laws, one can calculate a cake thickness (by applying Poiseuille's law to the pressure drop), and a specific permeability of the cake which can be considered as a qualitative index of its texture.

The example illustrated in the figure 1 was colleted during an experiment on the site of Bernay, during a turbidity spike. One can see that the turbidity increase is correlated to an increase of the cake thickness and a drop of the permeability of the cake. Interestingly, the thickness is not affected by the turbidity decrease whereas, the permeability of the cake increases after the turbidity spike is over.

The figure 2 illustrates the events related to the cake thickness

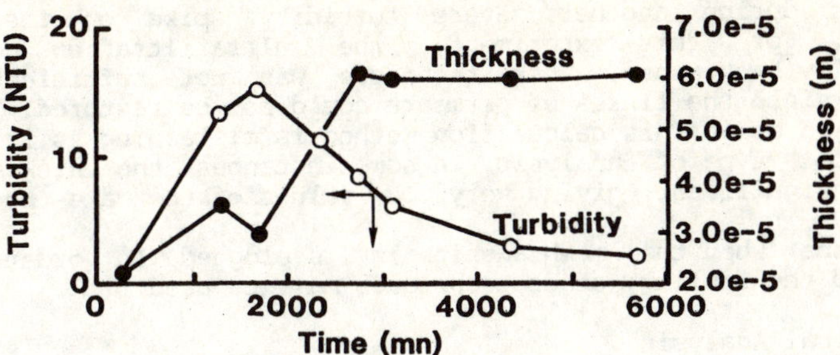

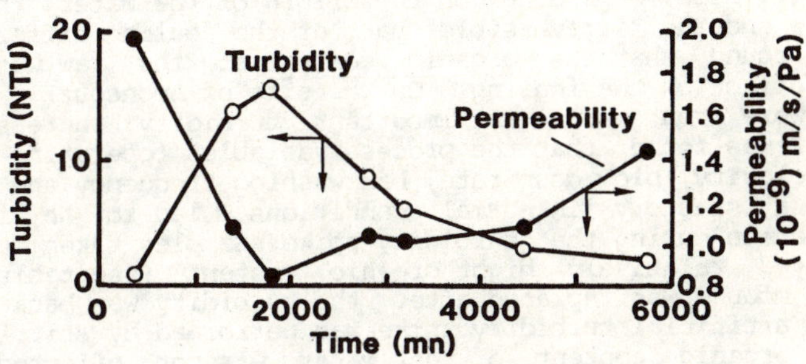

Figure 1. Cake thickness and permeability
during a turbidity spike at Bernay

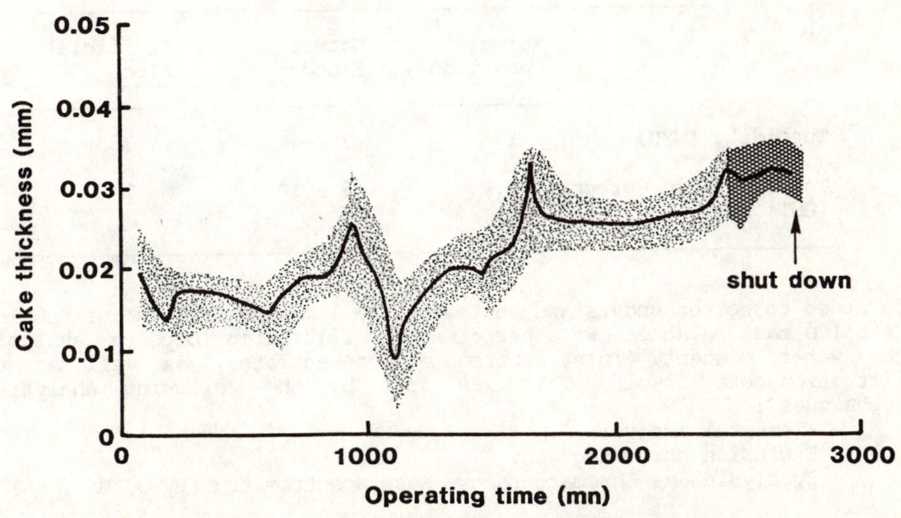

Figure 2. Cake thickness during a turbidity
spike at Amoncourt.   shaded areas represent the
extend of thickness variation due to backwashing

variations during another severe turbidity spike on the site of Amoncourt. For this experiment, the ultrafiltration module was periodically backwashed. This technique was not sufficient in this instance, since the fluxes of permeate could not be restored.

Applied as is this calculation method is misleading as it relies on a "black box" type of analysis. In some instances, the thickness of the cake is insignificant, giving very low values of the cake permeability.

This happens when the membrane itself is clogged. In order to better understand the fouling, other techniques must be used.

2.2 Chemical Analysis

Very early in this project, it was found that, applied on a given water, membranes of different compositions display different behaviour with respect to fouling. Eventhough all of them show a similar equilibrium flux, their ability to be cleaned by simple hydraulic means varies widely. Also, depending on the origin of the water, the fouling rate differs, and the "irreversible" part of the fouling differs as well.

It was found that the organic content of the raw water had a determining effect on the fouling. On the site of Amoncourt, filtration experiments were run while an important turbidity increase occured naturally. It was found that the process variables (operating pressure, tangential velocity, blow down rate, backwashing frequency and duration) which were satisfactory in normal conditions, led to an irreverible fouling situation during the turbidity spike. Samples taken during this event displayed relatively hight organic content (see table 1). The experimental module was replaced after the turbidity was back to normal level and an artificial turbidity spike was performed by stirring up the spring. The organic content of the water was not affected by this experiment, and the module did not show any sign of irreversible fouling.

**TABLE I. Raw water qualities on the site
of Amoncourt (Haute Saône, France)**

	Normal Condition	Natural Flood	Artificial Flood
Turbidity (NYU)	1	250	250
Total Organic Carbon (COT) mg/L	0.9	5 à 10	1

In order to better understand what were the limiting factors of fouling, detailed mass balances were performed on filtration loops by analysing the water content (inlet, permeate, concentrate), as well as the filtration cake itself. This was done by the following analytical techniques :
- Elemental analysis
- Exclusion chromatography
- Pyrolysis-Gas Chromatography- Mass Spectrometry (Py-GC-MS).

The elemental analysis of the deposits shows a predominance of iron, silicium, calcium and aluminium, interpreted as particulate material constituted of calcium carbonate, clays, and iron oxides. This is

TABLE II. Mineral analysis (mg/g of fouling materials)

Site / Element	NOGENT Psu	NOGENT DC	NOGENT Ceraver	BERNAY DC	BERNAY PP	ST CLAUDE DC	ST CLAUDE Ceraver	NEBIAS DC	VERNET DC
K	2,80	2,66	1,89	11,75	–	8,00	5,50	6,51	4,20
Zn	2,40	1,66	2,08	0,85	–	1,00	0,50	9,30	6,44
P	16,80	20,58	21,36	2,82	–	6,00	1,50	10,23	–
Li	0,04	<0,03	<0,01	<0,05	–	<0,05	<0,05	0,19	0,08
Ti	0,12	<0,03	0,09	0,47	–	0,30	0,35	0,09	0,34
Ba	1,98	1,13	0,32	0,42	–	2,12	1,55	2,42	0,73
B	1,00	1,03	1,09	2,02	–	1,55	0,60	1,02	0,53
Ag	0,28	1,09	0,15	1,55	–	40,00	2,80	1,11	0,25
Sr	0,24	0,13	0,28	0,05	–	0,10	0,10	0,09	0,06
La	<0,4	0,33	<0,2	–	–	<0,5	<0,5	0,53	<0,2
Mn	0,16	0,17	0,21	0,99	0,33	0,35	0,55	0,65	0,59
Fe	255,20	164,34	306,18	47,00	92,00	81,00	53,50	40,92	32,56
Mg	2,00	1,99	1,70	5,64	5,63	7,50	9,50	4,65	4,76
Ca	37,20	27,23	43,47	43,71	20,20	53,00	85,00	17,67	17,36
Al	(13,20)	(4,32)	(8,32)	(56,40)	(62,20)	(46,50)	(49,50)	(33,48)	(27,72)
Cu	3,60	1,16	1,45	0,23	–	0,05	0,10	3,44	4,76
Si	87,08	48,57	95,50	171,56	139,00	105,12	148,60	45,57	112,48
Na	2,00	1,99	0,94	4,23	–	3,00	1,50	3,72	2,24

DC   Cellulosic Derivatives
PP   Polypropylene
PAN   Polyacrylonitrile
PSu   Polysulfone

illustrated in the Table II.

- Exclusion Chromatography
   This technique splits the organic matter with respect to its apparent molecular weight (MW). In the tested waters, 3 categories appear
   - MW > 5000 daltons (G1)
   - 5000 > MW > 1500   (G3)
   - 1500 > MW   (G5)

   On most cases, the G1 fraction decreases from the feed stream to the permeate stream. This can be interpreted as either a concentration of the larger molecules in the concentrate or an entrapment of these compounds in the cake. There is no information enabling to discriminate these 2 hypothesis, since the Exclusion Chromatography is not applicable on the cake.

- Py-GC-MS
   This combined technique is applicable on either waters or cake. Examples of the qualitative partition of organics in the raw water, permeate and cake are given in the figure 3.
   This technique makes it possible to identify 4 categories of organic compounds :
   - Polysaccharides (PS)
   - Polyhydroxyaromatics (PHA)
   - Proteins (PROT)
   - Aminosugars (AS)
   On this example, it can be seen that the cake concentrates the polyhydroxyaromatics and the proteins, whereas the other categories are not affected by the filtration operation. This tendency is confirmed on the other sites.

2-3 <u>Electron microscopy and related techniques</u>
   In order to better control the fouling one must know where the fouling occurs. If the flux decay is due to the formation of a filtration cake, hydrolic techniques can be used, combined or not with chemical techniques. If on the other hand, the flux decay is due to pore filling, chemical techniques must be used.
   The observation of fouled membranes was performed using electron microscopy techniques :
   - Scanning Electron Microscopy (SEM)
   - Energy Dispersion System (EDS)
   - Secondary Ion Mass Spectrometry (SIMS)

- Scanning electron microscopy
   Many samples collected from either field tests or laboratory tests were examined using this technique. It makes it possible to assess the thickness and the texture of the filtration cake.
   The comparison if the thicknesses observed by SEM and hydrodynamic calculations is marred by 2 problems :

(i) The approximations made to use hydrodynamic laws.
(ii) The possible changes of structure and thickness due to the sample treatments performed prior to the observation.

   Interestingly, the texture of the cakes observed on samples coming from the field tests are similar to a kaolinite sample coming from laboratory tests, as shown by the figure 4 or the figure 5.
   In the first case, the clays particles are piled in the cake, and randomly arranged. In the second case, the organic compounds seem to

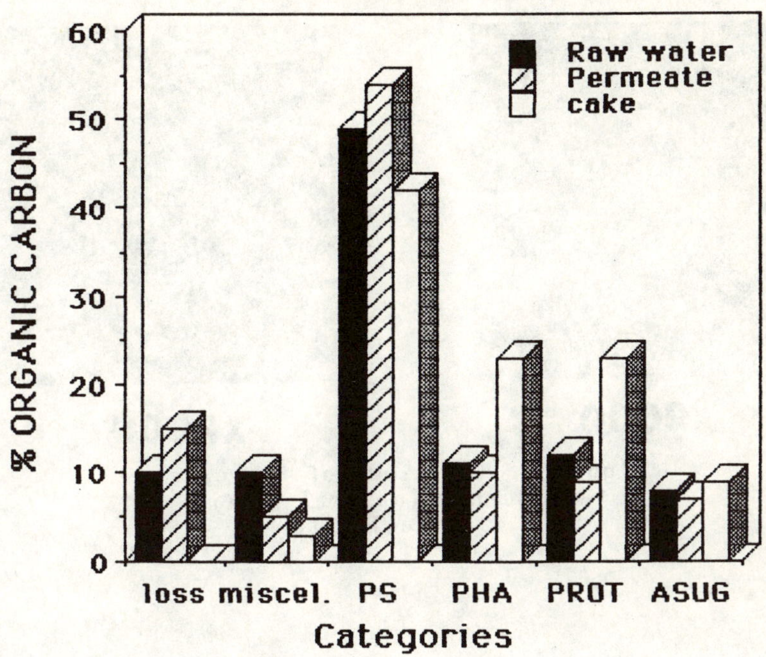

Figure 3. Comparison of organic compositions of raw water, permeate and cakes as analyzed by Py-GC-MS
PS : Polysaccharides
PHA: Polyhydroxyaromatics
PROT:Proteins
ASUG:Amino sugars

Figure 4. MEB views of the cake :
a : Bernay (natural water)
b : Kaolinite

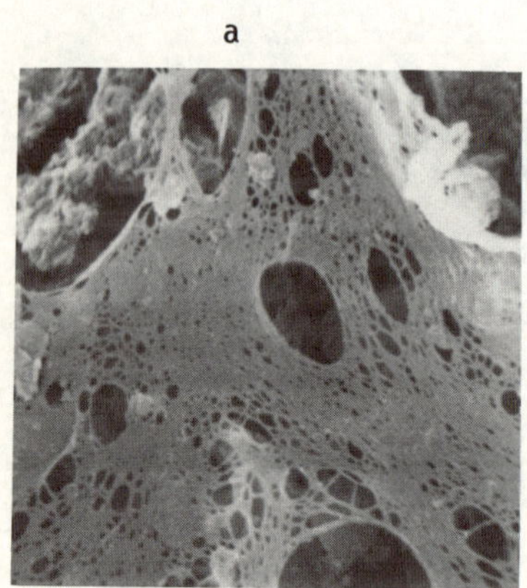
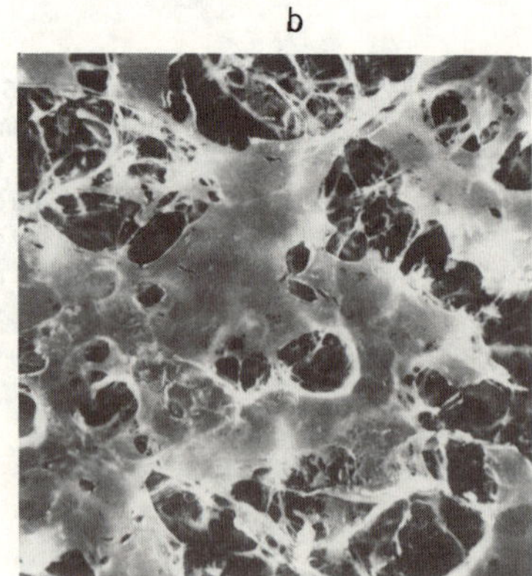

Figure 5. Organic deposits on membranes
a : Bernay = cake-water interface
b : Dextran deposit

concentrate in the cake in the form of a discontinuous net.

- Energy Dispersion System

This technique makes it possible to perform in situ analysis on a volume ranging from 5 to 15 10^{-18} m³. As for elemental analysis, silicium, aluminium and Iron predominate the overall analysis, thereby confirming the presence of clay and iron oxides in the cake. When hydrodynamically cleaning the membranes, it remains a residual cake, which was observed by SEM. The EDS showed that Silicium and Aluminium remain, whereas the Iron seems to be removed.

- Secondary Ion Mass Spectrometry

This technique confirmed the findings of EDS. Additionnally, the organic matter can be visualized by analyzing for nitrogen and phosphorus.

Comparing images of the different elements, it is found that Silicium and Aluminium are associated (clays), Iron is located everywhere in the cake, and the organics are located at the interface between the cake itself and the membrane.

2-4 Forced Fouling Experiments

The hypothesized role of organics in the fouling raised the question of the involved mechanisms. Aqueous organics can be associated to suspended material, in which case they may participate to the cohesion of the cake. Alternatively, dissolved organics can interact with the membrane material if they are not stopped by the cake. In all cases, the adsorption properties of aqueous organics on membrane materials are important and worth investigating.

In order to investigate these properties, the chromatographic technique said "step gradient" was used. This technique makes it possible to assess both the adsorption and the desorption of a solute on the membrane material. The figure 6 shows the basic principle of this technique. It can be seen that an actual membrane can be used, thereby allowing to account at once for the porosity of the material. Once the adsorptive properties were assessed, these with the strongest affinity for the membrane materials were used along with kaolinite to assess the relationship between adsorptivity and fouling potential.

- Adsorption-Desorption experiments

The solutes were chosen so as to best represent the natural aqueous organics, especially, these found in the cakes, or used in the manufacturing/conditionning of the membranes themselves. The list of the adsorbates used in these experiments is given in the table 3.

TABLE III. Adsorbates used in Adsorption-Desorption Experiments

Phenols	Proteins	Sugars	Miscallaneous
Phenol	Tyrosine	Glucose	Glucosamine
Salicylic Acid	Gly-Leu-Tyr	Dextrin (900)	Triton X100
	Lacbumine	Dextran (40000)	Tannic Acid

The tested membranes were made of the following materials :
- Cellulosic Derivatives (BRO23)
- Polysulfones (MO14, RPM 100)

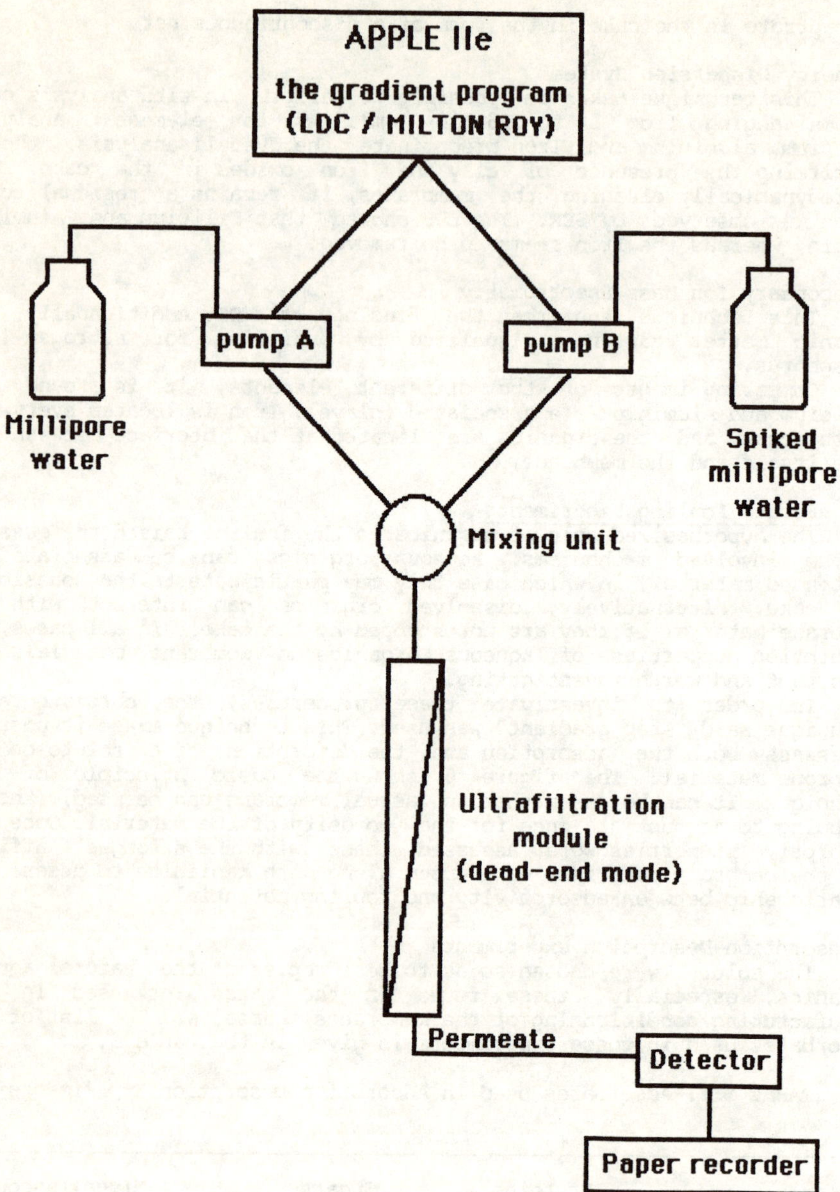

Figure 6. Schematics of the HPLC device used in adsorption/desorption experiments

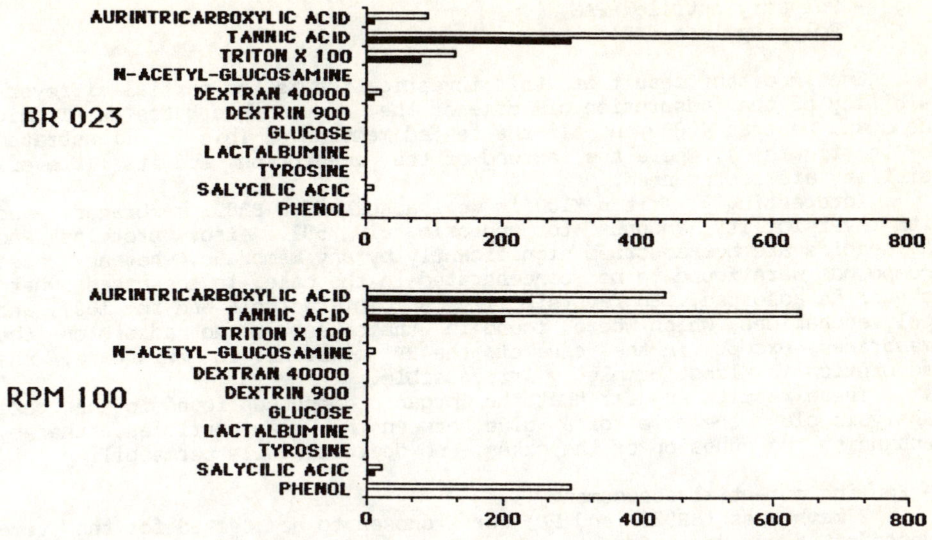

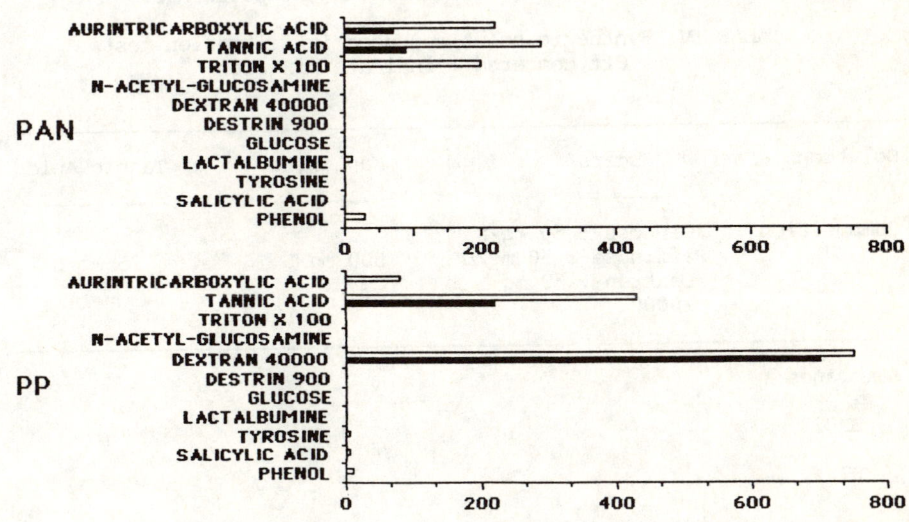

Figure 7. Adsorption/desorption results
on 4 selected membranes (see text)

- Polyacrylontrile (PAN)
- Polypropylene (PP)

One important result of this investigation is the partial irreversibility of the adsorption of some of the chosen absorbates as Tannic Acid and Dextran 40000 on all the tested membranes. This is illustrated by the figure 7, where the extend of the adsorption, and its irreversibility are represented.

Interestingly, Triton X100 is well adsorbed on BR023 membranes, and its reversibility amounts to approximately 50%. Also, proteins and aminoacids are not adsorbed significantly by any membrane, whereas these compounds were found to be concentrated in the cake. in the cases where phenol is adsorbed, its reversibility is complete (mo14 and rpm 100), and polysaccharides, which where found in the cakes, do not adsorb on the membranes, except in the case of the PP membrane. In this case, the adsorption is almost completly irreversible.

These results suggest that the organic compounds found in the cake analysis play the role of a glue between the clay particles, thereby enhancing the cohesion of the cakes, and decreasing its permeability.

- Fouling potential assessment
2 membranes (BR023 and PP) were chosen to be tested for the time dependent flux decay due to the presence of sugars, tannic acid and kaolinite in the raw water. The sugar mixture was constituted of glucose, raffinose and dextran 10000. All tests were performed without and with kaolinite.

The table 4 summarizes the tests which were performed.

TABLE IV. Synthetic solutions used in filtration tests.
Off centered + indicate mixtures

Solutions	Sugars	Kaolinite	Tannic Acid
Composition :	Glucose : 50 mg/L Raffinose : 50 mg/L Dextran : 50 mg/L 10000	500 mg/L	25 mg/L
Membranes :			
BR023	+	+	+
	+		+
PP	+	+	−
	+		−

Back washing experiments were performed only on BR023.

The general behaviour of the selected membranes with the chosen foulant can be summarized as follows :
- The presence of sugars or tannic acid in the water involves an increase of the flux decay rate. The fouling is irreversible, i.e. the permeate flux can not be restored by backflushing with the pemeate.
- The presence of kaolinite alone involves to a lesser extent a flux decay. The flux can be partially restored by backwashing. The SEM

examination of the membranes after these experiments showed a discontinuous cake.
- The simultaneous presence of kaolinite and organics displays intermediate results. The flux decay rate is less acute and the backwashing is more efficient than in the case of the presence of organics alone.

3- FOULING CONTROL

So far, it has been referred to backflushing in several instances. This very well known technique is the remede to "physical" fouling, i.e., when a filtration cake is formed during the filtration operation. It is then a matter of optimizing its frequency, duration and pressure/flowrate, according to the "intensity" and/or to the "rate" of fouling. In the case of drinking water filtration, this technique displays a relative efficiency but leaves a fouling residue which is responsible for long term flux decay. To tackle this problem, 2 categories of means are available :
- Precondition the membrane so that it is less sensitive to fouling. In mechanistic terms, the membrane must be less "adhesive" with respect to the foulants ;
- Stop the filtration operation and chemically cleans the membrane to rid it of entrapped foulants.

3-1 Membrane preconditionning

The idea behind preconditionning a membrane is to expose this membrane to a compound which will ease the permeate transfer and/or the backwashing. The specification of such a compound are the following :
   (i)   Chemically stable
   (ii)   Strong irreversible adsorption
   (iii) Non fouling (does not hinder the permeate transfer)
   (iv)   Hinders the adsorption of other compounds

Among the different compounds tested in the adsorption experiments, the Triton x100 is the closest to items (i) to (iii) of the former specifications. It was chosen as candidate to pretreat BR023 membranes, and to assess item (iv).

It was found that pretreatment with this compound at low concentration (5 mg/L) hindered further irreversible adsorption of Triton. Also, when exposed to tannic acid or dextran 40000 solutions, the pretreated membranes adsorb less than untreated membranes.

In the case of tannic acid, filtration tests show that untreated membranes display a strong flux decay (50% decay in 300 mn) whereas treated membranes have much better performances. This is illustrated by the figure 8. Dextran 10000, which hinder the permeation to a lesser extend, has also its fouling potential dereased by pretreatment of the membrane with triton X100.

This technique was used on the Seine river water, and showed no substantial effect. This still remains unexplained and will be repeated in the future for further investigation.

3-2 Chemical washing

Chemical washing involving the shut down of the filtration was investigated on modules either fouled during field tests or "artifically" fouled, using a tannic acid solution.

A screening procedure was performed using the washing conditions prescribed by the manufacturers.

3 classes of washing products were investigated :
- Oxydizing agents :   Hydrogen peroxide
                        Chlorine

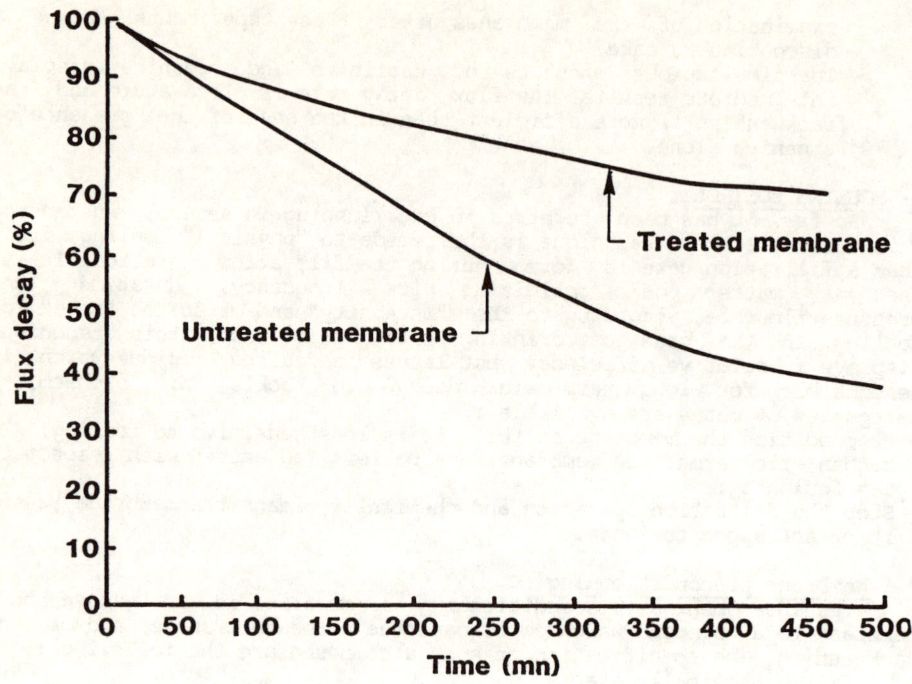

Figure 8. Flux decay as a function of
time for a tannic acid solution BRO23

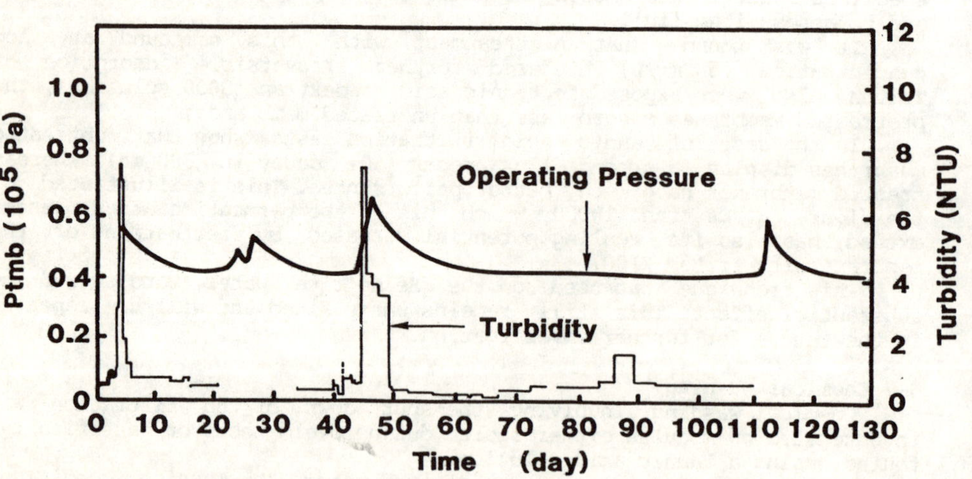

Figure 9. Constant flowrate, long term experiment at Douchy
Operating pressure (Ptmb) and Raw water turbidity
as a function of time

```
- Enzymes :                 Protease
                            Alpha Amylase
                            Hemicellulase
- Commercial Detergents : Ultrasils (Enkel)
                            Floclean (Pfizer)
                            MC's (Zenon)
```

Their efficacy was assessed by their ability to restore the permeate flux.

Among these products, only the oxydizing agents and the Ultrasil are efficient in restoring the permeate flux, for both conditions of fouling.

These 2 cleaning agents are currently being tested in order to optimize their conditions of use.

3-3 Operationnal compromise

In field tests, it was found that when backwashing with the permeate added with a mildoxidizing agent, the permeate flowrate could be maintained at reasonable levels over long periods of time. Furthermore, this technique makes it possible to wash a fouled membrane due to pollution spikes. This process is slow (several days) but does not require to stop the production. This is done without significant deterioration of the permeate quality.

This is illustrated by the figure 9, where a large scale module (50 m²) is in constant operation since June 1988. This module works under constant permeate flowrate (100 l m⁻²), and is periodically backwashed with chlorinated permeate (5 mg/L as chlorine).

4. CONCLUSIONS

- Membrane fouling in water treatment is a complicated problem which needs extended analytical studies to be solved ;
- The organic substances contained in the water play a determinant role in the cohesion of the filtration cake and possibly in the deterioration of membrane "long" term performances ;
- Laboratory scale "enhanced" fouling with model foulants showed the possible involvment of adsorption-desorption phenomena in the fouling. Pretreatement with selected substances showed a limited success in preventing flux decay with time ;
- Large scale pilot studies made it possible to design a hydraulic and chemical on-line washing method which maintain the filtration flux over several months without requiring off line chemical cleaning ;
- Chemical cleaning is still studied on membranes fouled in field tests as well as with model foulants.

INDUSTRIAL APPLICATIONS OF MICROFILTRATION

A. RENARD
TECH-SEP
Rue Penberton - St Maurice de Beynost
BP 347 . 01703 MIRIBEL CEDEX - FRANCE -

Summary

Tangential microfiltration is a technique related to other methods using membranes (Reverse osmosis, ultrafiltration etc...). This relatively new method tends to replace centrifugation and conventional filtration in the bio and pharmaceutical industries because of some inherent advantages : better yield, purification capability, safety and lower costs of operation.
After an overview of the basic principles governing the manufacture and the uses of membranes, a description of an industrial unit built by TECH-SEP in replacement of centrifugation will be given. However, microfiltration membranes and modules as well as the operational principles need to be improved to extend the uses of this technique. This is the subject of an important international research activity leaded by TECH-SEP within the frame of BRITE. The main features of this program are described in view to enlight the future trends in microfiltration.

1. INTRODUCTION

Tangential microfiltration is a method related to other techniques using membranes (reverse osmosis, ultrafiltration). Such a technique can be a good answer to problems encountered at the limits of classical equipments for liquid/solid separation. This is the case when the particles to be removed are micronic or sub-micronic and dispersed in a large volume of liquid phase. Centrifugation then needs a sophisticated and expensive equipment.

Conventional (dead-end) microfiltration suffers from a drastic reduction of flux (fouling). This can be overcome by a frequent and expansive change of the filtering medium (cartridges).

In tangential microfiltration, a shear stress is applied in the neighbourhood of the filtering medium in order to restrict or control the accumulation of particles on the filter. This is obtained by a flow of liquid along the membrane or by rotating the membrane in a still liquid.

Tangential microfiltration is a technique of phase separation applied to small particles where the Brownian movement becomes significant and the surface forces become predominant on the volume forces.

The improvement of performances by a tangential circulation of the fluids is done by an expense of energy wich has to be considered when one design a system.

This point is very important in the design of modules. However, the criteria used for reverse osmosis or ultrafiltration must be revised to answer the specificity of microfiltration :

- Mechanisms of fouling are different
- Fluxes of filtrate are higher

After an overview of the membranes and the operational parameter governing their use, an example of industrial application will be described. This example will show how to replace centrifuges by membranes in bioindustry.

Other examples, experienced by TECH-SEP can be found in the pharmaceutical and food industry (bacterial broths, fruit juices, dairy products).

Although tangential microfiltration is already operational in various industries, this technique can be improved in many ways :
- better knowledge of hydrodynamic and reduction of hold-up volume for a better control of fouling and higher fluxes of filtrate.
- better thermal resistance of membranes and modules to give very compact systems enduring sterilization by steam.

TECH-SEP has, since many years, contributed to the development of tangential microfiltration as an industrial technique. This development is to be accelerated by an international cooperation within the frame of the BRITE program. The main objectives and means of this action will be described in this paper.

2. MEMBRANES

The microfiltration industry developed rapidly in the 40's with the need for the bacterial analysis of water. This industry grew up independantly from the other membrane technologies and the convergence appeared only some years ago with close manufacturing routes and materials (e.g fluorinated polymers for UF and MF membranes).

Depending upon the technology used for manufacturing, MF membranes can be classified in four families :

2.1. Membranes from polymers

- Cellulose derivatives
- Acrylic copolymers, polypropylene
- Fluorinated polymers

These membranes are usually manufactured by the process know as "Phase inversion" and are under the form of thin films (about 100 micrometers) supported on an wowen or non wowen fabric.

2.2. Membranes from thermoplastics

These membranes are manufactured from polymeric powders by sintering or binding with porogenic agents. The main materials are PVC, polystyrene, polypropylene, polyamides. The membranes are manufactured under the form of plates and tubes and do no request to be supported.

2.3. Sieve membranes

Membranes of polycarbonate with a thickness of about 10 micrometers. They are manufactured by the process know as "track etching":dissolution of the zones sensitized by nuclear particles. These membranes have a low mechanical strength and their use is restricted to analytical applications.

2.4. Sintered membranes

Made by sintering of mineral powders on a carbon support (TECHSEP) or sintering of a powder to give a self supporting membrane (Alumina monolith).

The membranes from TECH-SEP, in sintered ceramic, are strongly asymmetric and have a high thermal, chemical and mechanical resistance.

The choice of the membranes is a function of many parameters (1), the first being the size of pores. But also very important are the adsorption properties which are related to fouling and flux. Thermal and chemical stability is also of primary importance in the design of a system.

3. OPERATIONAL PARAMETERS

The distinction between ultrafiltration and cross-flow microfiltration is arbitrary.
However, microfiltration has to be considered as a phase separation technique in opposition to reverse osmosis and ultrafiltration where soluble species (molecules and ions) are separated from a unique phase.

The filtration of particles using microfiltration membranes show the well known behavior of ultrafiltration.

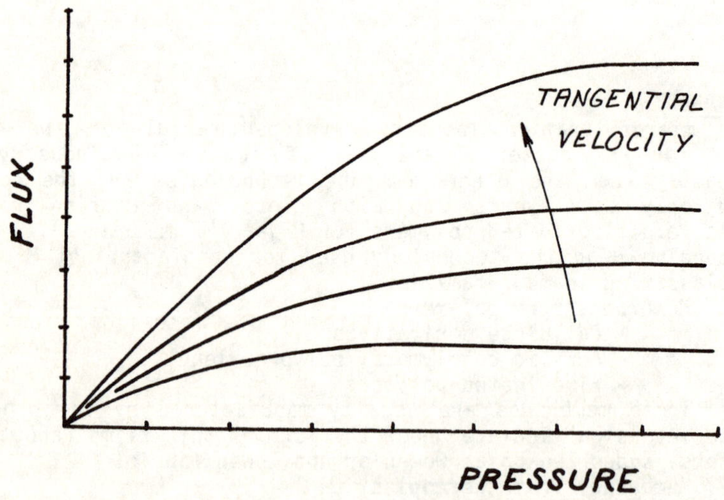

Figure 1 : The classical behavior of microfiltration membranes in clarification of fermentation broths.

The theory of ultrafiltration fails to predict the mass transfert coefficient as shown by the following figure.

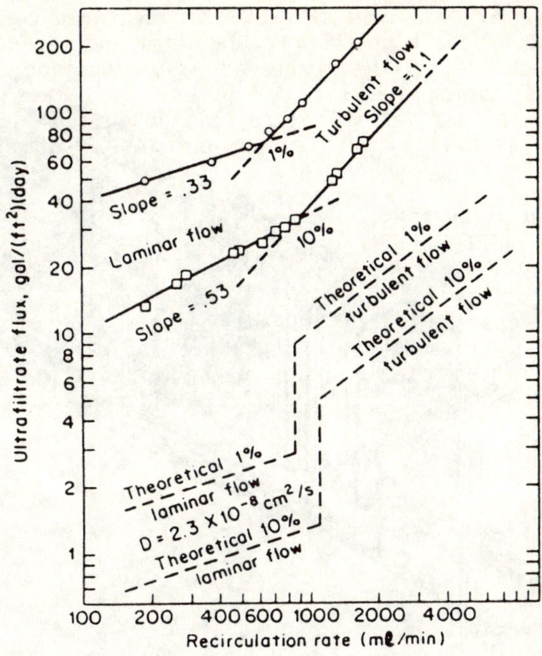

Figure 2 : Comparison of theoretical and experimental performance
curves for the filtration of styrene-butadiene latex (2).

Numerous attempts for a better fitting of cross-flow microfiltration
data are accessible in the litterature (3,4). However, there is still
a lack of a general theory because of the difficulty to obtain the
mass transfer values from experiments. In practice, the engineer is
faced to very complex mixtures and must act on the basis of empiric
experiments. This is especially true for the problems associated with
washing and regeneration of membranes.
An important effort has to be done for a better understanding of membrane fouling and regeneration. This cannot be dissociated from a proper adaptation of modules holding the membranes and defining the hydrodynamic conditions. With his various modules (cassettes, plate and
frame, tubular) and membranes, TECH-SEP is well placed to stay ahead
in the rush to industrial microfiltration.

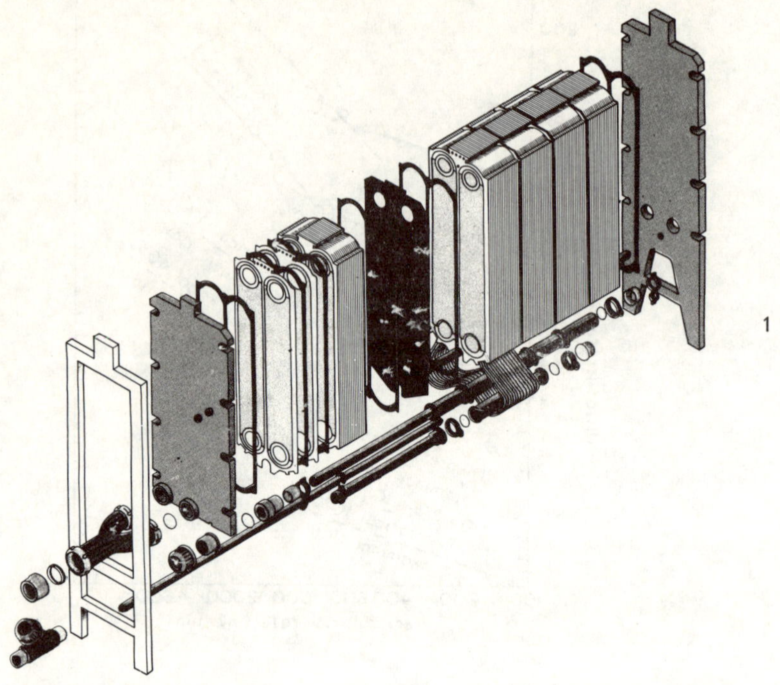

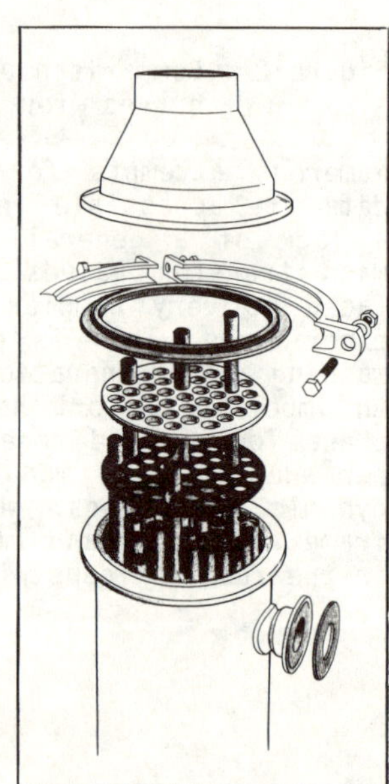

1 Plate and frame
2 Cassette
3 Tubular

Figure 3 : TECH. SEP Modules

4. **AN EXAMPLE OF INDUSTRIAL APPLICATION**

This application was developed in collaboration with Institut MERIEUX (Marcy l'Etoile, France) for the treatment of an hydroalcoholic solution of placental albumin. This solution results from an intermediate step in the purification process. At this step, albumin in already separated from gamma globulins and haemoglobin by two precipitations with alcohol. The volume of solution to be finally clarified is 32000 liters per day with the following composition :
Insoluble : 200 Kg (wet weight after centrifugation)
about 0,6 g/100 ml

Supernatant : Alcool 25 % - Traces of chloroform
NaCl 7,5 g/l
pH : 4,8   22°C
Proteins 4 g/l (97% of albumin).

The suspendend particles (insoluble proteins) have the following properties :
Concentration : 20 to 150-10^6 particles/ml
Size   : 0,5 to 60 micrometers (main part between 1 and 6 micrometers)
Zeta potential: + 23 mV

The traditional process used 9 centrifuges with closed bowls working for 10 hours per day.
The supernatant was then filtered an 1,0 micron on 1,0 micron cartridges prior ultrafiltration.
The high number of centrifuges is due to the limited operation time allowed by the process, the quantity of precipitate being relatively low.
Preliminary experiments were carried out on an laboratory system (UFP 2) in order to choose the adequate membrane. **IRIS (R) 3068** membrane was choosen.
Pore size : 0,8 micron
Mean flux on 5 hours : 200 l/h/m2
Easy regeneration with 0,1 N sodium hydroxide and sodium hypochlorite.
Perfectly clear filtrate
High life span (more than one year).
By recirculating the retentate, it is possible to concentrate 100 folds wich corresponds to a 60% load in insolubles. The flux fall to 25 l/h/m2 at the maximum concentration.
These good results allowed the total replacement of the centrifuges. Albumin can be separated on one tangential microfilter by the sequence concentration / diafiltration. The precipitate is rejected under the form of a slurry only 2,5 fold more diluted than the residue of centrifugation.
The installation for processing 32000 l a day of suspension was oversized of 30% from the results of laboratory and pilot experiments to keep a margin of safety. The unit is constituted of 8 UFP 10 modules, each with 9 square meters of IRIS (R) 3068 membrane. The system is fully automated and driven by a programmable process controller.
The advantages of such a unit are :
Quality and yield of production

Compliance with FDA good manufacturing practices
Closed system . Claning in place . Reproducibility.
Higher yield : suppression of losses washing of the precipitate without dismantling.

Economy
Savings on energy : 17 Kw against 133 Kw
Savings on labour : reduction by 50%
Less occupied surface and reduction of maintenance, reduction of penible operations and noise.
Better safety.

Figure 4 : Industrial unit of microfiltration

This operation was a very demonstrative example of the possibilities of microfiltration. Some problems are much more difficult to solve mainly for reasons of membrane fouling. Nevertheless, tangential micro-filtration is gaining importance in various fields of biotechnology and food industry (5).

At TECH-SEP, other similar operations are under way, especially with the dairy industry, to replace centrifugation by tangential micro-filtration.

5. FUTURE PROSPECTS AND THE BRITE PROGRAM

With the present state of the art, tangential microfiltration is gaining importance in various industries. However, the long term fluxes are of orders of magnitude lower than the fluxes for pure water.

Reducing the rate of fouling or removing it after its occurence is a major objective of research and development in this field.

Despite some successfull applications, tangential microfiltration is only becoming to compete with centrifugation.

TECH-SEP and partners have identified the field of biotechnology as one of the most important for a future development :
- Centrifuges are of current use despite their inconvenients : cost, safety, poor purification capability.
- High value products need closed systems with a very low dead volume to limit the losses.
- For reasons of contaminations and GMP, the system should be cleaned in place.
- The systems have to support in line steam sterilization - Mineral membranes can achieve this task but, by their design, they have an higher dead volume and this restricts their use to the processing of large volumes (e.g fermentation broths for production of medium value products : amino-acids, antibiotics etc...).

A major breakthrough in the process of high value fermentation products and pharmaceuticals can be achieved by combination of :

Organic membranes (allowing the design of very low hold-up volume systems) with high fluxes and low fouling properties. Surface properties of the membranes have to be tailored for adequate interactions with the filtered medium.

Ability of membranes and modules to endure repeated sterilization by steam. This will put the flexibility of organic membranes to the fields of applications where only steam can be used for sterilization.

The project will be subdivided in four major interacting tasks :
- Research on membranes : new materials for low fouling and high tempe- rature resistant microfiltration membranes.
- Fundamental studies on fouling with applications to the new materials.
- Improvement of hydrodynamic properties and thermal resistance of modules in view of application tests of the new membranes.
- Application tests for characterisation on a large scale of the proper- ties of the new modules and membranes.

TECH-SEP has set-up a program of research in association with two British partners :
- Advanced Protein Products will conduct the application tests.
- Looughborough university will carry out the studies on fouling.

Due to the growing importance of animal cell culture, applications tests will be mainly carried out on sterile clarification of cell culture media.

5.1. **Research on membranes**
The main purpose of this task is to make membranes for microfiltration showing the following properties :

Controlled and reproducible porosity in the range 0,05 to 1,5 micrometers.

Controlled surface properties. It is recognized that membranes have to be hydrophilic to limit the non-specific adsorption, first step of fouling.

Thermal stability : the target is to obtain membranes showing a high resistance to hydrothermal treatment (autoclave or in line steam sterilization).

On the basis of his experience, TECH-SEP will conduct the research along two lines :
- Manufacturing of membranes with blend and alloys of polymers.
- Grafting on surface by various chemical and physical methods.

Although the methods of grafting are well known in their princi- ples, they have given only a few industrial products. The purpose of this task is to define methods with a potentiality for industrializa- tion.

The main way to be explored is the use of blends of polymers already experienced, on a small scale, by TECH-SEP. The principle is to add, in a controlled extent, a polymer with a particular property in a polymer constituting the matrix of the future membrane.

By submitting the mixture to phase inversion in controlled conditions, one can obtain a segregation to the surface of the membrane during its formation. Various industrial and proprietary polymers (polyimides, high temperature resistant polyamides) will be experienced.

The new membranes have to be submitted to an extensive characterization before tests of applications.

The target is to be ready for industrialization by the end of the project. This work includes the following steps :

- Laboratory studies : physical chemistry of polymers and solutions. Analysis of the conditions requested for the formation of membranes. Analysis of surfaces : wettability, composition, Assessment of thermal stability.
- Pilot plant studies for the definition of pre-industrial parameters and production for further studies : characterization of fouling and applications.

5.2. Research on fouling

Ideally the new generation of membrane should have a predictable rejection and flux rate as well as a prolonged period of operations.

However, such aims will not be achieved unless the fouling characteristics of the membranes are fully understood and, preferably, fouling effects reduced to a minimum.

Flux rate is reduced for three reasons :
- Concentration polorization
- Fouling
- Physical change of membranes

Concentration polarization contributes to fouling and is more or less under control, depending on the operating conditions. This cannot be neglected in the study of fouling because this work, in the latter stages, deals with real, as opposed to model, systems.

The membranes produced by TECH-SEP will be extensively studied for their fouling properties.

This will need the construction of a special test rig and the development of analytical methods for quantitation of protein and lipids adsorption.

The new membranes will be evaluated for their fouling characteristics under static and dynamic conditions using model mixtures of proteins and lipids. These model mixtures will be designed to imitate real feed stocks such as milk, whey proteins, blood fractions.... but have the advantage of being far more consistent in composition.

Tests will be designed to determine the various fouling rates at different points during the concentration of a particular feed stock. The effects of processing conditions and concentration polarization on fouling rate will be experienced with various model mixtures.

The most promizing membranes will be tested using real industrial mixtures such as whey or blood plasma fractions. The analytical techniques will be adapted to accomodate the complexe and instable mixtures to be processed.

5.3. Research on modules

Although a range of new membranes showing a reduced fouling and a satisfactory thermal resistance is highly desirable, this objective is not sufficient for an industrial technology.

The performances of a membrane are strongly associated with the quality of the module in which they are assembled. Consequently the present modules, well adapted to RO and UF, have to be specifically improved to get the best from the new membranes. TECH-SEP has made the choice of the plate and frame system because of its versatility, ease of handling and cleaning.

The open channel design (no spacers between the membranes) associated with the plate and frame configuration makes this concept very attractive for cross-flow microfiltration in biotechnology where mediums to be processed can be viscous or with a high content in suspended materials.

Microfiltration needs a high tangential velocity and a rather low transmembrane pressure to minimize the clogging of membranes by mechanical compaction of particles on the surface. This implies a careful design of the module configuration for each set of given parameters (viscosity of the fluid to be processed, acceptable shear rate for the product etc....). The more accessible parameters of the configuration for a proper hydrodynamic are :
- Thickness of the liquid channel
- Arrangement of liquid channels in parallel or series depending upon
  the viscosity of the fluid.
If rules are well established for this type of module working in ultrafiltration, a lot of work is still to be done to define the optimal conditions of use in microfiltration.

Materials have also to be adapted for steam sterilization. Particularly important will be the study of heat transfert in such a complex assembly of various materials.

5.4. **Application tests**

The main competitor technologies are "Dead-End" Microfiltraand Centrifugation.

"Dead-End" Microfiltration is typically used to remove small amounts of contaminating particulates, from suspension. The method is used only for small batch sizes. Flux rates are low compared with cross-flow microfiltration. The filtration efficiency is variable, since the filtration action is frequently achieved by the bed of contaminating material, rather than by the filter itself. Flux rates drop off significantly if a temporary loss of pressure occurs. The method is expensive, since cartridges are normally discarded after each use. For example, filter cartridges suitable for sterilization of mammalian cell culture media additives cost up to 6000 FF. These cartridges are discarded after filtering approximately 250 l of foetal calf serum, or only 100 l of adult bovine serum. Porcine serum and sheep serum are even more difficult to filter by this method.

Centrifugation is more usual where suspended solid levels are high. Because of the small density difference between liquid and suspended solids, high g-forces are normally required. The formation of an aerosol containing the suspended material, with its attendant health hazards, is therefore a possibility. Microfiltration will enable cleaning in place without dismantling. It will also enable automation to be readily applied, and will enable direct connection to subsequent processes such as ultrafiltration.

Other technologies also exist in principle, although they are presently at early stage of development. These include electrodialysis(electro decantation), and two-phase aqueous separation.

A classic use for "Dead-End" microfiltration is that of sterilization to remove bacteria, particularly where heat-labile fluids are concerned. This application is becoming more important, due to the growing importance of mammalian cell culture for example hybridoma culture for monoclonal antibody production. Media additives include foetal calf serum (FCS), adult bovine serum (ABS), and immunoglobulin depleted (modified) bovine serum (MBS) plus various proteinaceous "Growth Factors". Since these come from natural sources, sterilisation by filtration is normally essential. Small pore microfilters (less than 0,05 um) could also be expected to remove most virus, and some mycoplasmas.

Natural materials (FCS, ABS, MBS) contain both cell nutrients and "growth factors". For this application it is important commercially that these should not be removed in the filtration, through membrane adhesion. In the present state of knowledge, this can only be confirmed by performing mammalian cell culture studies, and this has been built into the program.

A growing market for microfiltration is that of removal of cells and cell debris, and plant and mammalian cell culture media. For various reasons, it is often necessary that the cells remain intact. Assessment of cell damage, through cell viability studies, will be carried out.

There is also a growing market for haemoglobin, free of "ghosts" (red cell envelope) and non-haem proteins for use as a "blocking agent" in monoclonal antibody test kits.

The following areas have been identified as being of future importance and sufficiently representative to give generalisable information :
- Clarification and sterile filtration of animal sera.

Produced sera will be tested on specific mammalian cell lines to ensure that "growth factors" are not retained by the membranes. Comparisons with existing techniques will be made : cartridges, costs, labour, processing time, ease of cleaning, life span of membranes.
- Removal of red cell "ghosts".

This will be used as an example of clarification of a protein solution wich is well known to cause filtration problems.
- Removal of mammalian cell from culture media.

This is a very sensitive test to evaluate the stress endured by fragile organisms during concentration.

6. CONCLUSIONS

Tangential microfiltration is gaining importance in various industrial applications.

With the development of biotechnology, there is a growing demand for new and efficient processes able to perform difficult separations in favorable economic conditions.

Membranes technology has achieved considerable progress in the last decade and industrial applications are becoming a reality. In this paper, it was shown that microfiltration today can be incorporated in the processing of large volumes of complexe mixtures with an important economic benefit.

Nevertheless, fundamental and applied research have to be intensified to allow the extension of microfiltration to more complex and fragile mixture. A better understanding of the mechanisms of mass transfert and fouling, a better adaptation of modules and a better control of operational parameters are requested by the manufacturer, the process designer and the end user.

Taking those needs into account, TECH-SEP and partners have set-up a very important program of research and development. This program will make a new level of performance available for the bio, pharmaceutical and food industry.

REFERENCES

(1) MELTZER T.H. Filtration in the pharmaceutical industry
     MARCEL DEKKER, INC N.Y. 1987.
(2) PORTER M.C. "Membrane filtration" in handbook of separation Techniques for Chemical Engineers
     Ed. Schweitzer PA. Mc Graw Hill - 1979.
(3) BELFORT G, CHIN P, DZIEWUKSKI DM (1982) "A new gel polarisation model incorporating lateral migration for membrane fouling "in WORLD FILTRATION CONGRESS III DOWNINGTON PA, p 548.
(4) FANE AG (1983) "Factors Affecting Flux and Rejection in Ultrafiltration" J. Separ. Proc. Technol., 4 , 1 , p 15.
(5) US Patent 4.664.056

Prototype of an inorganic ultrafiltration membrane
for nuclear and petroleum applications

New charged ultrafiltration organo-mineral membrane for effluent recovery

Preparation and application of composite membranes

Membrane preparation by low temperature plasma :
application to the enhancement of membrane
hemocompatibility

Treatment of surface-water by electrodialysis

Emulsion membrane filtration applied to phenol
recovery

PROTOTYPE OF AN INORGANIC ULTRAFILTRATION MEMBRANE
FOR NUCLEAR AND PETROLEUM APPLICATIONS

P. BERGEZ[*], J.CHARPIN[*] and J.M.MARTINET[**]
* IRDI/DESICP/Département d'Etude des Lasers et de la Physico-Chimie
CEA. CEN/SACLAY, 91191 GIF/SUR/YVETTE Cedex - France
** IRDI/DESICP/Département de Séparation Isotopique
CEA. CEN/VALRHO, 30205 BAGNOLS/SUR/CEZE - France

Summary

The Commissariat à l'Energie Atomique has developed an inorganic membrane for specific applications in the nuclear and non-nuclear fields.
A membrane was elaborated, adapted to ultrafiltration of cooling water for PWR under nominal conditions; this membrane may also be used for specific petroleum applications.
The specifications of this membrane were: no modification up to 500°C, good behavior in corrosive environments such as solvents and high temperature water, absolute reliability to differences of pressure up to 100 bars.
The membrane we finalized is a composite material which consists of (i) a macroporous, metallic substrate (nickel or stainless steel) of approximately one millimeter thickness, which provides the mechanical properties and (ii) a mesoporous ultrafiltration layer (zirconia) of a few microns, which is sustained by the metallic filter.

The Commissariat à l'Energie Atomique has developed an inorganic membrane for specific applications in the nuclear and non-nuclear, mainly petroleum, fields.

The first problem to solve was the elaboration of a membrane adapted to ultrafiltration of cooling water for PWR under nominal conditions; this membrane may also be used for specific petroleum applications.

The specifications of utilization of this membrane are the following: no modification up to 500°C, good behavior in corrosive environments such as solvents and high temperature water, absolute reliability to differences of pressure up to 100 bars.

The membrane we finalized is the first one to fit these specifications. It is a composite material which consists of (i) a macroporous, metallic substrate (nickel or stainless steel) of approximately one millimeter thickness, which provides the mechanical properties and (ii) a mesoporous ultrafiltration layer (zirconia) of a few microns, which is sustained by the metallic filter.

1. MEMBRANE FOR ULTRAFILTRATION OF COOLING WATER OF A PWR

In the primary circuit of a PWR, water circulates at a pressure of around 155 bars and a temperature close to 300°C in order to cool the core and to transfer the heat from the core to the steam generator. The pressurized water has a very high speed and a high corrosive power because of the presence of boric acid used for modifying the reactivity of the core.

The corrosive pressurized water induces oxide particles (iron, nickel or cobalt) coming from its contact with the components of the circuit and carries them at a very high speed. A great part of these particles is under a colloidal form.

The device used to get rid of these particles consists of mechanical filters poorly adapted to extraction of colloids for which an ultrafiltration approach is much more relevant because of their spectrum of granulometry. FRAMATOME Company has conceived an ultrafiltration process (patent 83 15130) and entrusted CEA with finalization of adapted membranes.

Corrosion studies have led to the choice of membranes with metal substrates (stainless steel or nickel) and filtering layer made of zirconia. The granulometry analysis imposes a pore diameter of around 300 A for the zirconia layer the thickness of which is around 1 micron.

The prototypes have been tested at the Cadarache Nuclear Center on the BERTA loop, which reproduces the thermohydraulic and chemical conditions of a PWR, i.e.:

$$T : 280 - 350°C$$
$$p : 150 \text{ bars}$$
$$\Delta p : 6 - 7 \text{ bars}$$
$$pH : 7.5 - 8.5$$

The mean selectivity, or retention ratio, is high: 85 to 90%. On stainless steel-zirconia membranes, the stability of the permeated flux is outstanding: no shift during 50 days which is exceptionnal for this type of separation unit; on nickel-zirconia membranes, one observes a slight shift attributed to a slight chemical corrosion of nickel.

The flow per hour, which is already high ($2.5 \text{ m}^3/\text{m}^2$) under Δp of 6 bars, may be improved considering a decrease of the thickness i.e. decrease of the substrate roughness.

2. PETROLEUM APPLICATIONS OF THESE MEMBRANES

The properties of our membranes have been explored for petroleum applications which involve severe conditions of use.

Three applications have been considered:

1 - separation through liquid phase ultrafiltration of hydrocarbonated oil and solvent of de-asphaltation in the de-asphaltation process of oils containing asphaltenes (patent IPF-CEA 86/04827). The temperature is in the range of 80 to 400°C under a Δp of around 10 bars: taking into account technico-economical considerations the pore diameter is in the range of 50 to 300 A. Among the possible membranes, ours might fit.

2 – separation through ultrafiltration of catalysts for liquid phase hydrotreatment of heavy hydrocarbides (patent IFP-CEA 86/01879). This process leads to conditions of temperature in the range of 150 to 300°C with a Δp of 10 bars. Optimization of c onditions in function of pore diameter will be achieved in the frame of a EEC contract. Stainless steel-zirconia membranes seem to be well adapted.

3 – in the process of refining engine used oils, an optimized stainless steel or nickel-zirconia membrane could be convenient.

NEW CHARGED ULTRAFILTRATION ORGANO-MINERAL MEMBRANE FOR EFFLUENT RECOVERY

C.EYRAUD[*], C.BARDOT[*], P.BERGEZ[**] and J.CHARPIN[**]
* Laboratoire d'Automatique et de Génie des Procédés - URA D 1328
Université Claude Bernard
43,Bld du 11 Novembre 1918, 69622 VILLEURBANNE Cedex - France.
** IRDI/DESICP, Département d'Etude des Lasers et de la Physico-Chimie.
CEA. CEN/SACLAY, 91191 GIF/SUR/YVETTE Cedex - France

Summary

A concept of charged ultrafiltration membrane of organo-mineral nature has been finalized by deposition of a thin polymer layer previously functionnalized (suphonated polysulphone: PSS) on a porous permeable substrate made of ceramics or carbon. This type of membrane has two vocations:
- retention of solids, macromolecules, colloids by sifting effect
- retention of ionic solutes due to functionnalization of the separative layer by DONNAN effect; the lower the ion concentration, the greater the role of this effect.

1. INTRODUCTION

A concept of charged ultrafiltration membrane of organo-mineral nature has been finalized by deposition of a thin polymer layer previously functionnalized (suphonated polysulphone: PSS) on a porous permeable substrate made of ceramics or carbon.

This concept presents the following advantages:
- permeation flow much higher than those of reverse osmosis (around 20 to 40 times higher) while the pressure of use is lower (< 10 bars). This new process hence presents double savings in comparison with reverse osmosis.
- the structure of the ultrafiltration element with an organic polymer separative layer which is embedded in the pores of the porous substrate leads to a solid, not fragile element, with high resistance to pressure and good mechanical properties.

Elaboration of this membrane has been patented on june 20th 1986 (patent CEA 86.08947). In the USA, University of Kentucky and Mickley Associates company are working in this domain but as fas as we know it does not exist any research programm in France nor in Europe.

These membranes (called CUF: Charged Ultra-Filtration) are particularly adapted to de-mineralization of effluents.

2. <u>CHARGED ULTRAFILTRATION MEMBRANES IN SULPHONATED POLYSULFONE (PSS)</u>: <u>Elaboration and properties</u>

The tubular macroporous susbstrate (800 mm length and 15 or 6 mm diameter) can be either alumina or alumina-TiO_2 composite or carbon or carbon-ZrO_2 composite. The separative layer is formed by immersion-coagulation in an appropriate bath following the method by LOEB and SOURIRAJAN. By adjustment of the collodion composition, of the duration of drying, of composition and temperature of the coagulation bath, one can obtain ultrafilters the pore radii of which vary from 10 to 70 A with permabilities respectively of 0.2 to 0.9 $m^3 m^{-2} j^{-1} bar^{-1}$. The separative layer is schematically represented on the following figure:

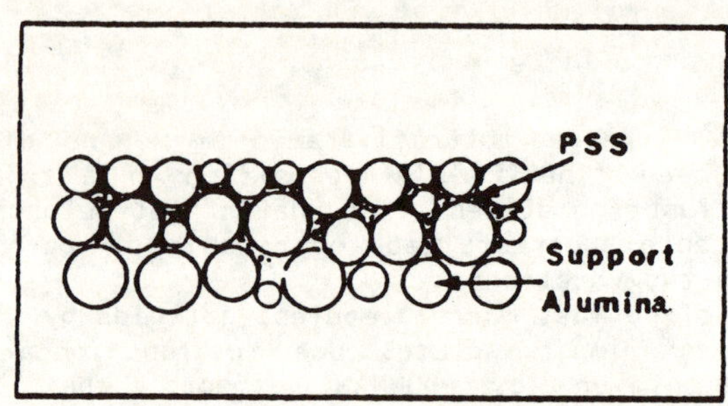

The efficiency of CUF membranes is evaluated by a NaCl or $NaNO_3$ separation test; the retention ratios $R = (1-C_p/C_o) \, 100$ are calculated from analysis of cation concentration of the permeated flux and the original solution.

The PSS membrane with fixed sulphonic groups is cationic for every pH and is caracterized by an improvement of its properties from pH = 8.5 with a more or less important decrease of permeability.

If sodium nitrate is added to a cobalt nitrate solution with increasing quantities, retention of cobalt is increased of more than 50% while sodium is less retained. The same phenomenon is observed replacing cobalt with manganese or lead, retentions are almost doubled. It is therefore necessary to invoke a second order DONNAN exclusion taking account distributions of counter-ions in the capillary and the screening effect due to highest charge compensating ions.

3. <u>APPLICATIONS</u>

These membranes are convenient for (i) recovery of effluents containing toxic (Hg, Pb, Cd...) or radioactive (water from nuclear plant pools) heavy metals, (ii) water softening, (iii) obtention of "electronic" quality waters, (iv) recovery of metals (v) retention of low molecular weight organic species (dyes, amino acids, ionic polymers).

A few cases have been considered:

- <u>sodium recovering</u> from water:

1 - for permeabilities in the range of 0.2 to 1 $m^3 m^{-2} j^{-1} bar^{-1}$, the concentration of sodium being constant, the retention ratio decreases from 90 to 60% when the permeability increases

2 - at constant flow the retention ratio decreases with concentration for example:
- 82% for 10 ppm Na,
- 50% for 50 ppm Na,
- 30% for 100 ppm Na.
Similar results have been observed with other cations.

- <u>decontamination of water from PWR pools</u>:

with traces of colloids and radioactive ions (^{110}Ag, ^{55}Mn, ^{60}Co, ^{122}Sb) in very low quantities (a few ppb), simulation tests have been achieved in inactive environment. Retention rations of 50 to 60% for cobalt and manganese have been determined, for a membrane of permability equal to 0.6 $m^3 m^{-2} j^{-1} bar^{-1}$ and for Co = 5 ppm.

PREPARATION AND APPLICATION OF COMPOSITE MEMBRANES

R. LEYSEN, Ph. VERMEIREN and W. DOYEN
Studiecentrum voor Kernenergie, SCK/CEN,
Department of Materials Development
Boeretang 200, 2400 MOL, BELGIUM

Summary

Composite membranes, containing large amounts of inorganic oxides or hydroxides, have been synthesized and characterized. Initially, these membranes have been developed for use in electrochemical systems, e.g. alkaline water electrolysis, alkaline batteries. A significant increase in performance has been obtained using these composite membranes in the alkaline water electrolysis, compared to the classical used asbestos diaphragms. In the last three years, other applications have been envisaged, e.g. ultrafiltration, biotechnology.
In the case of ultrafiltration, the addition of inorganic oxides or hydroxides increases the possibility of monitoring the pore size and pore size distribution. It also increases the hydrophilicity and as such the permeability, while a decrease in fouling has been noticed. In the field of biotechnology, these composite membranes seem to have good properties for enzyme-immobilization and so for use in bioreactors. Recently, these membranes have also been used for cell harvesting in order to prepare valuable biochemicals.

1. INTRODUCTION

Since the late seventies, the Belgian Nuclear Research Centre (SCK/CEN) has carried out research work aimed at the development of a new separator material for use in the alkaline water electrolysis. Most of this work has been sponsored by the Commission of the European Communities in the framework of their Indirect Energy Programme, sub-programme Hydrogen. The search for a new kind of separator, instead of asbestos, in the alkaline water electrolysis, has been initiated by the world energy crisis at the time.

Therefore, a new kind of separator has been developed on the basis of polysulfone, as the binding material, which was made hydrophilic by loading with inorganic material. Most suitable as inorganic material was found to be zirconium oxide for reasons of chemical stability and good wettability properties. This zirconium oxide-polysulfone membrane has now been commercialized for use in low temperature alkaline water electrolysis by the firm Hydrogen Systems N.V.

Parallel to this development of water electrolysis using film cast membranes, a research programme for the development of a novel type of electrode to be used in alkaline fuel cells was undertaken. In this case, polytetrafluoroethylene (Teflon ®) is used as the binding material which holds the platinized carbon particles together. The cold rolling technique is hereby used.

From these two programmes a lot of experience has been gained concerning the preparation and characterization of porous composite membranes. In this paper, we will discuss the preparation and the electrochemical as well as some other applications of these membranes.

2. PREPARATION OF COMPOSITE MEMBRANES

As indicated in the introduction the term "composite membranes" means a membrane composed of a polymeric matrix filled with inorganic materials, such as metal oxides or hydrated metal oxides. These oxides are homogeneously distributed in the polymeric matrix.

Most frequently, polysulfone or polyvinylidene fluoride is used as the polymeric material when applying the film casting technique for making these composite membranes. In this case, the casting dope consists of a suspension of inorganic material (e.g. oxide) into a polysulfone solution. The inorganic oxide is machined in such a way that a homogeneous distribution is reached in the suspension. The suspension is poured onto a substrate and the membrane is formed by a combination of thermally induced phase inversion (by evaporation) and normal phase inversion (by extraction) [1].

If polytetrafluoroethylene is used as the binding material, the cold rolling technique is applied for making thin foils. After thorough mixing of both the Teflon and the inorganic powder, a cake is being pressed. Afterwards, this cake is dry rolled until a thin foil of the desired thickness is obtained. In some cases a pore filling material is initially added to the mixture and removed afterwards. With this technique it is possible to obtain membranes with an overall porosity ranging between 10 and 90 %. Also multilayer membranes can be envisaged and have already been fabricated. Figure 1 shows a cross section of such a multilayer zirconium oxide/Teflon ® based membrane.

The thickness of the membranes varies mostly between 20 and 1000 µm. With the existing equipment, flat membranes can be manufactured up to a surface area of about 5,000 cm^2. However, scaling-up may easily be envisaged. The film casting technique can also be used to prepare these composite membranes as hollow fibres or as a coating inside a porous tube (tubular membranes). Figures 2 and 3 illustrate respectively a hollow fibre module and a tubular composite membrane.

3. APPLICATIONS

Initially, these composite membranes have been developed for electrochemical applications [2]. They have first been used in low temperature alkaline water electrolysis using a composite membrane composed of an inorganic ion exchanger and polysulfone. Later on a combination of zirconium oxide and polysulfone has been optimized for this purpose. Table I indicates the most important characteristics of this composite membrane.

TABLE I : Basic characteristics of the membrane, used for low temperature alkaline water electrolysis

- Composition	80 wt % ZrO$_2$, 20 wt % PSF + reinforcing polymer structure
- Thickness	650 µm
- Main pore size diameter (skin)	200 Å
- Electrical resistance	0.3 Ω cm^2, 20°C, 30 wt % KOH
- Maximum operating temperature	120°C
- Chemical stability	2 % weight loss after 10,000 hours, 120°C

FIGURE 1. Scanning Electron Micrograph of a cross-section of a multilayer membrane based on zirconium oxide and Teflon®.

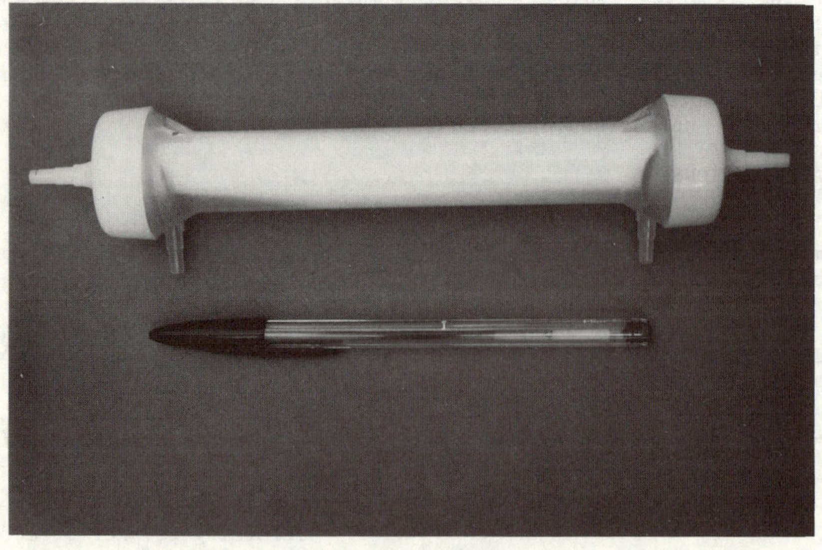

FIGURE 2. Photograph of a hollow fibre module composed of ± 100 composite fibres (Module was fabricated by Organon Teknika N.V.).

FIGURE 3. Photograph of a tubular membrane whereby the composite layer is at the inner side of a porous carbon tube.

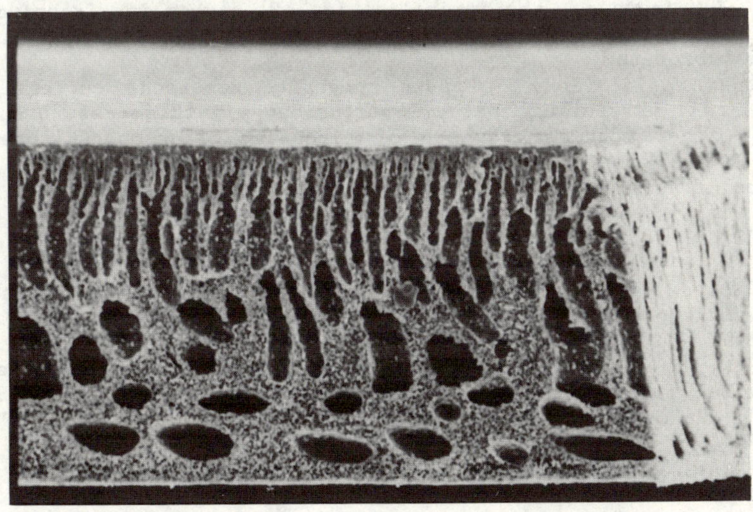

FIGURE 4. Scanning Electron Micrograph of a cross-section of an ultra-filtration membrane.

Besides low temperature water electrolysis, a membrane based on the combination of zirconium oxide-Teflon has been studied for the medium temperature alkaline water electrolysis. With this combination, it is possible to operate electrolysis at a temperature of 160°C. The same components have also been used to manufacture a diaphragm for the chlor-alkali electrolysis.

A similar kind of composite membrane as the one used in the low temperature alkaline water electrolysis has been developed as separator in alkaline batteries. Studies are going on with an industrial partner in order to evaluate the use of these membranes for this application.

Over the last few years, several non-electrochemical applications have been envisaged for these composite membranes. The studied applications are in the field of ultrafiltration and biotechnology.

For ultrafiltration, it was found that these composite membranes may offer some distinct advantages, as compared to 100 % polymeric membranes. Experiments have shown that the addition of the inorganic compound to the polysulfone polymer influences the pore size distribution. As a result of this, it was found that the retention curves of the inorganic loaded membranes are steeper as compared to some other UF-membranes. This may lead to better and more efficient separations. Figure 4 shows a Scanning Electron Micrograph of a cross section of such a membrane.

There is also experimental evidence that the presence of the metal oxide increases the permeate flux. Compared to 100 % polysulfone membranes, an increase in permeability by a factor of 2.5 could be measured. This is probably due to and increase in surface porosity and/or a decrease in "skin" thickness.

Besides the increase in permeability, an increased resistance against fouling was also noticed. This has been experienced with experiments for the separation of highly stable oil-water emulsions [3].

In the field of biotechnology, the intrinsic properties of the inorganic material are exploited for the immobilization of some bioactive species. For instance, the immobilization of enzymes in zirconium oxide-polysulfone membranes is presently considered for the development of an enzymatic bioreactor.

Another application for these composite membranes currently under study, is cell harvesting and cell perfusion. In this case, use is made of hollow fiber modules composed of zirconium oxide-polysulfone hollow fibers for the growth of cells in the extracapillary space or for the production of very specific biochemicals.

REFERENCES

[1] LEYSEN, R. and VANDENBORRE, H., Synthesis and characterization of polyantimonic acid membranes. Mat. Res. Bull., Vol. 15, pp.437-450 (1980).

[2] LEYSEN, R. et al., The use of heterogeneous membranes in electro-chemical systems, in "Synthetic Polymeric Membranes", Eds.B, Sedláček and I. Kahovec, pp.89-99, Published by W. de Gruyter, Berlin-New-York (1987).

[3] DOYEN, W. et al., The use of ZrO_2-based composite membranes for the separation of oil-water emulsions, in "Particle Technology in Relation to Filtration of Separation", Proceedings published by the Royal Flemish Society of Engineers (K.V.I.V.), Volume 11, pp.3.21-3.33, (1988).

MEMBRANE PREPARATION BY LOW TEMPERATURE PLASMA:
APPLICATION TO THE ENHANCEMENT OF MEMBRANE HEMOCOMPATIBILITY

G. CLAROTTI, F. SCHUE, J. SLEDZ, A. AIT BEN AOUMAR,
K. GECKELER *, A. ORSETTI ** and G. PALEIRAC ***

Université des Sciences et Techniques du Languedoc, Montpellier, France
* University of Tübingen, Tübingen, West Germany
** Centre Hôspitalier Régional Lapeyronie, Montpellier, France
*** Centre de Transfusion Sanguine, Montpellier, France

Summary

Plasma polymerization of the gases present in a low temperature
plasma allows to deposit on a substrate a thin film which properties
depend more on the experimental conditions and the atomic elements
introduced in the plasma than on the molecular structure of these
gases. This paper briefly reviews the different types of plasma
treatments allowing to modify a substrate surface without affecting
its bulk properties and summarizes the opportunities they offer in
membrane preparation. The advantages of the treatment of commercial
microfiltration low-cost membranes to confer them the required bio-
and hemocompatibility to be implanted in an organism are then
presented. Since the surface energy of a polymer seems to be one of
the key properties affecting its hemocompatibility, we have tried to
deposit films from a plasma containing a mixture of ethylene oxyde
(EOx) and perfluorohexane (PFH) in order to obtain very hydrophobic
(100% PFH), very hydrophylic (100% EOx) and intermediate coatings
that will allow us to study their hemocompatibility. The first
results obtained with different preliminary plasma treatments are
discussed. The treated substrates have been characterized by weight
variations, surface profilometry, SEM and ESCA analysis.

1. INTRODUCTION TO PLASMA TREATMENTS [1,2,3,7 and 12]

Low temperature plasma treatment is a unique and powerful new
technique that affects only the outermost layers of membranes exposed to a
"cold" plasma and therefore ensures that their bulk properties do not
change. This technique offers a wide range of possible plasma/membrane
combinations and the surfaces obtained are, mostly, very stable.

A plasma is a globally neutral, partially ionized, complex gas
composed of electrons, ions, free radicals, photons of various energies
and gas atoms or molecules both in in the ground state or higher excited
states. Low temperatures, or "cold", plasmas are characterized by a gas
temperature (Tg) lower than the electron temperature (Te) with Te/Tg
ranging from 10 to 100 when the electron energy varies from 1 to 10 eV.

Cold plasmas are usually obtained by submitting a low pressure (0.1
to 10 mm Hg) gas to a high frequency (13.56 MHz or microwaves) electric
discharge (10 to 100 watts). This can practically be done by letting the
gas flow through a tube or a bell-jar glass reactor containing two
parallel plates coupled to the power source (internal electrodes) or
through a glass tube surrounded by an electric coil (external electrodes).

Any gas contains a small number of electrons and ionized atoms at room temperature. The electric discharge increases the concentration of charged particles and confers them an energy sufficient to break any bond, therefore leading to the complex interacting excited species mixture that forms the plasma.

When a substrate is exposed to plasma, three types of basic reactions can occur and one can control which reaction will predominate by varying the conditions of the discharge so that a wide range of possible surface properties can be obtained with the same gas/reactor/substrate system.

In surface <u>Plasma Modification</u> (PM) by a non-polymerizable gas (He, $Ar, O_2, N_2, Air, Fluorocarbons$ with $F/C \geq 3$), the excited species transfer their energy to the surface, creating reactive sites such as radicals and excited species (surface activation) that can then react with atomic or molecular oxygen and nitrogen to form a surface layer containing hydroxyl, oxide peroxide and amine polar groups. These activated surfaces can also be subsequently exposed to reactive molecules or monomers that will then react with the reactive sites and graft or polymerize in a classical way (<u>Plasma Induced Grafting and Polymerization</u> (PIG and PIP)).

<u>Atom implantation</u> can also occur when an atom coming from the gas phase replaces an atom of the substrate such as in polyethylene fluorination (or teflonation): $-CH_2- + 2F^\bullet --> -CF_2- + 2H^\bullet$

Surface <u>Plasma Etching</u> (PE) occurs when very reactive gases $(Cl_2, F_2, O_2$ or Fluorocarbons) and/or high power plasmas and/or substrates containing many light and reactive groups ($-OH$, NH_2, or similar) interact leading to the volatilization of the surface of the membrane.

<u>Plasma Polymerization</u> (PP) requires a polymerizable gas (Hydrocarbons, Silicon organic compounds, Fluorocarbons with $F/C < 3$) that will be converted to reactive species in the gas phase. This species will then react with each-other and with the substrate surface and deposit a very thin (10 to 1000 Å) and cross-linked film that is tightly bound to the substrate surface by adhesion or covalent links.

The energy of the plasma being so high, monomers don't need to contain double and triple bonds or cyclic structures to polymerize. The gas polymerizes by an "atomic polymerization" process that leads to a polymer which structure is more determined by the experimental conditions and by the atomic elements introduced in the plasma than by the starting monomer structure : methane, ethane and ethylene, for example, all lead to a plasma polymerized film similar to polyethylene.

2. <u>PLASMA TREATMENT APPLIED TO MEMBRANE PREPARATION</u>

Since 1970, plasma treatment has been applied to insulate, protect or confer good adhesion properties to metal substrates or to etch semiconductor devices and other substrates, but recently many papers concerning their application to membranes have been published.[3 to 6]

The membrane surface modification obtained will depend on the plasma processing type/substrate morphology combination used.

<u>PM</u> can crosslink the surface of a homogeneous (nonporous) substrate and create a more selective membrane or a less permeable barrier. It can also vary the substrate hydrophylicity, therefore affecting the liquids diffusion rates and modifying the membrane wettability.

<u>PIG and PIP</u> can functionalize the membrane surface conferring specific properties much in the same way as classical surface grafting does.

PE can "clean" the membrane surface removing impurities and alter its permselectivity by varying the pore dimension.

PP deposits an ultrathin layer over the substrate, therefore modifying its surface properties as well as its permselectivity. The deposition of this film on the surface of a "classic" membrane can reduce its permeability and increase its selectivity, until the film covers all the pores, therefore forming a composite membrane similar to an asymmetric conventional membrane, but with an even thinner selective layer (as low as 100 Å !!)

There are two principal fields in which membranes prepared or modified by plasma treatment are used : Reverse Osmosis and Gas Separation.

In the former field, composite membranes obtained by PP of an allylamine film on conventional polysulfone membranes yielded excellent results such as an initial increase in water flux and salt rejection followed by a good stability independent of salt concentration and applied pressure [4]. Poly(acrylonitrile) membranes modified by an He plasma discharge have been reported working smoothly for 3 years in an electrocoloring bath, effluent treatment plant [SANO & al., 1984, cited by [5]].

Organosilicic plasmas were used to deposit gas separation thin films on porous polypropylene films [6]. It appeared that the nature of the starting monomer determined the properties of the membrane, the oxygen permeability increasing and the ratio $P(O_2)/P(N_2)$ decreasing with increasing oxydation state of Si in the deposited layer.

3. APPLICATION TO MEMBRANE HEMOCOMPATIBILITY ENHANCING

Plasma treatment suits particularly well biomedical material processing because it is a very "clean" method that can be applied to preformed material without risks of contamination by potentially toxic additives and catalysts or pathogenous microorganisms. Plasma treatment furnishes sterile, atoxic and ready-to-be-implanted materials.[7,8,12]

Our Montpellier team has a long experience in membrane surface modification applied to the reduction of membrane thrombogenicity for biomedical applications involving their contact with blood. With the collaboration of our German Colleagues from Tübingen, we are presently estimating the possibilities offered by different plasma treatment in order to obtain low thrombogenicity membranes. Our goal is to obtain hints on the feasability of a plasma treatment conferring a commercial microfiltration low-cost membrane, the required bio- and hemocompatibility to be implanted in an organism. For this we have chosen polysulfone films and membranes to be the substrates we wish to render hemocompatible.

Two type of gases will be used for PP treatments.
PP of a mixture of ethylene oxyde (EOx) and perfluorohexane (PFH) has been used by RATNER and HAQUE [11] to obtain coatings having surface energies varying over a wide range. By varying the ratios of the two gases they could obtain very hydrophobic (1 erg/cm² for 100% PFH plasma), very hydrophylic (45 erg/cm² for 100% EOx plasma) and intermediate coatings. Surface energy being one of the most important surface properties related with blood compatibility [9 and 10], we wish to submit the membranes to the same treatments to study how surface energy variations affect the blood reactivity towards them.

<u>PP of an organosiloxane</u> should cover the membrane with a poly(siloxane) coatings similar to poly(diméthyl-siloxane). According to CHAWLA [12] these films show remarkably low blood-platelet adhesion, which is the first step of blood clotting [9].

4. FIRST PLASMA TREATMENTS AND RESULTS

4.1 <u>Experimental set-up description</u>

The first plasma treatments were performed at the laboratory of Pr. H. SUHR in Tübingen whom whe thank for his precious help and advice.

We used a bell-jar reactor containing two parallel plates electrodes (3 cm apart) and powered by a radio frequency generator (13.56 MHz). Argon was used as a carrier gas to fix the operating pressure and, depending on the different treatments, mixed whith H_2 or EOx in percentages controlled by a needle valve (for EOx) or MKS electronic mass flow controllers. PFH was used as a liquid and evaporated into the reactor at a rate controlled by the operating pressure (0.3 or 0.4 Torr corresponding to total carrier gas flows of 5 or 10 sccm) and the temperature of a thermostated bath (between 230°K and 250°K).

Three types of substrates were arranged in different patterns on the lower electrode : Polysulfone asymmetric membranes and films were used to study the effects of the plasma treatment on the polymer surface while microscope glass slides and pieces of Silicium wafers were used to study the plasma polymerized deposit.

4.2 <u>Treated substrates characterization</u>

The substrates were weighed immediately before and after the treatment in order to approximate the Amount of plasma polymer deposited or of Substrate etched material (in mg/cm^2) as well as the Etching or Deposition Rate ($mg/cm^2/min$).

Surface Profilometry on the glass slides (TENCOR alpha-step) allowed us to measure the Thickness of the Deposited Layer as well as its Roughness and Density.

ESCA analysis (Pr. GÖPEL, Un. Tübingen, RFA) of the treated silicium wafers and polysulfone substrates was used to determine their Surface chemical Composition and Structure, and, therefore, the Composition and Structure of the deposited layers.

Scanning Electron Microscopy of the treated surfaces allowed us to have a glance at their aspect and to confirm the results of the previous analysis.

Infra-Red Analysis of the glass slides has also confirmed the results concerning the Surface chemical structure (Multiple Internal Reflectance mode) and the Thickness of the Deposited Layer by the study of the reflected Interference Fringes.

4.3 <u>Results</u>

Table I presents the results obtained with different plasma treatments). We can see that high plasma power and/or low PFH flow lead to plasma etching of the membranes (negative deposition rates), while the use of low power plasmas (< 30 W) and/or higher PFH flow allowed us to obtain deposited, fluorocarbon-type, films. The profilometry tests showed that the etched polysulfone substrates had a rough surface while polymer deposition allowed us to obtain a smooth and dense (scratch resistant) film. However, the reaction yield was low, more than 3 g of PFH being consumed for the deposition of less than 6 mg of plasma polymer.

Adding H_2 to the plasma system allowed us to obtain a similar deposition rate with lower plasma power and PFH flow, thus multiplying the reaction yield by five. But then, the deposited film was less smooth and SEM analysis showed it contained many irregular clusters protruding from the smooth surface.

The first ESCA analysis of the substrates surfaces also showed the presence in the film of oxygen possibly originating from a simultaneous polysulfone membrane-etching or, more probably, from a post-reaction of the long-life radicals present at the surface with air.

This presence is mostly undesired because oxygen is probably present in polar carbonyl, carboxyl or even peroxide groups that will increase the surface hydrophilicity, while we wish to have, in this case, a very hydrophobic surface. We have been able to reduce the oxygen content of the fluorocarbon-type coatings by a post-treatment done by first filling the reactor with Ar or PFH up to the atmospheric pressure and by letting then the radicals react with the gases for 30 or 60 minutes.

Addition of EOx to PFH and Argon doubled the deposition rate and increased further the reaction yield, but the polymerized deposits are very powdery and are easily scratched by the stylus used in the profilometry assay.

While the litterature [11] reports high deposition rates with EOx plasmas, we could not obtain a satisfying plasma polymerization with EOx, either alone or as a 50/50 mixture with Argon. This might be due to the fact that the use of EOx proved to be problematical because of the gas tendency to soften the polymer tubing and to swell the valves gaskets, thus affecting the control of the treatment. Therefore, use of EOx will require an experimental system containing only few polymer components (needle valve gaskets), while polymer tubing will be replaced by glass or metal tubing.

Further analysis of the treated substrates concerning their surface quantitative composition, their surface energy and hemocompatibility is under way. The modifications of the membranes filtering properties due to the plasma treatment will also be studied accurately.

<u>ACKNOWLEDGEMENTS</u>

*This work is supported by grant N°ST2*351 of the CEC "Stimulation" program.*

<u>REFERENCES</u>

[1] BOENIG H.V., Plasma science and technology, Cornell University Press, London (1982).

[2] YASUDA H., J. Polym. Sci., Macromolecular Reviews, <u>16</u>, 199-293, (1981)

[3] VARIOUS AUTHORS, Proceedings of the ACS division of polymeric materials : science and engineering, Vol. 56., American Chemical Society, (1987)

[4] MATSUZAWA Y. YASUDA H. and al., Ind. Eng. Chem. Prod. Res. Dev., 23, 153-163 & 163-169, (1984)

[5] YASUDA T., GAZICKI M. and YASUDA H., J. App. Polym. Sci., Applied Polymer Symposium, 38, 201, (1984)

[6] SAKATA J., YAMAMOTO M. and HIRAI M., J. App. Polym. Sci., 31, 1999, (1986)

[7] HOFFMANN A.S, Adv. Polym. Sci., 57, 141-157, (1984)

[8] HOFFMANN A.S. and GOMBOTZ W.R., CRC Critical Reviews in Biocompatibility, 4, pg.1, (1987)

[9] IKADA Y., Advance in Polymer Science, 57, 104, (1984)

[10] ANDRADE J.D. and HLADY V., Advance in Polymer Science, 79, 3, (1986)

[11] HAQUE Y. and RATNER B.D., J. App. Polym. Sci., 32, 4369, (1986)

[12] CHAWLA A .S., Plasma polymerization and plasma modification of surfaces for biomaterials applications for biomaterials applications, in Polymeric Biomaterials, pg. 231, E. PISHKIN and A.S. HOFFMANN Editors, M. Nijhoff Publications, Dordrecht (Holl.),(1986)

TABLE I : Plasma treatments and substrates properties

PLASMA TREATMENTS			SUBSTRATES CHARACTERISATION		
CARRIER GAS MIXTURE (Gas : Sccm)	PFH FLOW (Sccm)	PLASMA POWER (Watts)	DEPOSITION RATE (mg/cm²/min)	DEPOSIT THICKNESS (nm)*	SURFACE COMPOSITION (ESCA)
Ar : 10	0.7	15	- 30	-10^{5**}	------
Ar : 5	13.0	40	10	1000	C, F, H, O
Ar : 8 / H₂ : 2	2.6	15	11	1500	C, F, H, O
Ar : 8 / EOx : 2	1.0	30	20	3500	C, F, H, O
Ar : 5 / EOx : 5	---	20	≤ 1	≤ 100	C, H, O

 * After 30 minutes of treatment
 ** Estimated by weight loss
   (Gas flows are in Sccm : Standard Cubic Centimeter per Minutes)

TREATMENT OF SURFACE-WATER BY ELECTRODIALYSIS

R. ROOFTHOOFT (LABORELEC) and J. VAN WELY (IONICS - UK)

Summary

The treatment of surface water by electrodialysis has been studied at the Langerbrugge Power Sation (Belgium).
The principles of electrodialysis are explained. The main results of the tests are presented.
As the system is also economically interesting a plant for the treatment of surface water at 20 m³/h has been ordered. It will come in operation in the beginning of 1989.

1. Principles of electrodialysis
Electrodialysis is an electrochemical separation process in which ions are transferred through membranes from a less concentrated to a more concentrated solution as a result of the flow of direct electric current.
    Two types of membranes are used :
- anion transfer membrane : allows the passage of anions. It is electrically conductive and water impermeable under pressure ;
- cation transfer membrane : allows the passage of cations. It is also electrically conductive and water impermeable under pressure.
    Figure 1 illustrates what occurs when a DC-potential is applied. As shown in this figure alternating compartments of demineralised and concentrated solutions are formed in the membrane cell when a DC-potential is applied across the electrodes. The unit will yield two major and separate streams-demineralised and concentrate waters-and two minor streams from the electrode compartments. Several hundred of these demineralising and concentrating compartments are assembled into a membrane stack to obtain the desired flow rate.
    In order to avoid fouling reversal of polarity and flow streams is introduced at a regular interval. A turbulent flow through the stacks of membranes is ensured by a tortuous path spacer between the membranes. This greatly reduces the chance of settling of suspended solids.

2. Water treatment at Langerbrugge power station
a) Actual situation :
The make-up water for HP-boilers has to fulfil very severe specifications. In the Langerbrugge power station the make-up is prepared by demineralisation (ion-exchange) of potable water.
b) Planned modifications :
The price of potable water is increasing and a shortage can be foreseen in the near future.
It has been decided to build a new plant in order to treat the water of the canal between Ghent and Terneuzen. The aim is to produce a water of lower salinity than the potable water and to demineralise this water on the existing ion-exchange unit.

The pretreatment will be done by flocculation and settling followed by dual media and activated carbon filtration. Afterwards the pretreated water is going through an electrodialysis unit. The output of the system will be 20 m³/h.

3. <u>Pilot plant study</u>
In a pilot plant study two types of water have been treated :
a) well water with a very high content of organics (humic acids) ;
b) canal water (Ghent-Terneuzen).

The pretreatment which gives the best results is the flocculation with WAC (Al-polychlorosulphate) and filtration on dual media and activated carbon. Especially activated carbon is of great value to reduce the turbidity and the organics. The electrodialysis has been tested during more than 6 months. Table I gives the values of the main chemical parameters after treatment, before the ion-exchange.

4. <u>Economic aspects</u> :
A comparison of the cost of demineralised water has been made for the Langerbrugge power station.
In these calculations the following parameters are accounted for :
- cost of potable water
- cost of regenerants
- cost of pretreatment
- cost of electrodialysis
- investment for pretreatment and electrodialysis.

On basis of a plant which produces 240 000 m³/year, the pay-back time of a new installation (pretreatment + electrodialysis) is about three years. Calculated on a five year basis the return on investment is 15 % for the same installation.

<u>TABLE I</u>

	potable water	canal water	treated water
organics mgO_2/l	0,44	13,1	3
TOC mg/l		0,2 – 0,6	0,2 – 0,6
Ca mg/l	108	156	4
Mg mg/l	25	34	1,5
Na mg/l	65	400	27
Cl mg/l	185	600	28
SO_4 mg/l	80	150	48
TDS mg/l	503	1600	200
pH	7,23	7,42	6,6 – 7
conductivity μs/cm	585	2100	230
turbidity NTU	0,56	16	0,5
SiO_2 mg/l	13,9	10,7	10

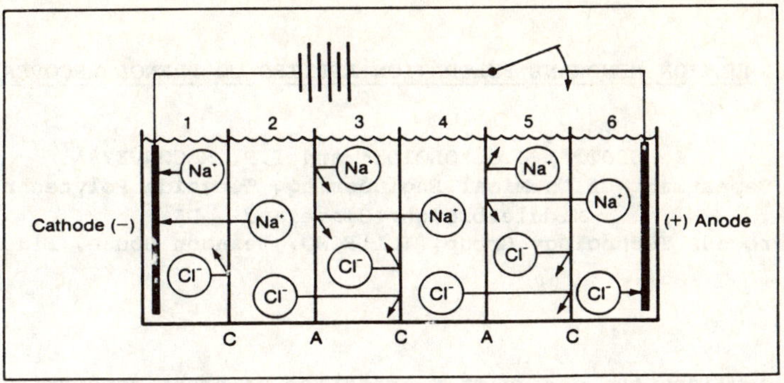

Fig. 1 : Principles of electrodialysis

EMULSION MEMBRANE FILTRATION APPLIED TO PHENOL RECOVERY

K. SCOTT*, J.O. OLOIDI* and I.F. McCONVEY**
* Department of Chemical Engineering, Teesside Polytechnic,
Middlesbrough, Cleveland. UK
** Process Technology Group, ICI FCMO, Hexagon House, Blackley

Summary

This work examines the design of a crossflow membrane separation process
for the recovery of phenol from dilute solutions. Emulsion membrane
filtration (E.M.F.) is, in the context of this work, a method whereby low
concentrations of organics are separated from aqueous solution using a
combination of solvent extraction and membrane filtration. Extraction is
carried out by using a low inventory of a highly specific solvent in the
form of a "stable" emulsion. This emulsion is difficult to break by
normal procedures and thus separation of the organic (solvent) phase from
the aqueous phase is achieved by membrane separation.

This work describes experimental results of the filtration of an
emulsion of tridecanol and water to be applied in the recovery of phenol.
A number of microfiltration membranes with and without surface
modifications are tested and compared.

1. THE EMULSION MEMBRANE FILTRATION PROCESS

The pilot-scale equipment is shown in Figure 1, it is composed of a 6
litre and a 20 litre QVF "mixing" vessels, a Stuart stainless steel
internal, 3-13.7 m head 11.4-54 1/minute capacity centrifugal pump. A
Heidolph PZR50 stirrer motor, which could give a maximum of 2000 rpm,
static mixer, flat photocell, rotameters, pressure gauges and the membrane
unit. The crossflow membrane unit is made of Trovidur PP-D plastic of 127
mm diameter cell, with 3 mm diameter channels for the inlet and outlet
streams. The membrane is of 30 cm^2 surface area. The permeate flows
through the membrane tangentially and out through three 1 mm holes in the
lower part. The unit is constructed in such a way that it could be
operated with other cells connected in series of parallel. All the
process line is of QVF pipe sections and reinforced PVC tubing.

The organic solvent used was tri-decanol supplied by ICI. It had the
following properties: flash point-118°C, density 847 kg/m^3, viscosity 48.3
mN/m and solubility in water of 0.01%. The emulsion of a specific
loading, in some cases with phenol, is made up in the vessel and stirred
continuously for about 1 hour. Further mixing is by the centrifugal pump,
static mixer and recycling back to the mixing vessel for another 30
minutes.

The applied pressure was 12 psi in all cases, excepted when the
effect of the operating pressure was investigated and all runs were
performed at room temperature (21-23°C). The runs varied from about 2
hours to six hours depending on the permeability. The cross flow velocity
was in the range 0.8-3 m/s.

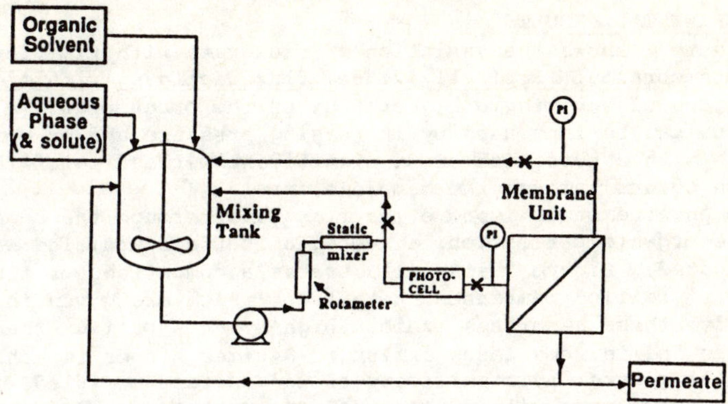

Figure 1 : Schematic E.M.F. separation unit

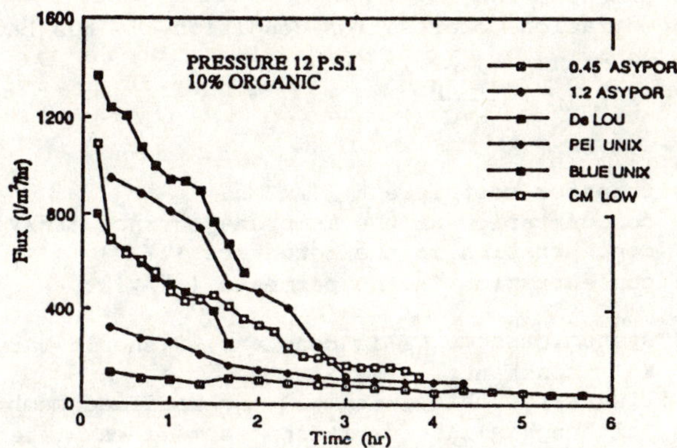

Figure 2 : Comparison of flux decline rate for surface modified membranes

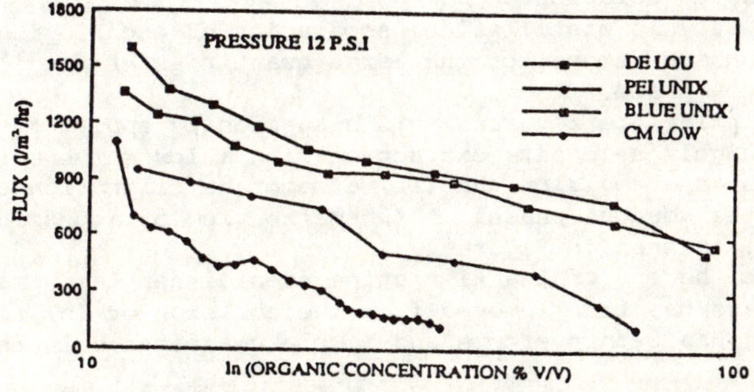

Figure 3 : Effect of varying oil concentration on flux

2. EXPERIMENTAL RESULTS

Figure 2 shows the variation of flux rate with time for commercial Asypor[R] membranes. Typically we see flux declining with time, with the larger pore size giving a higher flux of the order of 0.1 m^3 m^{-2} h^{-1}. This flux can be increased by increasing pressure and/or pore size, although with a pore size of 3μ significant permeation of the organic emulsion occurs through the membrane.

In an attempt to improve the flux rate through the membranes and to minimise organic permeation, a number of surface modified membranes were investigated. Figure 2 also illustrates a comparison of flux decline rate for these modified membranes, details of which are given in Table 1. Generally, these membranes exhibit higher rates of flux than the Asypor membranes and in some cases a flux of 6 times higher is achieved. This is due to the greater hydrophilicity of the surface modified, i.e. charged. Significantly, even with fluxes of 1 m^3 m^{-2} h^{-1}, negligible penetration of organic solvent through the membrane is achieved.

Figure 3 illustrates the flux decline rate for the surface modified membranes as a function of organic loading. The uncrossed linked membranes showed a higher flux than the crossed linked. The significant feature of this performance is that higher water separations are achieved with values in excess of 90%.

Under concentration polarisation condition, it has been shown that the flux is given by:

$$J = K \ln \frac{C_s - C_p}{C_r - C_p}$$

where J = flux $(1/m^3/h)$
K = mass transfer coefficient $(1/m^2/h)$
C_s = oil concentration at the membrane surface (%^{V}/v)
C_r = oil concentration in the retentate (%^{V}/v)
C_p = oil concentration in the permeate (%^{V}/v)

Thus with this system, assume C_s is constant, J should vary linearly with $\ln(C_r)$ if K is constant.

Figure 3 illustrates this, with all the modified membranes have similar slopes, although slight deviation from lineary are seen.

3. PROCESS APPLICATION

The concept of emulsion membrane filtration is seen as an alternative method for the separation of organic (or other) species produced by synthesis in dilute solution.

Typically, say in distillation, separation of organic is achieved by the energy intensive removal of the large quantities of the lower boiler aqueous "solvent" phase.

E.M.F. in principle reverses this situation by applying two single steps. (i) a highly selective extraction with a low solvent inventory in the form of a stable emulsion and (ii) a membrane filtration of a stable emulsion from the aqueous phase. A schematic view of a possible process adopting E.M.F. is seen in Figure 4.

Notably at the end of the filtration step either (a) almost all of the aqueous phase has been removed from the emulsion or (b) the stability of the emulsion has been overcome and simple coalescence is then possible.

4. CONCLUSIONS

The ability of microfiltration membranes to separate an aqueous phase from an emulsion with a despersed solvent has been demonstrated. The rate and degree of separation depends in particular on the property of the membrane, although separations in excess of 90% has been achieved.

In all cases a continuous drop in flux is seen and is part contributed to a fouling of the membrane probably due to a "contaminated" originating from the solvent. Further work will clearly be required to investigate this effect further and develop means of surface regeneration.

At this preliminary stage of the work we see the application of E.M.F. as a replacement to solvent extraction and/or distillation as a means of separation of synthesised organics produced in dilute solution from, say biochemical or electrochemical processes. In addition the technique may find application in effluent treatment and similar areas.

TABLE -1   Membranes used in experiments

Name	Pore Size (μm)	Modifications
Asypor (70% Cellulose Nitrate and 30% Cellulose Acetate)	0.2, 0.45, 1.2 and 3.0	--------
Nypor (Nylon)	0.2	---------
Tefpor (Teflon)	0.2	----------
De (Low) (Diethyl amino ether on regenerated Cellulose)	0.8 -1.0	crosslinked, + charge density 0.3 Milli eqv/g
CM (Low) (Carboxyl methyl on regenerated cellulose)	0.8 -1.0	crosslinked, - charge density 0.3 Milli eqv/g
Blue (Cibacron Blue Dye on regenerated cellulose)	0.8 -1.0	uncrosslinked Dye Density 8 μmol/cm^3
PEI (Polyethyleneimine on regenerated cellulose	0.8 -1.0	uncrosslinked, + charge 0.4 Milli eqv/g

CONCLUSIONS

<u>ROUND TABLE DISCUSSION</u>

<u>Chairman : Professor Meares</u>

The round table discussion began with a summary of each session presented by the corresponding chairman.

<u>SESSION I - Chairman : Professor J. Néel</u>

"Fundamentals of membrane systems and engineering aspects"

The first session brought together four papers of more general interest, surveying different membrane technologies and their various applications taking into account both the industrial point of view as well as economic aspects.

The first paper entitled :

"Review of nuclear and non-nuclear applications of membrane processes. Present problems and future R & D work"

presented an overview of the different membrane processes i.e. reverse osmosis, ultrafiltration, microfiltration and electrodialysis showing their respective advantages and disadvantages in comparison with conventional processes. In particular, this paper emphasised that the use of membranes for nuclear applications require radioresistant materials, e.g. inorganic membranes such as Carbosep (carbon tube internally coated with a zirconia or alumina membrane), although polymeric membranes can also give very good results. In addition, this paper aimed at reviewing the common problems of membrane processes in both nuclear and non-nuclear applications (permeate flow, fouling, clogging, mechanical and chemical resistance).

The second paper

"Production and characterisation of membranes.
How science interacts with practical know-how"

was oriented towards membrane characterisation emphasizing the comparability of the results achieved in different laboratories.
In this respect, this paper highlighted that the credibility of membrane processes in terms of reliability and efficiency can be increased and maintained if a very good interactive exchange of information exists between the membrane manufacturer and the user, particularly during the experimental testing and operation of the separation plant of concern.

For example, in order to maintain a constant permeate flow, back-flushing is worthwhile even for cross-flow microfiltration applying modules equipped with tubular or hollow-fibre membranes. However, normally back-flushing cannot substitute chemical cleaning everytime severe membrane fouling occurs.

The third paper

"Economical evaluation of the membrane processes"

underlined that even if most of the membrane industry is located in the U.S.A., the respective markets in Europe and in the U.S.A. are comparable. It is interesting to point out that the annual increase of worldwide sales of synthetic membranes is currently between 12 and 14% (only membrane modules and not the ancillary equipment).

This paper also showed that the European industry could successfully find openings in the US market by providing advanced technologies like pervaporation in which Europe is ahead.

The fourth presentation

"Studies on transport phenomena in hollow fibre membranes
by means of dimensionless groups"

concerned a more specialized topic where the influence of the dimensions of the hollow fibre membrane and the influence of gas viscosity on the performance of the membrane were studied and quantified. In this respect a dimensionless number has been developed which enables :

- the determination of the optimal fibre dimensions (e.g. length), when the process parameters and the fibre characteristics are known;

- the distinction between hollow fibre types, as well as a comparison between homogenous fibres and asymmetric fibres;

- a considerable simplification of the calculations of the yield of the corresponding membrane unit.

This development is not only applicable to perpendicular flow in hollow fibre bundles but also for other arrangements like co- and counter currents.

SESSION II - Chairman : Dr. H. Strathmann

Industrial applications (Part I)

The five papers of this session represented very well the spirit of the symposium where new ideas tackled traditional problems and where ways to solve them were put forward.

The first presentation

"Recent developments in testing new membrane systems at
AERE-Harwell for nuclear applications"

mainly dealt with the setting-up of a new device (Direct Membrane

Cleaning) based on the use of an electrical field to prevent irreversible fouling of the membrane. High voltage drops (20V) were used in this system depending mainly on the conductivity of the solution. The high current densities did not significantly modify the pH of the solution by decomposing water because the pulse time was only from 1 to 5 seconds. The membrane itself was used as a cathode. For specific application of DMC on Carbosep type membranes it was found that the irregular deposition on $ZrO2$ had a beneficial effect on the contact between liquid and graphite and thereby improved process performance. This new idea could have very broad applications. In principle, modules equipped with DMC should be commercially available within the next two years.

The second presentation

> "Current applications for drying acetic and formic acid
> and commercialised techniques"

outlined the possibilities of drying acetic and formic acid in comparison with conventional processes. Since no details of the membrane itself were given, it was difficult to comment on the membrane development.

The third and the fourth presentations

> "The studies on supported liquid membranes carried out
> at the ENEA–Casaccia Research Centre"

> and

> "Use of liquid membranes for treatment of nuclear effluents"

have demonstrated good prospects as alternative processes to liquid/liquid solvent extraction when high decontamination or separation factors are required. However subsequent scaling up of these processes needs a better understanding of the transfer mechanism of the elements and the chemical species through the membranes.

The last presentation

> "Characteristics and performances of new types of
> ultrafiltration membranes with chemically modified
> surfaces"

demonstrated a commercial approach for improving the properties of a membrane using an inexpensive route through modifying the surface of a membrane made of a basic polymer. No correlation has yet been drawn from this work between the degree of hydrophobicity and the fouling effect but it has been demonstrated by measurement of the wetting angle that a significant improvement in membrane flux can be achieved. A tentative explanation of this improvement is that absorption dominates the flux limitation as a result of increasing the membrane resistance (pore structure of the membrane structure covered). Nevertheless if the flow decreases, due to polarisation concentrations and increasing viscosity of the liquid very close to the membrane surface, the problem is different and has to be solved by other techniques.

From this session it became clear that new ideas are on the drawing board and in laboratories to solve current problems but there is still a lot of work to be done in order to obtain reliable working processes.

<u>SESSION III - Chairman : Dr. R. Gutman</u>

<u>Industrial applications (Part II)</u>

The nine papers of this session encompassed a number of other industrial applications from gas separation to the treatment of low level radioactive waste.

The first presentation

> "Membrane separation of CO_2 and H_2S
> from mixtures with gaseous hydrocarbons"

looked into the development of a new membrane type capable of exhibiting high selectivity whilst keeping an acceptable permeability for the purification of gaseous hydrocarbons.

Experimental investigations on this new process were completed with an economic assessment which highlighted the predominating contribution of the housing device in the total cost of the module. As a result, the utilisation of expensive membranes should not significantly affect the cost effectiveness of the process.

Despite the intrinsic interest of this research, industrial applications remain limited given the current state of the petroleum market.

For the specific purpose of

> "Natural gas transportation through pipelines the removal
> of CO_2 and H_2O by means of membranes"

was the subject of the second presentation on gaseous permeation.

In this respect, the application of pressure resistant membranes (up to 10 MPa) belonging to the aromatic polyimide family proved to be quite promising as far as dehydration is concerned. In addition, experimental investigations showed that by increasing the pressure the process selectivity decreased and that sweetening and dehydration of natural gas could not be achieved using the same membrane type.

The third presentation dealt with the

> "Optimisation of an UF pilot plant for the treatment of
> radioactive waste"

which is currently operated at Harwell for the site low level active liquid waste. This facility is intended to test a number of chemical precipitants as well as several UF/MF membranes (possibly equipped with DMC) for improving decontamination performances and volume reduction of the resulting sludges.

Experiments demonstrated the effectiveness of the DMC system for in situ membrane cleaning (flux steady state increased by 2 to 4) as well as its better performance over simple back-flushing or depressurisation of the UF loop.

The subject of the fourth presentation relates to the

> "Processing of waste water from the olive oil industry by using membrane technology".

This process will allow direct discharge of the waste water after treatment and at the same time recover by-products and energy. The prerequisite for a successful operation of the reverse osmosis step is the setting up of a suitable pretreatment of the oil waste waters (i.e. ultrafiltration).

On the economic view point this new process scheme is expected to lead to a significant reduction of the management cost of the olive oil waste waters compared to the thermic process by a factor of 5 to 7.

The fifth presentation reported on a

> "Study on membrane technology for the textile industry"

placing emphasis on membrane preparation and characterisation. The main purpose of this R & D activity was to process textile liquid waste in order to avoid on the one hand release of polluting effluents into the environment and on the other hand to enable recycling of the chemical products. As a result of the good efficiency of the waste pretreatment step, this process demonstrated its capability to continuously operate a R.O. unit without any detrimental membrane fouling.

As an alternative to conventional solid/liquid separation techniques, the

> "Application of electro-osmosis to radioactive waste processing"

was the subject of the sixth presentation. This process was shown to be ideally suited for the thickening of slurries resulting from gravity settling or crossflow microfiltration as it uses an electric field instead of pressure as the driving force across a microporus membrane. Compared to ultrafiltration, electro-osmotic dewatering was demonstrated to operate at a higher rate while achieving similar concentration factors. Although developed so far for concentrating active chemical precipitation sludges, this process appeared to have good prospects for non-nuclear applications.

The seventh presentation referred to another nuclear application of membrane processes :

> "Reverse-osmosis for the treatment of liquid effluents produced by nuclear power plants"

In order to improve the operation of the waste treatment system of nuclear power plants in terms of reduction of the waste product arisings and the releases of radioactive effluents, a reverse-osmosis unit was designed and constructed. Already tested at Savannah River, this process demonstrated a good potential for future application in the French nuclear power plants.

The

> "Analysis and the control of fouling during natural water
> treatment"

by ultrafiltration was the subject of the eighth paper.

Membrane fouling in water treatment is a complicated problem because the permeate flowrate should be maintained for a long period of time at the lowest possible energy requirement.

Extended analytical studies have been carried out in view of understanding and solving the problem. It appeared that organic substances contained in the water could play a determinant role in the deterioration of membrane long term performances. To cope with this, an hydraulic and chemical on-line washing method has been designed which maintains the filtration flux over several months.

The last paper dealt with the

> "Industrial applications of tangential microfiltration".

This process which is replacing "Dead-end" microfiltration and centrifugation is gaining importance in biotechnology which demands new and efficient processes capable of performing difficult separations in favourable conditions. Nevertheless in spite of research intensification over the last decade, a better understanding of the mechanism of mass transfer and fouling as well as new module designs are required to extend the use of tangential microfiltration to more complex and fragile mixtures.

MAIN ACHIEVEMENTS OF THE ROUND TABLE DISCUSSION

The main objective of the Commission is to support new technologies in order to improve the industrial competitiveness in Europe.

To achieve these objectives, the Commission must orient its programmes towards the development of new applications. Indeed the tendancy actually is to develop a good membrane and to find afterwards the right application, which is the wrong way. The correct one should be the development of a membrane for a particular need. Nevertheless it is not the role of the Commission to find or to identify new potential applications but through a very good interactive communication between the CEC, the membrane producers and industrial people confronted with separation problems to define where membrane technology might help. It is necessary to know if we have a market pull or a technology push.

Professor Smolders insisted on the fact that Europe must have a membrane science strategy. In reality, the market is mainly shared by the USA and Japan. The only chance for Europe to establish a place in this market is to develop new membrane processes or maintain our advance in some membrane systems like pervaporation. For the university, it is very frustrating to see that a big research project can be stopped in a very short time due to a market study producing a result indicating that industrial breakthrough is not expected before ten years.

The Commission's representative, Mr van der Eijk (Head of Division "BRITE"), reminded the audience that the main objectives of the research programmes of the Commission are :

- Increasing European competitiveness,
- Creating a cooperation between Member States, industrial sectors, scientists and engineers.

In the case of the BRITE programme which supports industrial applied research, we have to reach industrial applications. This remark supports one point raised by the Committee of the BRITE evaluation which noted that "Technology is not always seen as a strategy variable by companies management". The management of technology often appears to be isolated from the production and commercial functions of the companies involved. Thus demonstrating what we need in Europe is an industrial technological research strategy.

Nevertheless other programmes at the Commission can cover more fundamental research not supported by industry, like SCIENCE.

Dr. Madsen supported Prof. Smolder's comment by insisting on the fact that very few scientists are working in polymer science. Some countries are pushing very hard in this direction e.g. China, Australia. As example of the different behaviour : in many symposiums in Europe benefits of the membrane distillation system have been presented but never the development of a real industrial system. On the other hand, a membrane distillation plant will be built in China.

This argument was reinforced by the intervention of J-L. Bersillon which allowed the following conclusions to be drawn :

We must speak of membrane technology and no more of membrane science in order to manufacture at an industrial level and at low cost a good reliable membrane operating in a well designed module for particular applications.

Another important point which is very often a problem in the new technology is the standardization (harmonization) of data expression. Actually, each membrane manufacture has his own standards and his own methods of characterization which makes it impossible to compare to membrane characteristics for specific applications.

Two points could be undertaken by the CEC :

- the normalisation of the characterization methods and of the data expression,
- evaluation of the data collection.

The Commission is aware of the normalisation in R & D areas. One of its main objectives is to define European norms resulting from the various R & D programmes. Another programme, BCR (Bureau Communautaire de Référence, Communautary Bureau of Reference) is also dealing with this kind of problem. (The development of standards is a vast worldwide activity which could almost be classified as an industry in itself : Majorie F. Hill).

GENERAL CONCLUSIONS

The lessons to be drawn from this symposium are many and varied, concerning purely technical subjects and economic problems.

Firstly, it is important to point out that at the level of exchange of information between specialists in membranes drawn from very different disciplines and sectors of activity, the symposium proved to be particularly beneficial in establishing fruitful contacts between researchers from universities and industrialists. It is to be hoped that these first exchanges will soon give rise to concrete actions.

As regards technology, this symposium has shown that there is much in common between nuclear and non-nuclear applications, and that solutions developed in one of these areas can advantageously be used in the other, as was illustrated by the presentations on composite membranes, on membranes modified by appropriate surface treatment and on membranes incorporating an electrochemical direct cleaning device.

All the same, it should be noted that non-nuclear industry is oriented principally towards polymer membranes while the nuclear industry prefers inorganic membranes because of the obvious problems of radiation resistance.

Growing interest was also demonstrated in techniques for dehydration of chemical precipitates, where electro-osmosis has been shown to be particularly flexible.

In the implementation of membrane processes, there are further analogies between nuclear and non-nuclear applications linked principally to problem of fouling. Regarding this, it is interesting to note that for the specific case of water treatment, organic materials have been identified as being one of the major causes of blockage of the membranes pores.

Obviously the nature and composition of the solutions to be treated play an important role in judging the effectiveness of membrane process such as reverse osmosis. Still, it has become apparent that for the food and textile industries a judicious combination of chemical pre-treatment, choice of membrane, and process flow-sheet can more easily overcome the fouling problem.

In the area of purification of hydrocarbons, gaseous permeation is without doubt destined to play an important role in the future. Confronted most often with requirements for membranes with ever higher levels of permeability and selectivity combined with resistance to high pressures, gaseous permeation wholly merits the title of new technology.

Finally, as a method of purifying liquid effluents, mention must be made of supported liquid membranes which, because they allow the use of novel chelating agents, open new prospects towards chemical separation techniques for highly specific applications.

<u>Economically speaking</u>, the impact of membrane processes depends on the type of application.
In the nuclear field, the principal advantage of processes such as reverse osmosis and ultrafiltration resides with the reduction in volume of radioactive waste to be handled, in as much as these processes do not themselves generate additional radioactive effluents.

For non-nuclear applications, there is no universal criterion for selection of membrane processes, each case having to be treated separately (reduction of transport costs, exploitation costs, improvement of efficiency, recycling of chemical products, reduction of pollution...).

The main brake on the development of new membrane processes rests nevertheless with the size of the markets for the different applications considered.

However, for the coming years it seems that the developing markets will be in the gas, food, biochemical, and, to a lesser extent nuclear industries. It is with this in mind that research priority should be in membrane processes.

Community R & D in membrane technology should concentrate on :

- identifying the advantages and disadvantages of membrane processes compared to other separation techniques for a selection of specific applications.

- giving impetus to collaboration and exchange of information between scientific staff, researchers, and industrialists from different disciplines.

- promoting European cooperation with the aim to reinforce the competitiveness of European industry.

LIST OF PARTICIPANTS

AIMAR, P.
C.N.R.S.
Laboratoire de Chimie Physique
Route de Narbonne 118
F - 31062 TOULOUSE CEDEX

AMORIM, M.T.
Universidade do Minho
Palacio de Vila Flor
av. D. Alfonso Henriques
P - 4800 GUIMARAES

ARRIBAS, J.J.
University of Valladolid
Facultad de Ciencias
Dept. Fisica Aplicada II
E - 47071 VALLADOLID

AUDINOS, R.
Université Paul Sabatier
Lab. de Chimie-Physique
& Electrochimie
Route de Narbonne 118
F - 31400 TOULOUSE

AVENI, A.
Istituto Ricerche Breda
Via F. De Blasio
I - 70123 BARI

BARDOT, C.
Université Claude Bernard
Lab. d'Automatique
& de Génie des Procédés
Bat. 305
F - 69622 VILLEURBANNE CEDEX

BARNIER, H.
CEA/CEN Cadarache
IRDI/DERDCA
B.P. 1
F - 13108 ST. PAUL-LEZ-DURANCE CEDEX

BEACH, G.
Curtis Systems AG, Zug, Switzerland
c/o WEKO GmbH
Postfach 1260
D - 6239 KRIFTEL

BERGEZ, P.
CEA/CEN Saclay
IRDI - DLPC
F - 91191 GIF-SUR-YVETTE CEDEX

BERSILLON, J.L.
Lyonnaise des Eaux
Central Laboratory
rue du Président Wilson 38
F - 78230 LE PECQ

BOKELUND, H.
C.E.C.
Joint Research Centre
Karlsruhe Establishment
Postfach 2340
D - 7500 KARLSRUHE

BOTTINO, A.
Istituto di Chimica Industriale
Università di Genova
Corso Europa 30
I - 16132 GENOVA

BRIGGS, A.
UKAEA - AERE
Harwell Laboratory
Harwell, Didcot
UK - OXON, OX11 ORA

BRUSCHKE, H.
G.F.T.
Gerberstrasse 48
D - 6650 HOMBURG/S

BUYS, H.C.W.M.
T.N.O.
Division for Tech. for Society
P.O. Box 108
NL - 3700 AC ZEIST

CAPANNELLI, G.
Istituto di Chimica Industriale
Università di Messina
Contrada Papardo
I - 98100 MESSINA

CARBONARO, A.
ENICHEM S.P.A.
New Division Development
Via Medici del Vascello 26
I - 20138 MILANO

CASAS Y GARCIA, J.
CEA/CEN Cadarache
SEDFMA
B.P. 1
F - 13108 ST. PAUL-LEZ-DURANCE CEDEX

CAZEMIER, P.G.
Delair Droogtechniek
en Luchtbehandeling
Oude Kerkstraat 2
NL - 4844 AA ETTEN-LEUR

CECILLE, L.
C.E.C.
DG Science, Research and Development
rue de la Loi 200
B - 1049 BRUSSELS

CHARPIN, J.
CEA/CEN Saclay
IRDI - DLPC
F - 91191 GIF-SUR-YVETTE CEDEX

CHARTIER, P.
Saint-Gobain Recherche
Quai Lucien Lefranc 39
F - 93303 AUBERVILLIERS

CHIESA, G.
Castagnetti S.P.A.
R & D
Via Acqui 86
I - 10090 RIVOLI (TORINO)

CLAROTTI, G.
Université des Sciences
et Techniques du Languedoc
Lab. de Chimie Macromoléculaire
Place Eugène Bataillon
F - 34060 MONTPELLIER CEDEX

CLIFTON, M.
C.N.R.S.
Laboratoire de Chimie-Physique
& Electrochimie
Université Paul Sabatier
Route de Narbonne 118
F - 31062 TOULOUSE CEDEX

CONBOY, T.M.
British Nuclear Fuels plc.
Hinton House
H. 311
Risley
Warrington
UK - CHESHIRE WA3 6AS

COOPER, J.B.
British Petroleum Chem. Intern.
Sattend
UK - HULL HU12 8DS

COT, L.
Ecole Nationale Sup. de Chimie
rue de l'Ecole Normale 8
F - 34075 MONTPELLIER CEDEX

COURTOIS, Ch.
CEA/CEN Cadarache
DRDD/SDFMA
B.P. 1
F - 13108 ST. PAUL-LEZ-DURANCE

COWIESON, D.
Anotec Separations Ltd.
Wilomere Road
Daventry Road Industrial Estate
UK - BANBURY, OXON OX16 7JU

CUEILLE, G.
Tech-Sep
Groupe Rhone-Poulenc
St. Maurice de Beynost
BP 347
F - 01703 MIRIBEL CEDEX

CUMMING, I.W.
UKAEA - AERE
Harwell Laboratory
Chemical Engineering Division
Building 353
Harwell, Didcot
UK - OXON OX11 ORA

DEAN, R.
Consultech
Heathrow Copse 3
Baughurst
UK - BASINSTOKE, HANTS RG26 5JG

DEL FELICE, G.M.
SNIA Fibre
Centro Sperimentale
Via Friuli 55
I - 20031 CESANO MADERNO

DESCHAMPS, A.
Institut Français du Pétrole
avenue de Bois Préau
F - 92502 RUEIL MALMAISON

DESGRANCHAMPS, G.
ELF-Aquitaine
Groupement de Recherches de Lacq
BP 34 Lacq
F - 64170 ARTIX

DHEIN, R.
Bayer AG
ZF-FKH
R 79
Rheinuferstrasse 7
D - 4150 KREFELD-URDINGEN

DICK, R.
Institut National de Recherche
Chimique Appliquée (IRCHA)
BP 1
F - 91710 VERT-LE-PETIT

DOYEN, W.
Studie Centrum voor Kenernergie
SCK/CEN
Boeretang 200
B - 2400 MOL

DOZOL, J.F.
CEA/CEN Cadarache
SEDFMA
B.P. 1
F - 13108 ST. PAUL-LEZ-DURANCE CEDEX

DRIOLI, E.
Università di Calabria
Dipartemento di Chemica
I - 87030 ARCAVACATA DI RENDE (CS)

DUGGAN, J.J.
Air Products and Chemicals
Hersham Place
UK - WALTON-ON-THAMES, SURREY

EFREMENKOV, V.
International Atomic Energy Agency
Wagramerstrasse 5
A - 1400 VIENNA

FINZI, S.
C.E.C.
DG Science, Research and Development
rue de la Loi 200
B - 1049 BRUSSELS

FUTSELAAR, H.
University of Twente
Postbus 217
NL - 7500 AE ENSCHEDE

GAETA, S.
Separem Membrane
Via Per Oropa 118
I - 13051 BIELLA

GARCIA ARROYO, A.
C.E.C.
DG Science, Research and Dev.
rue de la Loi 200
B - 1049 BRUSSELS

GASPARINI, G.M.
ENEA-CRE Casaccia
COMB-MEPIS-Ciclo a valle
S.P. Anguillarese 301
I - 00100 ROMA AD

GAUGER, J.
Cerac S.A.
Engineering Materials Dept.
Chemin des Larges Pièces
CH - 1024 ECUBLENS

GESQUIERE, R.
Metal Working Services S.A.
avenue des Martins Pêcheurs 21
B - 1640 RHODE ST. GENESE

GEUS, E.R.
Delft Technical University
Scheihundige Technologie
Vahgroep Organische Chemie
Julianalaan 136
NL - 2628 BL DELFT

GINZEL, K.D.
Hoechst Aktiengesellschaft
Hauptlaboratorium 830
Postfach 80 03 20
D - FRANKFURT/M. 80

GOOSSENS, W.
Studie Centrum voor Kenenergie
SCK/CEN
Market Studies & Project Dev.
Boerctang 200
B - 2400 MOL

GORISSEN, H.
AKZO Engineering B.V.
P.O. Box 9300
NL – 6800 SB ARNHEM

GUIBAUD, J.
S.C.T.
Process Engineering
B.P. 1
F – 65460 BAZET

GUTMAN, R.
Pall Europe Ltd.
Europa House
Havant Street
UK – PORTSMOUTH PO1 3PD

HACHE, J.
Société Bertin et Cie
B.P. 3
F – 78373 PLAISIR CEDEX

HALLSTROM, B.
Lund University
Div. of Food Engineering
P.O. Box 124
S – 221 00 LUND

HAUCHECORNE, K.
Solvay & Cie
Laboratoire Central
rue de Ransbeek 310
B – 1.120 BRUXELLES

HENRIKSEN, P.
Technical University of Denmark
Instituttet for Kemiteknik
Dept. of Chemical Engineering
DtH, Bygning 229
DK – 2800 LYNGBY

HERMIA, J.
Université Catholique de Louvain
Faculté des Sciences Appliquées
Unité des Procédés
Voie Minckelers 1
B – 1348 LOUVAIN-LA-NEUVE

HERNANDEZ GIMENEZ, A.
University of Valladolid
Faculdad de Ciencias
Dept. Fisica Aplicada II
E – 47071 VALLADOLID

HOZUMI SHIMIZU
Nitto Belgium N.V.
Industrie Park-Zuid
Eikelaarstraat 22
B – 3600 GENK

HUET, Y.
Framatome
Tour Fiat
Place de la Coupole 1
Cedex 16
F – 92084 PARIS-LA-DEFENSE

IVENS, R.
British Nuclear Fuels plc.
Hinton House
Risley
Warrington
UK – CHESHIRE WA3 6AS

JAFFRIN, M.
Univ. de Techn. de Compiègne
Dept. Génie Biologique
B.P. 649
F – 60206 COMPIEGNE CEDEX

JANSSEN-VAN ROSMALEN, R.
N.V. Nederlandse Gasunie
P.O. Box 19
NL – 9700 MA GRONINGEN

KAMPS, P.
University of Twente
P.O. Box 217
NL – 7500 AE ENSCHEDE

KNIBBS, R.
UKAEA – AERE
Harwell Laboratory
Chemical Engineering Division
Harwell, Didcot
UK – OXON OX11 ORA

KONTTURI, K.
Helsinki Univ. of Technology
Lab. of Physical Chemistry
& Electrochemistry
Kemistintic 1A
SF – 02150 ESPOO

KOOPS, G.H.
University of Twente
P.O. Box 217
NL – 7500 AE ENSCHEDE

KRISTENSEN, S.
De Danske Sukkerfabrikker Filtration
Stavangervej 10
P.O. Box 149
DK - 4900 NAKSKOV

KUSTERS, P.
Delair Droogtechniek
en Luchtbehandeling
Oude Kerkstraat 2
NL - 4844 AA ETTEN-LEUR

LAIDLER, D.
ICI Chemicals & Polymers Ltd
P.O. Box 8
The Heath
UK - RUNCORN WA7 4QD

LAMBRECHT, J.
Industrial Consultants
Max Rüttgers Strasse 29
D - 8021 IRSCHENHAUSEN

LEFELLE, R.
Ets Dewarin Fils & Co.
Peignage d'Auchel
B.P. 5
ZI 1
F - 62260 AUCHEL

LETESSON, P.
Université Catholique de Louvain
Faculté des Sciences Appliquées
Unité des Procédés
Voie Minckelers 1
B - 1348 LOUVAIN-LA-NEUVE

LEWIS, H.G.
British Nuclear Fuels plc.
Hinton House
Risley
UK - WARRINGTON, CHESHIRE WA3 6AS

LEYSEN, B.
Studie Centrum voor Kernenergie
SCK/CEN
Material Development
Boeretang 200
B - 2400 MOL

LUTIN, F.
Université Paris Val de Marne
Laboratoire Energétique
Electro-Chimique et Biochimique
rue de l'Isly 3
F - 75008 PARIS

MADSEN, R.
D.D.S.
Langebroggde 5
Box 17
DK - 1001 COPENHAGEN

MARCHAL, P.
Ets Dewarin Fils & Co.
Peignage d'Auchel
B.P. 5
ZI 1
F - 62260 AUCHEL

MASSIGNAN, L.
Centro Ricerche Bonomo
Contrada Castel del Monte
I - 70031 ANDRIA (BA)

MAUREL, A.
CEA/CEN Cadarache
IRDI/DERDCA
B.P. 1
F - 13108 ST. PAUL-LEZ-DURANCE

MCGEEHIN, P.
Compton Consultants
School Road
UK - NEWBURY, BERKS RG16 0Q4

MEARES, P.
University of Exeter
Dept. of Chemical Engineering
North Park Road
UK - EXETER EX4 4QF

MEJNDERSMA, G.W.
DSM Research/PT
P.O. Box 18
NL - 6160 MD GELEEN

MIETTON PEUCHOT, M.
IGEPA
rue Marcel Pagnol
F - 47510 FOULAYRONNES

MILISIC, V.
Institut de la Filtration et
des Techniques Séparatives
rue M. Pagnol
F - 47510 FOULAYRONNES

MIRANDA, T.M.
Universidade do Minho
Palacio de Vila Flor
av. D. Alfonso Henriques
P - 4800 GUIMARAES

MULDER, M.
University of Twente
P.O. Box 217
NL - 7500 AE ENSCHEDE

NAFPLIOTIS, L.
Synfina-Oleofina
R & D Biochemistry
rue J. de Lalaing 4
B - 1040 BRUXELLES

NEEL, J.
Ecole Nationale Supérieure
des Industries Chimiques
rue Grandville 1
B.P. 451
F - 54001 NANCY CEDEX

NEWELL, P.A.
DOW Rheinmünster GmbH
Industriestrasse 1
D - 7587 Rheinmünster

OHLROGGE, K.
GKSS - Forschungszentrum
Institut für Chemie
Max Planck Strasse
Postfach 11 00
D - 2054 GEESTHACHT

ORLOWSKI, S.
C.E.C.
DG Science, Research and Development
rue de la Loi 200
B - 1049 BRUSSELS

PARKER, J.
Tech. Sep. Groupe Rhone-Poulenc
St. Maurice de Beynost
B.P. 347
F - 01703 MIRIBEL CEDEX

PATERSON, R.
University of Glasgow
Dept. of Chemistry
UK - GLASGOW G12 8QQ

PEINEMANN, K.V.
GKSS - Forschungszentrum
Institut für Chemie
Max Planck Strasse
Postfach 11 00
D - 2054 GEESTHACHT

PEPPER, D.
PCI Membrane Systems Ltd
Laverstoke Mill
Whitchurch
UK - HANTS RG28 7NR

PFENNINGER, H.
Ciba-Geigy `AG
K - 674.712
CH - 4002 BASEL

PILLODS, F.
S.G.N.
rue du Héron 1
Montigny le Bretonneux
F - 78182 ST. QUENTIN-EN-YVELINES

PINERI, M.
CEA/CEN Grenoble
DRF-G/Service de Physique
Groupe Physico-Chimie Moléculaire
85 X
F - 38041 GRENOBLE CEDEX

PINHO, M.N.
Universidade Técnica de Lisboa
Instituto Superior Técnico
Dep. de Engenharia Quimica
P - 1096 LISBOA CODEX

PLANCHON, M.
Solvay & Cie S.A.
Direction Générale des Rech.
rue de Ransbeek 310
B - 1120 BRUXELLES

REDEKER, B.
Universität Dortmund
CT/MV
Postfach 50 05 00
D - 4600 DORTMUND 50

RENARD, A.
Tech. Sep.
Membranes Departement
St. Maurice de Beynost
B.P. 347
F - 01703 MIRIBEL CEDEX

RIOS, G.
Université de Montpellier II
(U.S.T.L.)
Centre de Génie & Techn. Aliment.
Ingéniérie des Procédés
Place E. Bataillon
F - 34060 MONTPELLIER CEDEX 1

ROBINSON, K.
UKAEA –AERE
Harwell Laboratory
Harwell, Didcot
UK – OXON OX11 ORA

ROOFTHOOFT, R.
Laborelec
rue de Rhode 125
B – 1630 LINKEBEEK

ROSENFELD, J.L.J.
Shell Research B.V.
(K.S.L.A.)
P.O. Box 3003
Badhuisweg 3
NL – 1031 CM AMSTERDAM

RUMEAU, M.
Université de Montpellier II
(U.S.T.L.)
Place E. Bataillon
F – 34060 MONTPELLIER CEDEX 1

SANDER, U.
Lurgi GmbH
R & D New Technologies
Gwinnerstrasse 27/33
D – 6000 FRANKFURT/MAIN 60

SCHNEIDER, K.
Akzo
Research Laboratories Obernburg
D – 8753 OBERNBURG

SCHNEIDER, M.
Hoechst AG
Verfahrenstechnik
Postfach 80 03 20
D – 5230 FRANKFURT/MAIN 80

SCHUURMANS, H.J.A.
Shell Research B.V.
(K.S.L.A.)
P.O. Box 3003
Badhuisweg 3
NL – 1031 CM AMSTERDAM

SCOWEN, P.
Pall Process Filtration Ltd.
Europa House
Havant Street
UK – PORTSMOUTH PO1 3PD

SEHN, P.
DOW Rheinmünster GmbH
Industriestrasse 1
D – 7587 RHEINMUNSTER

SIMON, R.
C.E.C.
DG Science, Research and Dev.
rue de la Loi 200
B – 1049 BRUSSELS

SINGH, J.
UKAEA – AERE
Springfields Laboratory
Salwick, Preston
UK – LANCS PR4 ORR

SKELTON, R.
Koch Intern. (UK) Ltd.
Friars Mill
Friars Terrace
UK – STAFFORD ST17 4AU

SMOLDERS, K.
University of Twente
P.O. Box 217
NL – 7500 AE ENSCHEDE

STAIRMAND, J.W.
UKAEA – AERE
Springfield Laboratory
Salwick, Preston
UK – LANCS PR4 ORR

STAM, H.
AKZO Engineering B.V.
P.O. Box 9300
NL – 6800 SB ARNHEM

STRATHMANN, H.
Fraunhofer Gesellschaft
Nobelstrasse 12
D – 7000 FRANKFURT 80

TELLIER, J.
ELF – Aquitaine
Groupement de Rech. de Lacq
B.P. 34 Lacq
F – 64170 ARTIX

TEN ASBROEK, W.G.H.
AKZO Engineering B.V.
Salt & Basic Chemical Div.
P.O. Box 25
NL – 7550 GC HENGELO –OV–

TER MEULEN, B.Ph.
MT - TNO
Dept. of Chemical Engineering
P.O. Box 342
NL - 7300 AH APELDOORN

TOUSSAINT, J.C.
C.E.C.
DG Science, Research and Development
rue de la Loi 200
B - 1049 BRUSSELS

TURNER, A.D.
UKAEA - AERE
Harwell Laboratory
Materials Development Division
Building 429
UK - HARWELL, DIDCOT, OXON OX11 ORA

VALENTINI, C.
Eniricerche S.P.A.
Via Ramarini 32
I - 00015 MONTEROTONDO (ROMA)

VALIENTE, M.
Universidad Autonoma de Barcelona
Quimica Analitica
E - 08193 BELLATERRA (BARCELONA)

VAN DER EIJK, W.
C.E.C.
DG Science, Research and Development
rue de la Loi 200
B - 1049 BRUSSELS

VAN DER WAAL, M.J.
University of Twente
Postbus 217
NL - 7500 AE ENSCHEDE

VAN GAANS, C.
Billiton Research B.V.
P.O. Box 40
NL - 6800 AA ARNHEM

VAN PIJKEREN, G.
N.V. Nederlandse Gasunie
P.O. Box 19
NL - 9700 MA GRONINGEN

VANDERSTUYFT, J.
Sogefiltres S.A.
avenue Tedesco 5 - 7
B - 1160 BRUXELLES

VERHOOG, H,M.
Hoogovens Groep
P.O. Box 10000
NL - 1970 CA IJMUIDEN

VLIEGEN, J.
Metallurgie Hoboken-Overpelt
A. Greinerstraat 14
B - 2710 HOBOKEN

WALLNER, H.
Biochemie GmbH Kundl
Dr. Bachmann Strasse
A - 6250 KUNDL

WILSON, F.G.
Fairey Filtration Dev. Centre
Fareham Industrial Park
UK - FAREHAM HANTS PO16 8XG

YIANNI, J.
Biocompatibles Ltd.
Brunel Science Park
Kingstron Lane
UK - UXBRIDGE, MIDDX UB8 3PQ

YUKSEL, L.M.
Henkel KGaA
R-TV Verfahrenstechnik
Postfach 1100
D - 4000 DUSSELDORF 1

INDEX OF AUTHORS